THE ELEMENTS: SYMBOLS, ATOMIC NUMBERS, AND MOLAR MASSES

Element	Symbol	Atomic number	Molar mass, g/mol
Actinium	Ac	89	227.03
Aluminum	Al	13	26.98
Americium	Am	95	241.06
Antimony	Sb	51	121.75
Argon	Ar	18	39.95
Arsenic	As	33	74.92
Astatine	At	85	210
Barium	Ba	56	137.34
Berkelium	Bk	97	249.08
Beryllium	Be	4	9.01
Bismuth	Bi	83	208.98
Boron	B	5	10.81
Bromine	Br	35	79.91
Cadmium	Cd	48	112.40
Calcium	Ca	20	40.08
Californium	Cf	98	251.08
Carbon	C	6	12.01
Cerium	Ce	58	140.12
Cesium	Cs	55	132.91
Chlorine	Cl	17	35.45
Chromium	Cr	24	52.01
Cobalt	Co	27	58.93
Copper	Cu	29	63.54
Curium	Cm	96	247.07
Dysprosium	Dy	66	162.50
Einsteinium	Es	99	254.09
Erbium	Er	68	167.26
Europium	Eu	63	151.96
Fermium	Fm	100	257.10
Fluorine	F	9	19.00
Francium	Fr	87	223
Gadolinium	Gd	64	157.25
Gallium	Ga	31	69.72
Germanium	Ge	32	72.59
Gold	Au	79	196.97

Element	Symbol	Atomic number	Molar mass, g/mol
Hafnium	Hf	72	178.49
Helium	He	2	4.00
Holmium	Ho	67	164.93
Hydrogen	H	1	1.008
Indium	In	49	114.82
Iodine	I	53	126.90
Iridium	Ir	77	192.2
Iron	Fe	26	55.85
Krypton	Kr	36	83.80
Lanthanum	La	57	138.91
Lawrencium	Lr	103	257
Lead	Pb	82	207.19
Lithium	Li	3	6.94
Lutetium	Lu	71	174.97
Magnesium	Mg	12	24.31
Manganese	Mn	25	54.94
Mendelevium	Md	101	258.10
Mercury	Hg	80	200.59
Molybdenum	Mo	42	95.94
Neodymium	Nd	60	144.24
Neon	Ne	10	20.18
Neptunium	Np	93	237.05
Nickel	Ni	28	58.71
Niobium	Nb	41	92.91
Nitrogen	N	7	14.01
Nobelium	No	102	255
Osmium	Os	76	190.2
Oxygen	O	8	16.00
Palladium	Pd	46	106.4
Phosphorus	P	15	30.97
Platinum	Pt	78	195.09
Plutonium	Pu	94	239.05
Polonium	Po	84	210
Potassium	K	19	39.10
Praseodymium	Pr	59	140.91

Element	Symbol	Atomic number	Molar mass, g/mol
Promethium	Pm	61	146.92
Protactinium	Pa	91	231.04
Radium	Ra	88	226.03
Radon	Rn	86	222
Rhenium	Re	75	186.2
Rhodium	Rh	45	102.91
Rubidium	Rb	37	85.47
Ruthenium	Ru	44	101.07
Samarium	Sm	62	150.35
Scandium	Sc	21	44.96
Selenium	Se	34	78.96
Silcon	Si	14	28.09
Silver	Ag	47	107.87
Sodium	Na	11	22.99
Strontium	Sr	38	87.62
Sulfur	S	16	32.06
Tantalum	Ta	73	180.95
Technetium	Tc	43	98.91
Tellurium	Te	52	127.60
Terbium	Tb	65	158.92
Thallium	Tl	81	204.37
Thorium	Th	90	232.04
Thulium	Tm	69	
Tin	Sn	50	
Titanium	Ti	22	
Tungsten	W	74	
Uranium	U	92	
Vanadium	V	23	
Xenon	Xe	54	
Ytterbium	Yb	70	
Yttrium	Y	39	
Zinc	Zn	30	
Zirconium	Zr	40	

Catalytic Chemistry

THE WILEY SERIES IN CHEMICAL ENGINEERING

Catalytic Chemistry

Bruce C. Gates

University of Delaware

John Wiley & Sons, Inc.

New York · Chichester · Brisbane · Toronto · Singapore

Acquisitions Editor	Bradford Wiley, II
Designer	Laura Nicholls
Production Supervisor	Elizabeth Austin
Manufacturing Manager	Lorraine Fumoso
Copy Editing Manager	Deborah Herbert
Illustration	Ishaya Monokoff

Library of Congress Cataloging in Publication Data:

Gates, Bruce C.
　　Catalytic chemistry / Bruce C. Gates.
　　　　p.　cm.
　　Includes bibliographical references and index.
　　ISBN 0-471-51761-5 (cloth)
　　1. Catalysis.　I. Title.
QD505.G38　1991
541.3′95—dc20　　　　　　　　　　　　　　　　91-4192
　　　　　　　　　　　　　　　　　　　　　　　　CIP

Printed in the United States of America

10 9 8 7 6 5 4 3 2 1

For
Jutta

Preface

Most of the chemical reactions in industry and biology are catalytic, and many chemists and chemical engineers work to understand and apply catalysis. Catalysis is involved at some stage of the processing of a large fraction of the goods manufactured in the United States; the value of these products approaches a trillion dollars annually, more than the gross national products of all but a few countries in the world. Catalysis is the key to the efficiency of chemical conversions. In an age of increasingly limited energy and raw materials and concern for the environment, it is needed more and more. The technological needs are matched by the scientific opportunities: new techniques are bringing rapid progress in the understanding of molecular details of the workings of catalysts.

Notwithstanding these needs and opportunities, catalysis remains a neglected subject in chemical education. Students encounter it only as unconnected fragments in chemistry (organic, inorganic, and physical), biochemistry, and chemical engineering. The objective of this book is to integrate the fragments for students at the advanced undergraduate level.

Catalysis is a coherent subject, unified by concepts of chemical structure and reactivity, kinetics, and transport phenomena. It is developed in the following sequence:

Chapter 1. Introduction, with definitions and illustrations of catalysis and a brief review of kinetics.

Chapter 2. Catalysis in solutions, with emphasis on acid–base and organometallic catalysis. The ideas and mechanisms presented here are a foundation for the remainder of the book.

Chapter 3. Catalysis by enzymes. The amazing efficiency of each of nature's catalysts is related to the complementarity between its intricate structure and the structure of its particular reactant.

Chapter 4. Catalysis in and on synthetic polymers, ranging from swellable, solutionlike gels to rigid, porous solids with large internal surfaces. This chapter provides a transition from catalysis in solution to catalysis involving solids and surfaces.

Chapter 5. Catalysis within the molecular-scale cages of zeolites and other molecular sieves. The dramatic effects of shape selectivity at times allow a reversal of the reactivity patterns observed in solution catalysis.

Chapter 6. Catalysis on surfaces of inorganic solids, the most important industrial catalysts. This chapter is built on an introduction to the surface science of single crystals and extends to more complicated catalysts, including supported metal complexes, supported metals, metal oxides, and metal sulfides.

The principles are illustrated with catalytic cycles, reaction mechanisms, catalyst structures, and kinetics. The reader is led, with common threads of chemistry, from homogeneous solutions, to the confines of polymeric phases, enzyme pockets, cages, and surfaces. Catalysis is influenced by distributions of reactants and catalysts between phases and by geometric effects associated with groups bonded to catalyst molecules and surfaces. The complications and design possibilities of combinations of catalytic groups and the ideas of multifunctional catalysis are introduced from the beginning, paving the way to the enzymes and the complex surfaces of the solid catalysts used in industrial processes.

The central ideas are chemical, and the physical effects of diffusion, mass transfer, and heat transfer are integrated briefly within a context dictated by the chemistry. Industrial reactions and catalysts are emphasized to illustrate many of the principles, with tables and graphs of data included to provide concreteness and material for examples and problems. The selection of reactions and catalysts is intended to provide a sense of the great accomplishments represented by industrial catalytic processes, including alkylation, hydrogenation, hydroformylation, carbonylation, stereospecific polymerization, ammonia synthesis, ethylene oxidation, propylene ammoxidation, and cracking, reforming, and hydroprocessing of petroleum.

Students who have a foundation of undergraduate organic and physical chemistry and a beginner's familiarity with inorganic chemistry should be able to understand the book. Examples throughout the text and problems at the end of each chapter reinforce and test the student's grasp of the concepts.

This book was developed from notes used in one-semester courses at the University of Delaware (attended primarily by fourth-year students in chemical engineering) and the University of Munich (attended by students in physical chemistry). The Delaware catalysis course is paired with a course in kinetics and chemical reaction engineering, providing most chemical engineering students with their first integration of organic and inorganic chemistry into traditional chemical engineering. The course offers chemistry students an introduction to applied chemistry and some of the concepts of chemical engineering.

I thank many people for comments and suggestions, especially Professor G. C. A. Schuit. Other colleagues who read the manuscript in near-final form and offered helpful criticisms are R. H. Grubbs, Y. Iwasawa, H. Knözinger,

G. A. Mills, R. Prins, L. D. Schmidt, G. A. Somorjai, and R. J. P. Williams. Thanks are also due to Jeanne Grille, Lorraine Holton, Joan Pengilly, Frances Tilley and Cecilia Viering for typing the manuscript, and to Bradford Wiley, II, chemical engineering editor at John Wiley & Sons, Inc., and to its production staff. The book was written at the University of Delaware and the Institute of Physical Chemistry of the University of Munich. The Deutsche Forschungsgemeinschaft, the Fulbright Commission in Bonn, and the University of Delaware Center for Advanced Study provided fellowships to support the writing.

Bruce C. Gates
Newark, Delaware

About the Author

Bruce C. Gates is the H. Rodney Sharp Professor of Chemical Engineering and Professor of Chemistry at the University of Delaware and has frequently been a visiting professor at the Institute of Physical Chemistry of the University of Munich. His research group is active in the investigation of catalytic hydroprocessing and catalysis by zeolites, superacids, and supported metals and organometallics. Professor Gates is a coauthor of *Chemistry of Catalytic Processes* (McGraw-Hill, 1979) and coeditor of *Metal Clusters in Catalysis* (Elsevier, 1986) and *Surface Organometallic Chemistry: Molecular Approaches to Surface Catalysis* (Kluwer, 1988).

Contents

List of Notation

a	thermodynamic activity; area per unit volume
A	Arrhenius preexponential factor
A	acetone; acid; alcohol; reactant
Ac	acetyl, CH_3CO
B	base; reactant
C	catalyst; catalyst precursor
C	concentration
CN	coordination number
d	diameter
D	diffusion coefficient
DVB	divinylbenzene
E	enzyme
E	energy; dimensionless term defined in text where used
ESCA	electron spectroscopy for chemical analysis (XPS)
f	fraction
G	constant in Brønsted relation; Gibbs free energy; mass flow rate
G_M	flow rate, in moles per unit time per unit cross-sectional area
h	heat transfer coefficient; Planck's constant
h_0	defined immediately below
H_0	Hammett acidity function ($-\log h_0$)
I	intermediate
IB	isobutylene
In	initiator
j_D	dimensionless j factor defined in text where used
k	reaction rate constant
$\underset{\sim}{k}$	Boltzmann's constant
k_G	mass transfer coefficient, gas phase
k_L	mass transfer coefficient, liquid phase
K	equilibrium constant; adsorption equilibrium constant; dimensionless term defined in text where used

L	levo configuration
L	ligand
L	length (half-thickness of slab)
LEED	low-energy electron diffraction
M	metal; monomer
Me	methyl
n	integer
N	dimensionless number; principal quantum number; rate of surface bombardment
NMR	nuclear magnetic resonance
P	phenol; polymer
P	partial pressure; pressure; probability
Ph	phenyl
Q^+	cation
R	alkyl group; reactant
R	distance; gas constant; radius
r	distance; radius; reaction rate
S	selectivity; surface area
s	second(s)
T	temperature
t	time
V	volume
v	velocity
W	water
X	halogen; intermediate
XPS	X-ray photoelectron spectroscopy
Y	mole fraction
Y^-	anion
Z	polymer
z	distance

Subscripts and Superscripts

A	alcohol
act	activation
ads	adsorbed; adsorption
B	bulk
coll	collision
cat	catalytic
dis	disproportionation
E	enzyme
eff	effective
f	forward
g	gas
G	gas
H	heptane
isom	isomerization
I	intermediate

IB	isobutylene
K	Knudsen
L	liquid; length (of slab)
M	Michaelis
m	monolayer
mol	molecular
O	olefin
o	value for base case; value at saturation
obs	observed
P	particle; phenol
r	reaction; reverse
Re	Reynolds
s	surface
Sc	Schmidt
t	total
T	toluene
v	vacant
W	water
‡	transition state

Greek letters

α	constant $(0 < \alpha < 1)$; fraction
β	dimensionless term defined in text where used
γ	activity coefficient
ϵ	void fraction of particle or of fixed bed of particles
η	effectiveness factor
θ	fraction of surface sites
λ	thermal conductivity
μ	viscosity
ν	stoichiometric coefficient
ρ	density
σ	interfacial tension
τ	tortuosity
ϕ	contact angle; Thiele modulus
Φ	modified Thiele modulus

1

Introduction

Nitrogen and hydrogen flow through a tube at high temperature and pressure, but they do not react to give ammonia, although the chemical equilibrium is favorable. When particles of iron are placed in the tube, however, the gases coming in contact with them are converted rapidly into ammonia. Millions of kilograms of this chemical are synthesized every year in such tubular reactors at a cost of a few cents per kilogram. Ammonia is a raw material for nitrate fertilizers needed to feed the world's population. Fertilizers are assimilated by the cells of plants in complex sequences of biological reactions. Further metabolic reactions take place as animals consume the plants, reconstructing their contents to provide energy and molecular building blocks for growth. All these biological reactions proceed under the direction of naturally occurring macromolecules called enzymes; without them, the processes of life could not take place.

What is an enzyme, and what does it have in common with the iron in the ammonia synthesis reactor? Each is a **catalyst**—a substance that accelerates a chemical reaction but is not consumed in the reaction and does not affect its equilibrium. Catalysts cause reactions to proceed faster than they would otherwise, and they can be used over and over again. Catalysts are the keys to the efficiency of almost all biochemical processes and most industrial chemical processes.

How do catalysts work? This book is an attempt to answer the question, but it is already evident from the book's length that there is no simple answer. Nonetheless, it is fitting to begin with an oversimplified answer and to improve upon it along the way: a catalyst provides a new and easier pathway for reactant molecules to be converted into product molecules. By analogy, consider instead of reactant molecules a group of climbers crossing a mountain range, breaking ground slowly over the steep crest. Alternatively, they might be guided rapidly along a less direct pathway, experiencing gentle ups and downs but circumventing the crest and crossing a low pass to reach the opposite valley in a short time. According to this rough analogy, the guidance along this new pathway—

more roundabout and complicated but faster—is provided by the catalyst. The catalyst participates in the reaction by somehow combining with the reactant molecules so that they rearrange into products (and regenerate the catalyst) more rapidly than they could if the catalyst were not present.

A catalyst is by definition a substance that increases the rate of approach to equilibrium of a chemical reaction without being substantially consumed in the reaction.[1] A catalyst usually works by forming chemical bonds to one or more reactants and thereby facilitating their conversion—it does not significantly affect the reaction equilibrium.

The definition of a catalyst rests on the idea of reaction rate, and therefore the subject of reaction kinetics is central, providing the quantitative framework. The qualitative chemical explanations of catalysis take the form of reaction mechanisms. These are models of reactions accounting for the overall stoichiometry, identifying the sequence of elementary reaction steps, and (insofar as possible) explaining, in terms of chemical bond strength and geometry, the interactions of the catalyst with reactants.

There are threads of continuity in the book that are provided by the families of catalytic reactions proceeding by similar mechanisms; the mechanisms commonly involve the transfer of ions (often hydrogen ions), electrons, and/or radicals. Catalytic reactions take place in various phases: in solutions, within the solutionlike confines of micelles and the molecular-scale pockets of large enzyme molecules, within polymer gels, within the molecular-scale cages of crystalline solids such as molecular-sieve zeolites, and on the surfaces of solids. This list of phases in which catalysis occurs forms a progression from the simplest toward the most difficult to characterize in terms of exact chemical structure. It also forms the outline of this book. A given catalytic reaction can take place in any of these phases, and the details of the mechanism will often be similar in all of them. Consequently, the next chapter (Chapter 2) concerns catalysis in homogeneous solutions because it is most thoroughly understood. The succeeding chapters build on this platform to explain catalysis in other phases.

A full account of how a catalyst works requires a description of how reactant molecules are transported to the catalyst and of how product molecules are transported away. Because the reactants and products are often concentrated in a phase separate from that holding the catalyst, it is necessary to consider transport between phases by diffusion and convection and to deter-

---

[1] This definition is close to that given by Wilhelm Ostwald in about 1895: A catalyst is a substance that changes the rate of a chemical reaction without itself appearing in the products. Ostwald's definition allows for negative catalysis, whereby the catalyst slows the reaction. It does not allow any conversion of the catalyst.

Sixty years earlier Jakob Berzelius had coined the term "catalysis" when he recognized that changes in compositions of numerous substances that came in contact with small amounts of various "ferments," liquids, or solids could be classified by a single concept. But his definition turned on undefined "forces" and did not stand the test of time. Many years earlier, alchemists had been aware of the actions of some of these "ferments" and other substances. Their awareness probably encouraged the futile search for a philosopher's stone and a way to turn base metals into gold.

mine how these transport processes affect the rates of catalytic reactions. For example, in the ammonia synthesis, the reactants H_2 and N_2 are concentrated in the gas phase flowing through the reactor, but the catalytic reaction occurs on the surface of the iron particles. Therefore, the rate of the reaction depends not only on what happens on the surface but also on how fast the H_2 and N_2 are transported to the surface and how fast the ammonia is transported away. The reaction occurs on the surface in the steady state at just the rate of transport of the reactants to the surface and the rate of transport of the products from the surface.

Some of the essential ideas defining catalysis can be learned from pre-liminary consideration of an example. The reaction of ethylene and butadiene in solution to give 1,4-hexadiene is of industrial importance in the manufacture of the polymer nylon. The reaction has been suggested to proceed via the cycle of elementary steps shown in Fig. 1-1. The species entering the cycle are H^+, NiL_4 (a complex of Ni with unspecified ligands L), ethylene (symbolized by the double bond, $=$), and butadiene (symbolized by $\diagup\!\!\diagdown\!\!\diagup$). The species leaving the cycle are the hydrocarbon products, namely, three isomers of hexadiene. At this point it is not important what structures are formed from the reactants and the nickel complex. What is important is that the catalyst is formed from the reactants and the nickel complex and then converted cyclically through the various forms shown in the figure. The nickel is not consumed in the reaction once it has entered the cycle, but is repeatedly converted from one form to another in the cycle as the reactants are converted into products.

Catalysis always involves a cycle of reaction steps, and the catalyst is converted from one form to the next, ideally without being consumed in the overall process. The occurrence of a cyclic reaction sequence is a requirement for catalysis, and catalysis may even be defined as such an occurrence.

One might be tempted to designate NiL_4, the compound that enters the cycle, as the catalyst for the 1,4-hexadiene synthesis reaction. However, this choice would be arbitrary; any of the nickel compounds shown in the figure could with equal correctness be designated as the catalyst. The ambiguity in catalyst designation is typical, and the identification of one unique catalytic species is usually not possible. Reference to one species as the catalyst is imprecise but common because it is convenient.

Figure 1-1 shows that there is not just one cycle of reactions that converts reactants into products, but several parallel cycles. Many others, some involving nickel and some not, could also be imagined. They are not included because they are not kinetically significant; reaction via these imagined cycles is too slow to matter. A catalyst for a particular reaction always provides at least one pathway for conversion of reactants that is kinetically significant. The catalyst virtually always provides this pathway as a consequence of its becoming chemically bonded to reactants.

An **inhibitor** slows down a catalytic reaction; a *competitive inhibitor* slows down the reaction by competing with the reactants in bonding to the catalyst. A very strong inhibitor, one that bonds so strongly that it virtually excludes the reactants from bonding with the catalyst, is called a **poison.**

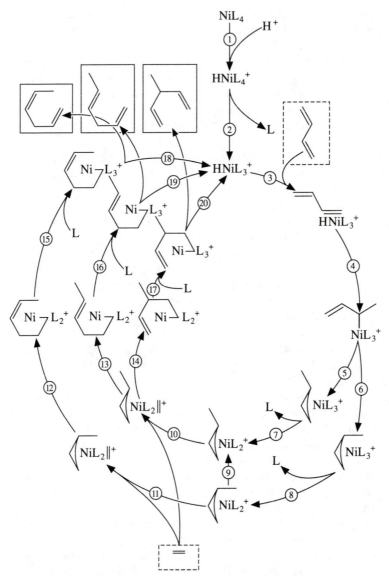

Figure 1-1
Catalytic cycles for the conversion of ethylene and butadiene into 1,4-hexadiene [1].

A quantitative measure of how fast a catalyst works is its **activity,** which is usually defined as the reaction rate (or a reaction rate constant) for conversion of reactants into products. Often products are formed in addition to those that are desired, and a catalyst has an activity for each particular reaction. A ratio of these catalytic activities is referred to as a **selectivity,** which is a measure of the catalyst's ability to direct the conversion to the desired products. For a large-scale application, the selectivity of a catalyst may be even more important than its activity. In practice, catalysts are inevitably involved in side reactions that lead to their conversion into inactive forms (those not appearing in the catalytic cycle). Therefore, a catalyst is also chosen on the basis of its **stability**; the greater the stability, the lower the rate at which the catalyst loses it activity or selectivity or both. A deactivated catalyst may be treated to bring back its activity; its **regenerability** in such a treatment is a measure (often not precisely defined) of how well its activity can be brought back.

The definitions of activity, selectivity, and stability all rest on reaction kinetics, and it is important at the outset to be familiar with some elementary ideas of catalytic reaction kinetics. For illustration, consider the simple catalytic reaction of Fig. 1-2. The overall reaction is an isomerization, $R \rightarrow P$, where R and P respectively represent the hypothetical reactant and product, shown in the cycle as $\lceil R \rceil$ and $\boxed{P}$.[2] The catalyst cycles between two forms: C and a complex incorporating the reactant, RC. (For simplicity, the form of catalyst initially entering the cycle is no longer shown.) Suppose that the following elementary reaction steps take place in the cycle:

$$R + C \xrightarrow{\ k_1\ } RC \tag{1-1}$$

$$RC \xrightarrow{\ k_2\ } R + C \tag{1-2}$$

$$RC \xrightarrow{\ k_3\ } P + C \tag{1-3}$$

This sequence of three steps indicates that the intermediate RC is formed from C and R and decomposes either to give back C and R, or to give C and the isomerized product P. Since each reaction is assumed to be an *elementary step* (or simply a step, i.e., a reaction taking place on the molecular level as written in the stoichiometric equation), the following rate expressions can be written directly from the stoichiometric equations:

$$r_1 = k_1 C_R C_C \tag{1-4}$$

$$r_2 = k_2 C_{RC} \tag{1-5}$$

$$r_3 = k_3 C_{RC} \tag{1-6}$$

Here r is the rate, C is concentration, and the ks are exponentially dependent on absolute temperature according to the Arrhenius relation $k = Ae^{-E_{act}/RT}$. If the reactions take place in an isothermal batch reactor with a negligible change in volume of reactants, then the reaction rates can be written as rates of change of concentration, for example,

---

[2] In most of the figures, reactants entering a cycle are indicated with a dashed rectangle, $\lceil \ \rceil$, and products leaving the cycle are indicated with a closed rectangle, $\square$.

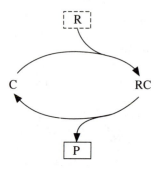

Figure 1-2
Hypothetical catalytic cycle for the isomerization R → P.

$$r_1 = -\frac{dC_R}{dt} = k_1 C_R C_C \tag{1-7}$$

$$r_3 = \frac{dC_P}{dt} = k_3 C_{RC} \tag{1-8}$$

The analysis is carried further by application of one of the frequently used simplifying assumptions of chemical kinetics, called the steady-state approximation. Assume that the concentration of the intermediate RC, after a short induction period, achieves a nearly time-independent value, that is, $dC_{RC}/dt \approx 0$. This approximation is most nearly correct when the intermediate is highly reactive and short-lived, which implies that C_{RC} is small.

A mass balance on RC (i.e., a statement of the principle of conservation of mass) gives the following:

$$\frac{dC_{RC}}{dt} = k_1 C_R C_C - k_2 C_{RC} - k_3 C_{RC} \tag{1-9}$$

Invoking the steady-state approximation and rearranging the equation leads to

$$C_{RC} = \frac{k_1 C_R C_C}{k_2 + k_3} \tag{1-10}$$

and, from Eq. 1-8,

$$r_3 = \frac{k_1 k_3 C_R C_C}{k_2 + k_3} = k' C_R C_C \tag{1-11}$$

This result shows that for the simple catalytic cycle of Fig. 1-2, the reaction is first order in the reactant concentration and first order in the catalyst concentration. These dependencies are especially simple and easy to remember. A convenient measure of the catalytic activity is the second-order rate constant k', which is identified in Eq. 1-11 as a function of the rate constants of the three elementary steps. Alternatively, the catalytic activity could be defined as the rate of product formation r_3, but this activity is not so useful, being dependent not only on temperature but also on the concentrations of reactant and catalyst; it is often convenient to use the rate, however, since less experimental effort is required to determine a rate than a rate equation and, thereby, a rate constant.

The example just considered is one of the simplest imaginable catalytic cycles, and the reaction kinetics is correspondingly simple. Consideration of a slightly more complicated cycle (Fig. 1-3) leads to kinetics more nearly representative of many catalytic reactions. The elementary steps are the following:

$$R_1 + C \xrightarrow{k_{12}} R_1C \tag{1-12}$$

$$R_1C \xrightarrow{k_{13}} R_1 + C \tag{1-13}$$

$$R_1C + R_2 \xrightarrow{k_{14}} P + C \tag{1-14}$$

A mass balance on R_1C gives the following:

$$\frac{dC_{R_1C}}{dt} = k_{12}C_{R_1}C_C - k_{13}C_{R_1C} - k_{14}C_{R_1C}C_{R_2} \tag{1-15}$$

Application of the steady-state approximation as before gives

$$C_{R_1C} = \frac{k_{12}C_{R_1}C_C}{k_{13} + k_{14}C_{R_2}} \tag{1-16}$$

and the rate of product formation is

$$r = r_{14} = \frac{dC_P}{dt} = \frac{k_{12}k_{14}C_{R_1}C_{R_2}C_C}{k_{13} + k_{14}C_{R_2}} \tag{1-17}$$

It is convenient to cast the rate equation in another form, one chosen to be more useful for the interpretation of typical experimental results. Often rates are determined in a series of experiments, whereby the concentrations of reactants and catalyst are varied one at a time in a systematic way. (A diluent may

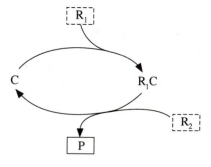

Figure 1-3
Hypothetical catalytic cycle for the reaction $R_1 + R_2$ → P. Note similarity to the cycle of Fig. 1-2.

be used in the reactant mixture to allow this variation.) For simplicity, rates are determined by extrapolation to initial conditions, so that initial rates are determined as a function of initial concentrations—this strategy eliminates complicating effects of reaction products on rates, such as inhibition. (Products can be present initially in known concentrations to allow determination of the dependence of rate on product concentrations.) The initial concentrations, designated by the subscript 0, are the following, provided that the formation of R_1C is virtually instantaneous:

$$C_{C_0} = C_{C_t} - C_{R_1C} \tag{1-18}$$

where C_{C_t} is the concentration of all forms of the catalyst (the total concentration), calculated from the amount of added catalyst and the initial solution volume. Furthermore,

$$C_{R_1} = C_{R_{10}} - C_{R_1C} \tag{1-19}$$

and

$$C_{R_2} = C_{R_{20}} \tag{1-20}$$

When these equations are combined with the mass balance, Eq. 1-15, and when the steady-state approximation is invoked and the term in C_{RC} is ignored (because C_{RC} is a relatively small term in comparison with the other concentrations), then the following equation for the initial rate results:

$$r_0 = \frac{k_{12}k_{14}C_{R_{10}}C_{R_{20}}C_{C_0}}{k_{13} + k_{12}(C_{R_{10}} + C_{C_0}) + k_{14}C_{R_{20}}} \tag{1-21}$$

The catalyst concentration C_{C_t} is usually small in comparison with the reactant concentration C_{R_1}, and so, to a good approximation,

$$r_0 = \frac{k_{12}k_{14}C_{R_{10}}C_{R_{20}}C_{C_0}}{k_{13} + k_{12}C_{R_{10}} + k_{14}C_{R_{20}}} \tag{1-22}$$

One important special case occurs in the limit as $k_{14}C_{R_{20}}$ becomes much greater than $(k_{13} + k_{12}C_{R_{10}})$, which implies that the intermediate RC is converted into product with relative rapidity—it is drained off into product almost as fast as it is formed, and its concentration attains only a low value. Correspondingly, Eq. 1-22 simplifies to

$$r_0 = k_{12}C_{R_{10}}C_{C_0} \tag{1-23}$$

In this simple limiting case, the form of the rate equation is the same as that of the preceding example, Eq. 1-11.

The second important limiting case occurs if the above inequality is reversed, that is, if $k_{14}C_{R20} \ll (k_{13} + k_{12}C_{R10})$:

$$r_0 = \frac{k_{12}k_{14}C_{R10}C_{R20}C_{C0}}{k_{13} + k_{12}C_{R10}} \qquad (1\text{-}24)$$

Again, the rate is proportional to the concentration of reactant R_2 and to the concentration of catalyst; but the dependence of rate on the concentration of R_1 is more complex, as is illustrated in Fig. 1-4. At low concentrations of R_1, the rate is proportional to this concentration; but at increasingly high concentrations, the rate approaches a limiting or saturation value. This is called *saturation kinetics* and is observed frequently in catalysis; it is usually referred to as Michaelis–Menten kinetics, after the two investigators who derived the expression for enzyme-catalyzed reactions in 1913 [2].

Alternatively, the kinetics illustrated in Fig. 1-4 could be explained as follows: At low concentrations of R_1, the catalyst is predominantly in the form of C rather than R_1C; and adding R_1 increases the concentration of R_1C in proportion to the concentration of added R_1. At higher concentrations of R_1, the balance shifts and R_1C predominates over C. In the limit, virtually all the catalyst is present as R_1C, so that addition of R_1 does not increase the concentration of R_1C; that is, the catalyst is "saturated" with reactant, and so the addition of R_1 does not increase the rate. In principle, the reverse situation could occur, in which the reactant could be saturated by catalyst and the reaction rate could become independent of catalyst concentration. This situation is observed infrequently because a large excess of reactant over catalyst is almost always used.

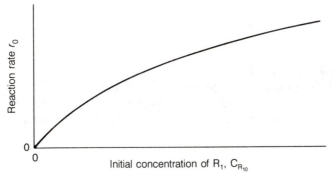

Figure 1-4
The illustration of saturation kinetics: the reaction rate is proportional to the reactant concentration at low values and independent of the reactant concentration at high values.

EXAMPLE 1-1   Catalytic Cycle for Ozone Depletion
Problem
Chlorofluorocarbons are used as refrigerants and in aerosols. Released on a large scale into the atmosphere, these chemicals are causing destruction of the ozone layer that protects us by absorbing ultraviolet radiation. The elementary steps in ozone destruction involve chlorine atoms (formed from the chlorofluorocarbon) and oxygen atoms:

$$Cl^{.} + O_3 \rightarrow O_2 + ClO \tag{1-25}$$

$$ClO + O^{.} \rightarrow O_2 + Cl^{.} \tag{1-26}$$

What is the catalytic cycle? What is the form of the rate equation?

Solution
The cycle is simply

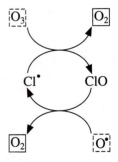

If the steady-state approximation is made, it follows that the rate of step 1-25 is equal to the rate of step 1-26, which is equal to the overall rate of the reaction:

$$r = k_{25} C_{Cl^{.}} C_{O_3} \tag{1-27}$$

■

Sometimes in the analysis of the kinetics of a sequence of steps, it is assumed that one of the steps is **rate determining**. This assumption is equivalent to stating that all the steps in the sequence other than the rate-determining step are so fast that they can be considered to be in virtual equilibrium. For example, in the sequence

$$A \rightleftharpoons B \tag{1-28}$$

$$B \rightarrow C \tag{1-29}$$

if the first step is in virtual equilibrium and the second is rate determining, then an equilibrium expression is written for the first:

$$\frac{C_B}{C_A} = K_{28} \tag{1-30}$$

where K_{28} is the concentration equilibrium constant for the reaction of Eq. 1-28. The rate of the sequence $A \rightarrow B \rightarrow C$ is then

$$r_{29} = k_{29}C_B = k_{29}K_{28}C_A \tag{1-31}$$

The key assumption is that the concentration of B is determined by its equilibrium with A and that all the resistance to the transformation is in the conversion of B to C. This assumption is more restrictive than the assumption that B is a highly reactive intermediate that rapidly attains a steady-state concentration, as illustrated by the following example.

EXAMPLE 1-2 A Simple Catalytic Cycle with a Rate-Determining Step
Problem
Consider the catalytic cycle represented by Eqs. 1-1 through 1-3. Assume that the third step is rate determining and find the rate equation; compare it with the rate equation based on the assumption of the steady-state approximation for the intermediate RC.

Solution
The rate-determining step (often called the *slow step*) is the third one; therefore, the rate is

$$r = r_3 = k_3 C_{RC} \tag{1-6}$$

The concentration of RC can be related to the concentrations of R and C by the equilibrium relation $C_{RC} = K_1 C_R C_C$, and the rate is therefore

$$r = k_3 K_1 C_R C_C \tag{1-32}$$

Since Eq. 1-2 represents the reverse of Eq. 1-1, the equilibrium constant for the reaction of R with C is $K_1 = k_1/k_2$ (can you show this?). Therefore, the rate equation can be written as

$$r = \frac{k_1 k_3 C_R C_C}{k_2} \tag{1-33}$$

This is just the form taken by Eq. 1-11, based on the steady-state approximation, in the limit as k_3 becomes very small in comparison with k_2. What is the physical meaning of this limiting case? ∎

Rates of catalytic reactions vary by many orders of magnitude, some being too low to be measured and some too high. But a good rule of thumb suggested by Weisz [3] is that a useful catalytic reactor must transform at least of the order of 10^{-6} mol/s for every cubic centimeter of volume occupied by catalyst. It is convenient to remember this as a rate of the order of 1 mole per liter per hour.

REFERENCES

1. Tolman, C. A., *J. Am. Chem. Soc.,***92,** 6777 (1970).
2. Michaelis, L., and Menten, M. L., *Biochem. Z.,* **49,** 333 (1913).
3. Weisz, P. B., *CHEMTECH,* **3,** 498 (1973).

FURTHER READING

To understand catalysis, the reader should know the principles of reaction kinetics. The introductory accounts given in standard physical chemistry textbooks are recommended for review. More complete treatments are given in *Kinetics of Chemical Processes*, by M. Boudart (Prentice Hall, 1968), which gives a condensed statement of principles; and in *Kinetics and Mechanism*, by J. W. Moore and R. G. Pearson (Wiley-Interscience, 1981), which gives a readable account of reactions in solution and includes many examples. The former is a good statement of concepts such as the steady-state approximation and the rate-determining step.

The subject of reaction mechanism is covered in modern textbooks of organic and inorganic chemistry. Further details at an introductory level are given, for example, in the paperbound *Organic Reaction Mechanisms*, by R. Breslow (Benjamin, 1969), and in *Mechanisms of Inorganic Reactions in Solution—An Introduction*, by D. Benson (McGraw-Hill, 1968).

The early history of catalysis is summarized in the book *Kurze Geschichte der Katalyse in Praxis und Theorie*, by A. Mittasch (Julius Springer, Berlin, 1939).

PROBLEMS

1-1   Consider the following steps in autoxidation of a hydrocarbon RH:

Initiation

$$In_2 \rightarrow 2In^{\cdot} \qquad (1\text{-}34)$$

$$In^{\cdot} + RH \rightarrow InH + R^{\cdot} \qquad (1\text{-}35)$$

Propagation

$$R^{\cdot} + O_2 \rightarrow ROO^{\cdot} \qquad (1\text{-}36)$$

$$ROO^{\cdot} + RH \rightarrow ROOH + R^{\cdot} \qquad (1\text{-}37)$$

Termination

$$R^. + ROO^. \rightarrow ROOR \qquad (1\text{-}38)$$

Use the steady-state approximation and establish a rate equation for the rate of oxidation of RH. Represent the process in terms of a cycle.

1-2 Consider the following hypothetical sequence of steps:

Initiation

$$ROOH + Co^{2+} \rightarrow RO^. + Co^{3+} + OH^- \qquad (1\text{-}39)$$

$$ROOH + Co^{3+} \rightarrow RO_2^. + Co^{2+} + H^+ \qquad (1\text{-}40)$$

Propagation

$$R^. + O_2 \rightarrow RO_2^. \qquad (1\text{-}41)$$

$$RO_2^. + RH \rightarrow R^. + ROOH \qquad (1\text{-}42)$$

Termination

$$2RO_2^. \rightarrow \text{stable products} \qquad (1\text{-}43)$$

Use the steady-state approximation (assume that the concentration of each free-radical species is independent of time) and determine a rate equation for the formation of ROOH in terms of concentrations of stable species. Represent the process by using the convention of cycles illustrated in the chapter.

1-3 The hydrolysis of urea is catalyzed by the enzyme urease. From the following data (obtained at constant urease concentration) [Kistiakowski, G. B., and Rosenberg, A. J., *J. Am. Chem. Soc.*, **74**, 502 (1952)], determine an appropriate rate equation and evaluate the parameters.

Concentration of Urea, mol/L	Initial Reaction Rate, mol/(L·s)
0.00032	0.130
0.00065	0.226
0.00129	0.362
0.00327	0.600
0.00830	0.846
0.0167	0.975
0.0333	1.03

1-4 When the product of a reaction is a catalyst for that reaction, the reaction proceeds autocatalytically. Derive an equation for the conversion in the irreversible reaction

$$R \rightarrow P \tag{1-44}$$

assuming that P is a catalyst and that the reaction takes place at constant temperature and volume in a batch reactor. Assume that k is the first-order rate constant for the uncatalyzed reaction ($r = kC_R$) and k_{cat} a second-order rate constant for the catalyzed reaction ($r = k_{cat} C_R C_P$). What is the shape of the conversion-vs.-time plot?

1-5 Consider the following sequence of steps in a catalytic cycle:

$$R_1 + C \rightleftharpoons R_1C \text{ fast} \tag{1-45}$$

$$R_1C + R_2 \rightleftharpoons P + C \text{ slow} \tag{1-46}$$

where the first step is in virtual equilibrium and the second, therefore, slow (or rate determining). What is the form of the overall kinetics and how does it compare with that obtained when the less restrictive assumption of the steady-state approximation is made?

1-6 When the product of a catalytic reaction bonds to the catalyst in competition with the reactant, the reaction is said to be slowed down by product inhibition. Consider the following sequence of steps illustrating this phenomenon:

$$R + C \rightarrow RC \tag{1-47}$$

$$RC \rightarrow R + C \tag{1-48}$$

$$RC \rightarrow P + C \tag{1-49}$$

$$P + C \rightarrow PC \tag{1-50}$$

$$PC \rightarrow P + C \tag{1-51}$$

Derive the kinetics for this sequence, assuming, first, that the third step is rate determining (slow) and, second, that the steady-state approximation is valid for C_{RC}. How do the equations express the product inhibition?

Catalysis in Solutions

2.1
INTRODUCTION

Catalysis takes place in solutions, on surfaces, and in environments that are microscopic phases, such as molecular-scale cages. Catalysis in solutions is the best understood, and therefore the logical starting point, because of the uniformity of the catalytic species and of the environment around the species participating in a reaction. The structures of reacting species, possibly including short-lived intermediates, may be determined by methods such as infrared, ultraviolet, and nuclear magnetic resonance (NMR) spectroscopy, and the dependence of rates of reaction on the concentrations of reactants and catalysts may be relatively easy to determine.

To understand how catalysts function in solutions, it is helpful to consider the organization of molecules in solutions. The fluids with the simplest structure are gases; in gases of low density, the interactions between molecules are negligible except when molecules in random motion collide. A collision between two molecules may result in no more than a transfer of energy between them, but occasionally it may lead to the forming or breaking of chemical bonds or both.

Collision frequencies are predicted by the kinetic theory of gases, which forms the basis for the simple collision theory of reaction rates: the rate of reaction r is assumed to be equal to a fraction of the frequency of collisions:

$$r = Pr_{\text{coll}}e^{-E_{\text{act}}/RT} \tag{2-1}$$

This equation incorporates the familiar exponential (Arrhenius) dependence of rate on absolute temperature. The exponential term indicates that only a fraction of the collisions involve molecules with sufficient energy to allow reaction. The term E_{act} is the activation energy, the energy required for the reacting molecules to pass over the energy barrier characteristic of the reaction. The fraction P expresses a geometric requirement for reaction: even if the colliding molecules have sufficient energy, reaction will occur only if they are aligned properly. The collision theory is of little predictive value except for some simple gas phase reactions.

A more realistic theory of reaction kinetics than the collision theory is the transition state theory originated by Henry Eyring and co-workers. In this theory, it is assumed that the reactants form a fleeting intermediate referred to as a transition state or activated complex (denoted by the symbol ‡), which reacts to give the products. It is assumed that there is a single potential energy surface on which the components will move as the reactants and products are interconverted. There is one path on this surface that requires the least energy for the transformation; this is the path over the lowest pass on the energy surface. The position along this path is called the *reaction coordinate*. The position of highest energy on the reaction coordinate is that of the transition state. The transition state is assumed to be in equilibrium with the reactants, and the rate of reaction is taken to be proportional to the concentration of this highly reactive (and hypothetical) intermediate. The rate constant, according to the theory in its simplest form, is universally $\underline{k}T/h$ (where $\underline{k}$ is Boltzmann's constant, T is the absolute temperature, and h is Planck's constant), the frequency of an unstable vibrational mode in the transition state. When one has a good idea of the structure of the transition state, one can use estimates of thermodynamic properties to determine the equilibrium constant for the reaction by which it is formed from the reactants, and thereby estimate the rate. For most reactions of any significant complexity, the theory is too simple and the identification of the transition state too uncertain to allow good quantitative estimates of reaction rates. Except for the simplest reactions (e.g., of small gas phase molecules and radicals), the only reliable way to determine rates of reactions is to measure them.

As a gas phase becomes more dense, ultimately approaching the character of a liquid, the interactions between molecules become more frequent and complex, but the motions are still rapid and essentially unrestricted. No longer is it sufficient to consider only two-body collisions. The motions of molecules are no longer random, and the fluid becomes weakly structured. The structure is far from the ordered array of a crystalline solid, however; in rough terms it may correspond to a clustering of molecules. For example, a solution of ions in a solvent may be described as a clustering of solvent molecules forming solvation shells around the ions, as sketched in Fig. 2-1. The shell is in dynamic equilibrium with the surrounding liquid, and solvent molecules rapidly enter and leave the neighborhood of an ion.

Chemical reactions in a liquid involve ions and molecules as reactants and catalysts. Since solvent–solute interactions are often strong, solvents often influence catalysis strongly, affecting the interactions between reactants and catalysts. Consequently, rates of catalytic reactions can be regulated by the choice of a solvent.

The most common solvent is water. Water is unique because the molecules are small and strongly polar, and water therefore dissolves many ionic

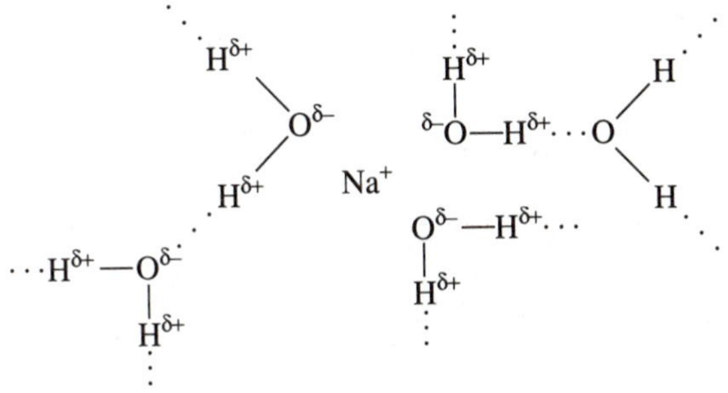

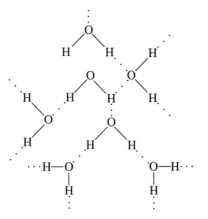

Figure 2-1
A cluster of solvent water molecules around an Na⁺ ion aqueous solution.

and polar compounds. The structure of liquid water is determined by the hydrogen bonds between the molecules, as illustrated in Fig. 2-2. Hydrogen bonds are not strong (the magnitude of the enthalpy of formation of a hydrogen bond is about 5 kcal/mol), but each water molecule can participate in several such bonds simultaneously. The hydrogen bonds in pure water are like those in aqueous solutions of alcohols and carboxylic acids, among other compounds (Fig. 2-3).

Solutions containing ions may also contain ion pairs and weakly organized environments of polar molecules surrounding positively and negatively charged ions. Nonionic solutions also have structure. For example, polar molecules have an affinity for other polar molecules; hydrogen-bonded structures like those of Fig. 2-3 are typical. The protons in these structures are highly mobile, which is of great importance in catalysis proceeding via cycles of proton transfer steps. When polar solute molecules such as CH_3CO_2H are present in nonpolar solvents, the hydrogen bonding between them is pronounced, since the solute molecules themselves are the only candidates for polar interactions. Sometimes the affinities of molecules for each other are so strong that they form a separate phase, which may be an ordinary liquid but which also may be a microphase, such as a micelle (described in Section 2.7 below).

Acid–base catalysis is the most thoroughly developed part of solution

Figure 2-2
Hydrogen-bonded structure of water.

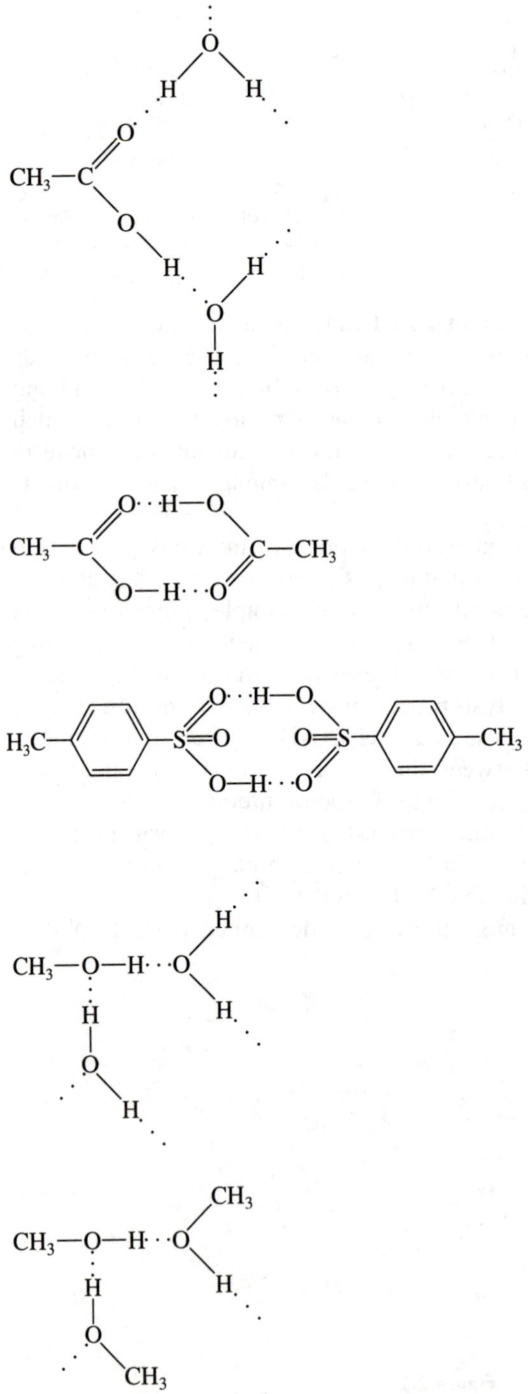

Figure 2-3
Some structures existing in liquids which involve hydrogen bonds.

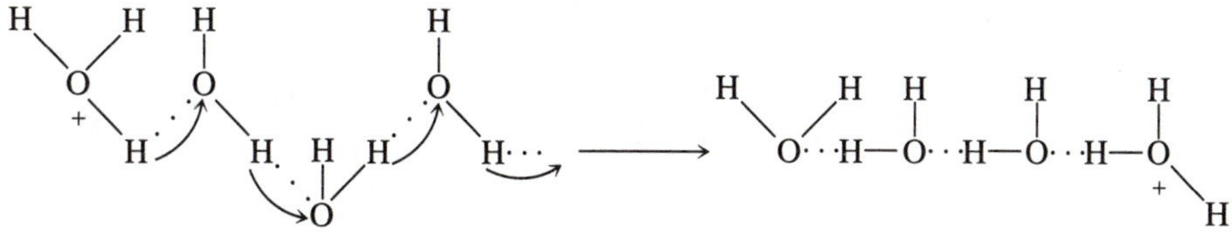

Figure 2-4
A schematic representation of proton transfer in a hydrogen-bonded chain
of water molecules.

catalysis and therefore a major subtopic of this chapter. The essential steps of
acid–base catalysis are often proton transfers. The protons in hydrogen-bonded
structures are highly mobile because the protonic mass is so low. Consequently,
there is a rapid interconversion of related structures as a proton moves; in
water, for example, a proton moves from the O atom of one molecule toward
the O atom of a neighbor, and this motion is compensated by the motion of
other protons to and from the corresponding O atoms, as shown in Fig. 2-4.
The rapid proton jumping accounts for the electrical conductivity of water and
of ice. Many reactions catalyzed by acids and bases in water proceed via such
proton transfers, and the catalytic cycles often involve water. Many of the
transformations of acid–base catalysis are analogous to the structural rearrangements shown in Fig. 2-4.

2.2
ACID–BASE CATALYSIS

2.2.1 Catalysis in the Gas Phase

Catalysis by acids and bases proceeds via cycles involving proton transfer
reactions, and solvent molecules like water often play a role. Reactions can
be investigated in the absence of complicating solvent effects if the catalyst
and reactants are maintained in a gas phase. An example of acid catalysis in
the gas phase is the dehydration of alcohol in the presence of HBr molecules.
The reaction has been suggested to proceed via the cycle shown in Fig. 2-5.
Both bond breaking and bond forming involve the motion of protons. The
catalyst donates a proton to the reactant at one position and retrieves a proton
from another position. As a consequence of the proton donation, the alcohol
becomes more reactive for the splitting out of a water molecule and a proton.

2.2.2 Catalysis in Dilute Aqueous Solution

Gas phase reactions catalyzed by acids and bases are unusual. Reactions catalyzed by acids and bases in aqueous solution, however, are common. The
first quantitative demonstration of catalytic action [2] was based on measurements of rates of sucrose inversion in the presence of dilute aqueous mineral
acids. In 1850 Wilhelmy [2] reported the rate to be proportional to the con-

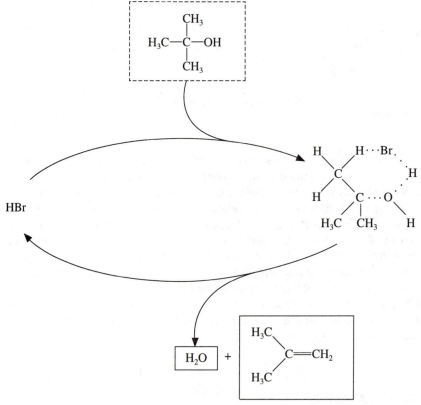

Figure 2-5
A postulated catalytic cycle for the dehydration of *t*-butyl alcohol in the gas
phase in the presence of HBr [1].

centration of acid and independent of which acid was used, provided that it
was a dilute strong acid, that is, one fully dissociated in water. These early
catalytic results were important in the elucidation of the nature of acids and
of their dissociation in water, and in beginning to place catalysis on an emerging
foundation of reaction kinetics. The results demonstrated that the sucrose inversion reaction was catalyzed by protons in the aqueous solution, which were
produced equally from all the strong acids.

Since Wilhelmy, many reactions have been found to be catalyzed by dilute
aqueous solutions of strong acids; these include ester hydrolysis,

$$R{-}\overset{\displaystyle O}{\underset{\displaystyle OR'}{C}} \ +\ H_2O \rightarrow R{-}\overset{\displaystyle O}{\underset{\displaystyle OH}{C}} \ +\ R'OH \qquad (2\text{-}2)$$

and its reverse, esterification, as well as the dehydration of tertiary alcohols, which was mentioned above to exemplify catalysis by HBr in the gas phase,

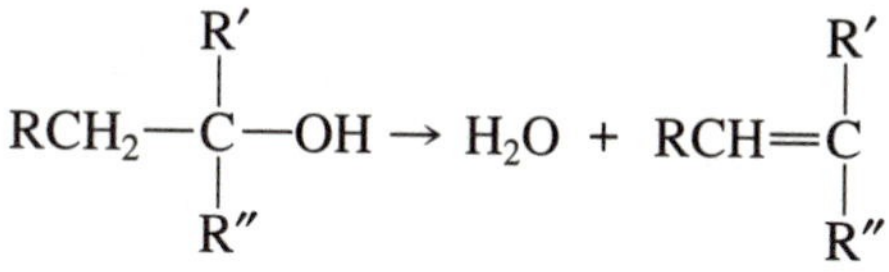

$$RCH_2\!-\!\overset{\displaystyle R'}{\underset{\displaystyle R''}{C}}\!-\!OH \rightarrow H_2O \ + \ RCH\!=\!\overset{\displaystyle R'}{\underset{\displaystyle R''}{C}} \tag{2-3}$$

There are many related reactions involving organic compounds containing N, O, and S atoms, all of which are good proton acceptors.

The mechanism that has been inferred for such a reaction is illustrated in Fig. 2-6. The reactants are

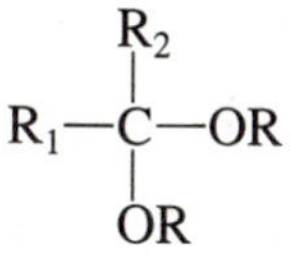

$$R_1\!-\!\overset{\displaystyle R_2}{\underset{\displaystyle OR}{C}}\!-\!OR$$

and water. The protons are consumed in one step of the catalytic cycle and released in another. The mechanism shown here is an oversimplification, because the hydration shells around the species in solution are not represented. The proton, for example, is actually present in H_3O^+, $H_5O_2^+$, $H_7O_3^+$, and related species in equilibrium with the hydrogen-bonded solution.

As is typical of these reactions, the first step is a protonation of the reactant, which is a base. The reaction in the absence of an acid would require the displacement of an alkoxide, which is a poor leaving group. The protonation of the oxygen provides a much better leaving group, that is, an alcohol (Fig. 2-6). In other words, the alkoxide is a much stronger base than an alcohol and therefore a poorer leaving group.

The reaction mechanism is consistent with the observed kinetics, namely,

$$r = k_4 C_{H^+} C_{\text{acetal}} \tag{2-4}$$

This result can be shown easily: the decomposition of the protonated acetal is slow and may be assumed to be rate determining, which implies that the preceding step in the cycle may be considered to be in virtual equilibrium. Therefore,

$$K_5 = \frac{C_{\text{protonated acetal}}}{C_{H^+} C_{\text{acetal}}} \tag{2-5}$$

From the rate-determining step, the rate of the overall cycle is determined to be

$$r = k_6 C_{\text{protonated acetal}} \tag{2-6}$$

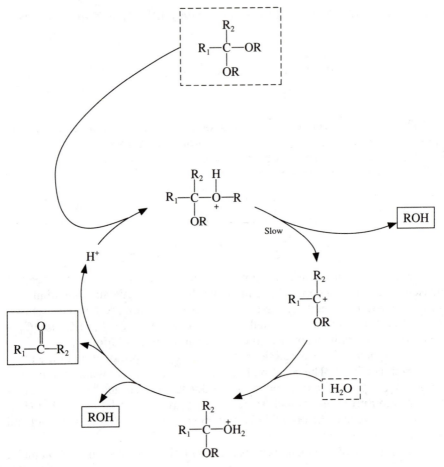

Figure 2-6
Catalytic cycle for acetal hydrolysis in aqueous acid solution [3, 4].

where k_6 is the rate constant for this slow step. Therefore,

$$r = k_6 K_5 C_{H^+} C_{acetal} \qquad (2\text{-}7)$$

which has the form of the observed kinetics.

The kinetics has been inferred here from the known mechanism, but a much more difficult and common challenge is to infer a mechanism from experimentally determined kinetics—and complementary results. The mechanism is almost never determined unequivocally by the kinetics. This point is illustrated by the present example. If the rate-determining step were, for instance, the protonation of the acetal, then the rate equation would take the same form as that above,

$$r = k_8 C_{H^+} C_{acetal} \qquad (2\text{-}8)$$

where k_8 is the rate constant for the acetal protonation step. More than one possible mechanism of the catalytic reaction is consistent with the form of the empirical rate equation for the overall catalytic cycle. The conclusion is general: determination of reaction mechanisms requires much more than determination of kinetics.

The rate equation mentioned above is of a form observed for many reactions of good proton acceptors catalyzed by dilute aqueous solutions:

$$r = kC_{H^+}C_{reactant} \tag{2-9}$$

Often, however, the kinetics becomes more complicated as the solution becomes more concentrated in acid. For example, data are shown in Fig. 2-7 for the hydrolysis of benzamide in solutions of HCl and solutions of H_2SO_4. At low acid concentrations, the curves are linear and coincide (as would be expected from Eq. 2-9); but at high concentrations, they deviate from each other, and the reaction rate passes through a maximum.

These results can be explained as follows [5]. The reaction cycle includes the sequence

$$
R\!-\!\underset{NH_2}{\overset{O}{\underset{\|}{C}}} + H^+ \rightleftharpoons (R\!-\!\overset{OH}{\underset{|}{C}}\!-\!NH_2)^+ \xrightarrow{H_2O} \text{hydrolysis products} \tag{2-10}
$$

At low acid concentrations addition of acid leads to an increased concentration of the protonated intermediate and to a higher overall reaction rate. But at higher concentrations, the predominant effect of added acid is to decrease the

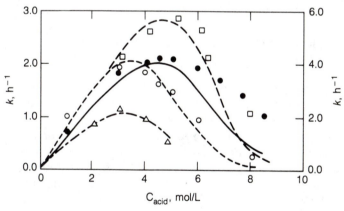

Left-hand scale:
$\left\{\begin{array}{l}\text{benzamide in HCl} \bullet \\ \text{benzamide in } H_2SO_4 \circ \\ p\text{-methoxybenzamide in } H_2SO_4 \triangle\end{array}\right.$

Right-hand scale: p-nitrobenzamide in H_2SO_4 □

Figure 2-7
The effect of acid concentration on the rate constant for benzamide hydrolysis [5].

reactivity of water, which is a reactant in the second step; consequently, the rate then decreases with increased acid concentration.

EXAMPLE 2-1   The Mechanism of the Phenol–Acetone Condensation Reaction to Give Bisphenol A: the Role of a Catalyst Promoter

Problem

A commercial process for the manufacture of bisphenol A (2,2-bis-4-hy-droxyphenylpropane), an important intermediate in the manufacture of epoxy resins and polycarbonates, involves the acid-catalyzed conden-sation reaction of phenol and acetone. The reaction kinetics has been reported to be the following (where A refers to acetone and P to phenol) [6]:

$$r = kC_{H^+}C_AC_P \tag{2-11}$$

and, alternatively [7],

$$r = kC_{H^+}C_AC_P^2 \tag{2-12}$$

Propose a reaction mechanism to account for the kinetics.

It has also been shown that compounds such as mercaptans are promoters for the acid-catalyzed reaction. A **promoter** is something which, although lacking catalytic activity itself, increases the activity (or possibly the selectivity or stability) of a catalyst when used in combination with it. Thioglycolic acid (HS—CH$_2$—CO$_2$H) (T) is a promoter in the bisphenol A synthesis, the kinetics being as follows [8]:

$$r = kC_{H^+}C_AC_P[k_1 + k_2C_T] \tag{2-13}$$

Propose an explanation for the role of the promoter.

Solution

The reaction is expected to proceed via protonation of a suitable base, the likely candidate being acetone. The protonated acetone is expected to be involved in a typical electrophilic aromatic substitution reaction with phenol. Such reactions are common in organic chemistry, for ex-ample, in the bromination or alkylation of aromatic compounds. The prod-uct is a tertiary alcohol, which can be easily protonated and dehydrated, and the product can be involved in another aromatic substitution. This sequence is summarized in the cycle of Fig. 2-8. The kinetics represented by Eq. 2-11 is accounted for if the step marked ① is rate determining, and the kinetics represented by Eq. 2-12 is accounted for if the step marked ② is rate determining. Can you demonstrate this?

Equation 2-13 implies that in the presence of the thioglycolic acid promoter, there are two independent catalytic cycles proceeding at sig-nificant rates. One of these (indicated by the first term on the right-hand

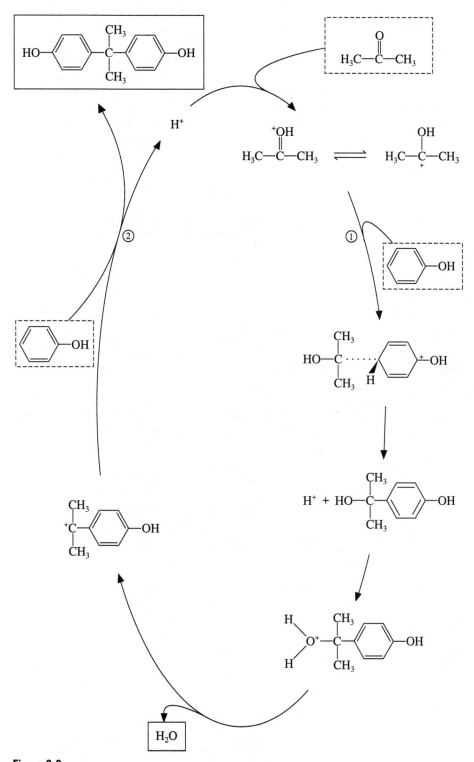

Figure 2-8
Catalytic cycle for the synthesis of bisphenol A from phenol and acetone.

side of Eq. 2-13) can be identified as the cycle of Fig. 2-8. The other cycle involves the promoter, and it is expected that the promoter functions by bonding to the reactants and/or catalyst. A good explanation of the promoter's role follows from the suggestion that it reacts to form more stable ionic intermediates than the carbenium ions shown in Fig. 2-8.

The following species, for example, might be formed:

$$\underset{H_3C-\overset{+}{C}-CH_3}{\overset{\overset{\displaystyle SCH_2CO_2H}{|}}{}}$$

and it is easy to write a catalytic cycle comparable to that of Fig. 2-8. What is this cycle, and what rate-determining step explains the second term on the right-hand side of Eq. 2-13?   ∎

2.2.3 General and Specific Acid and Base Catalysis

Thus far in this discussion of catalysis by aqueous solutions of acids, only hydronium ions (including H_3O^+, $H_5O_2^+$, $H_7O_3^+$, etc., but written as H_3O^+ for simplicity) have been explicitly identified as proton donor species. Often, however, other species in solution, such as undissociated acids, HA, can as well be the proton donors. The data shown in Fig. 2-9 illustrate the point: in the pH range from 6.5 to 8, the slope of the line is -1, which implies that the observed pseudo first-order rate constant k_{obs} is proportional to the concentration of hydronium ions. In the pH range from about 8 to 10, the slope of the line is nearly zero, which implies that the rate is independent of the concentrations of H_3O^+ and OH^-; separate results show that the rate is proportional to the concentration of a neutral species in this range. In the pH range from about 10 to 12, the rate is proportional to the concentration of hydroxide ions.

The data of this example are represented by a rate equation of the fol-

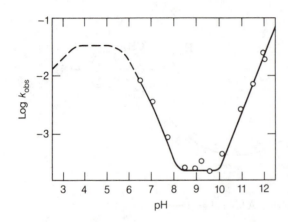

Figure 2-9
Dependence of the observed rate constant for oximation of acetone on pH at 25°C. The rate equation is $r = k_{obs} C_{acetone}$ [9].

lowing form:

$$r = k_{obs}C_R = [k_o + k_{H_3O^+}C_{H_3O^+} + k_{OH^-}C_{OH^-} + k_{HA}C_{HA} + k_{A^-}C_{A^-}]C_R \quad (2\text{-}14)$$

provided that the fourth and fifth terms inside the brackets are negligible. (The term k_o accounts for catalysis by water, which, when present in large excess, has a virtually constant concentration.) In the three pH ranges mentioned above, the second, first, and third terms, respectively, are predominant. There are many other examples consistent with this representation.

This development leads to several definitions: When the proton donor is H_3O^+ (or, more precisely, hydrated protons), the catalysis is referred to as **specific acid catalysis**. Most of the examples considered so far fall in this category. (What are the exceptions?) When the proton donors include other species, such as HA and H_2O, **general acid catalysis** occurs. Analogously, when the proton acceptor is OH^-, **specific base catalysis** occurs, and when the proton acceptors include other species, such as B and H_2O, **general base catalysis** occurs.

The occurrence of catalysis by undissociated acids (or bases) leads to a central question: How does the catalytic activity of an acid (or base) depend on the structure and properties of the acid (or base)? Such a question can be posed for many classes of catalysts besides acids and bases: How does the activity of a catalyst (or, more precisely, the reactivity of a species in a catalytic cycle) depend on some measurable properties of the catalyst? This question turns up frequently in the following pages because the answers provide a basis for predicting catalyst performance and selecting better catalysts.

Roughly speaking, the question can be answered as follows for general acid catalysis: the stronger the acid, the more active the catalyst. In other words, the greater the dissociation constant K_a,

$$K_a = \frac{a_{H^+}a_{A^-}}{a_{HA}} \quad (2\text{-}15)$$

the more active the catalyst HA. Often this statement is quantitatively expressed by the Brønsted equation

$$k_{HA} = G_{HA}K_a^{\alpha} \quad (2\text{-}16)$$

where G_{HA} is a constant characteristic of the acid, and α is a constant with a value between 0 and 1. Equation 2-16 indicates that a plot of log k_{HA} vs. log K_a should be linear with a slope of α; an example of such a Brønsted plot is shown in Fig. 2-10.

To repeat, general acid catalysis is often represented by the Brønsted relation which, for a series of catalysts (undissociated acids), relates the catalytic activity to a property (the acid strength):

$$\log k = \alpha \log K_a + \text{const} \quad (2\text{-}17)$$

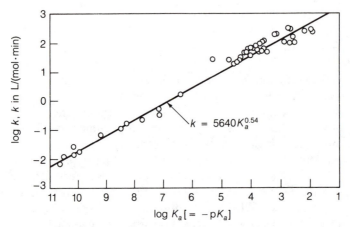

Figure 2-10
Brønsted plot for general acid catalysis of the dehydration of acetaldehyde hydrate. The catalysts include carboxylic acids and phenols. The rate equation is $r = kC_{HA} C_{acetaldehyde\ hydrate}$ [10].

where k is a second-order rate constant. In contrast, in the simplest examples of specific acid catalysis, the activity is the same for all the (strong) acids, since at a given concentration they all give the same concentration of H_3O^+. There is no equation for specific acid catalysis that is analogous to Eq. 2-17. Perhaps, the most nearly comparable expression is

$$\log k_{obs} = \log C_{H_3O^+} + \text{const} \tag{2-18}$$

where k_{obs} is a pseudo first-order rate constant.

The Brønsted relationship is often a good approximation in general acid catalysis and is therefore a good predictive tool, but it is empirical; it rests upon no sound "proof." A plausibility argument makes the Brønsted relationship more readily understandable, as follows:

Consider the general-acid-catalyzed reaction taking place by the following simple sequence:

$$HA + R \underset{}{\overset{K_{19}}{\rightleftharpoons}} RH^+ + A^- \tag{2-19}$$

$$RH^+ + A^- \underset{}{\overset{K_{20}}{\rightleftharpoons}} HA + I \tag{2-20}$$

$$I \longrightarrow P + H_2O \tag{2-21}$$

where the proton donor species is the undissociated acid HA, the reactant is R, the intermediate is I, and the products are P and H_2O.

If the second step is rate determining, then

$$r = k_{20}C_{RH^+}C_{A^-} = k_{20}K_{19}C_R C_{HA} \tag{2-22}$$

The observed rate constant, $k_{obs} = k_{20}K_{19}$, can be expressed in more familiar terms when a simplifying assumption is made, one frequently encountered in gas phase chemical kinetics but not often for liquid phase reactions of complicated molecules such as those of this example. The Polanyi relationship is assumed, which states that for elementary steps of a family of reactions, all involving geometrically similar reactants (so that steric effects are about the same for each), there is a simple relationship between the activation energies and the energies of reaction, as illustrated in Fig. 2-11:

$$E_{act} = \text{constant} - \alpha' q \tag{2-23}$$

where α' is a constant between 0 and 1.

This equation states a relationship between kinetics (the activation energy of reaction, E_{act}) and thermodynamics (the energy of reaction, q). The Polanyi relationship is only an occasionally valid generalization resting on nothing firmer than an empirical foundation.

The Brønsted relationship is developed for the sequence of elementary steps 2-19 through 2-21 by application of the Polanyi relationship to the rate-determining step:

$$k_{20} = \text{const } K_{20}^{\alpha'} \tag{2-24}$$

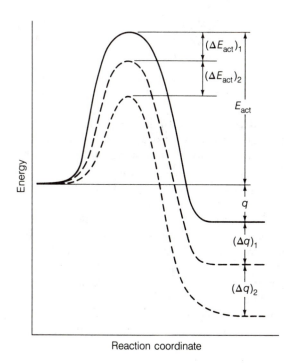

Reaction coordinate

Figure 2-11
Illustration of the Polanyi relationship for a family of elementary steps, all of which encounter the same steric restrictions.

(How does this statement follow from Eq. 2-23?) Substituting into Eq. 2-22 gives

$$r = \text{const } K_{19}K_{20}^{\alpha'}C_R C_{HA} \tag{2-25}$$

The equilibrium constants K_{19} and K_{20} are related to the acid dissociation constant K_a as follows:

$$K_{19} = \left(\frac{C_A - C_{RH^+}}{C_{HA}C_R}\right)_{\text{equil}} = \left(\frac{C_{H_3O^+}C_{A^-}}{C_{H_2O}C_{HA}}\right)_{\text{equil}} \left(\frac{C_{RH^+}C_{H_2O}}{C_R C_{H_3O^+}}\right)_{\text{equil}} = K_a \times \text{const} \tag{2-26}$$

This equality follows because the term in parentheses on the far right is a property of the reactant but not of the catalyst, and the goal of the development is a relationship between rate constant and acid strength for a series of acids catalyzing a particular reaction. Similarly,

$$K_{20} = \frac{\left(\dfrac{C_I C_{H_3O^+}}{C_{RH^+}C_{H_2O}}\right)_{\text{equil}}}{\left(\dfrac{C_{H_3O^+}C_{A^-}}{C_{H_2O}C_{HA}}\right)_{\text{equil}}} = \frac{\text{const}}{K_a} \tag{2-27}$$

Substituting into Eq. 2-22 gives the Brønsted relationship:

$$r = \text{const } K_a^{1-\alpha'}C_R C_{HA} \tag{2-28}$$

or

$$k = \text{const } K_a^{\alpha} \tag{2-29}$$

where $\alpha = 1 - \alpha'$.

Experimental results show that the Brønsted relationship is a good approximation for many catalytic reactions involving oxygen- or nitrogen-containing acids and bases as reactants or catalysts in aqueous solutions. The Brønsted relationship is one of a series of so-called linear free-energy relationships encountered in physical organic chemistry. They all relate reactivity (a rate constant) to an equilibrium constant.

When undissociated acids are used as catalysts in nonpolar (aprotic) solvents, linear Brønsted plots of log k vs. log K_a in the aprotic solution may also be found. But the kinetics of such a reaction is often complex, since a catalyst molecule cannot form hydrogen bonds with the solvent and therefore has a strong tendency to form hydrogen bonds with other catalyst molecules in competition with reactant molecules. The acid is therefore a proton donor and a proton acceptor in the same solution (cf. Fig. 2-3). The catalyst acts as a reaction inhibitor, even though it is still responsible for accelerating the reaction by protonating the reactant and thereby making it more reactive.

Other polar compounds, for example, water, can also play such multiple roles in acid–base catalysis, especially when they are present in low concentrations. Water can be a catalyst, serving as a proton donor (Eq. 2-14); it can be an inhibitor, accepting protons from an undissociated acid HA in competition with reactants; and it often solvates reactants and intermediates and assists in the proton transfer, accepting a proton from an acid molecule and more or less simultaneously donating a proton to a reactant—this process is like the one depicted in Fig. 2-4 for proton transport through water.

2.2.4 Catalysis in Concentrated Strong Acid Solutions

These points just begin to indicate the complexity of the kinetics of acid catalysis associated with the variety of interactions between proton donors and acceptors such as water and polar reactant molecules. Catalysis in highly concentrated strong-acid solutions is especially complex, in part because there are various proton donors and they are not all identified. It is not easy to measure the concentrations of the proton donors. Some quite general results have emerged, however, from correlations of catalytic activity with empirical measures of the solution's acidity. The *Hammett acidity function*, defined as a measure of the tendency of a solution to donate a proton to a neutral base, correlates with the catalytic activity of the solution in many examples of acid catalysis that involve protonation of neutral reactant molecules, as is illustrated in the following paragraphs.

In dilute aqueous solutions of strong acids, the proton donor strength of the solution is measured simply by the concentration of hydrated protons, given by the pH. This measure is not adequate for highly concentrated acid solutions in which the solution is no longer predominantly water, since the proton donors include numerous species other than hydrated protons. A more general measure of the proton-donor strength of strongly acidic solutions was developed by Hammett and Deyrup [11]. Solutions are characterized by their tendency to protonate neutral basic indicators (B):

$$H^+ + B \rightleftharpoons BH^+ \qquad (2\text{-}30)$$

The acid dissociation constant of BH^+ is, by definition,

$$K_a = \frac{a_{H^+}a_B}{a_{BH^+}} = \frac{a_{H^+}\gamma_B}{\gamma_{BH^+}} \frac{C_B}{C_{BH^+}} \qquad (2\text{-}31)$$

A ratio of the concentrations C_B/C_{BH^+} can be measured spectrophotometrically for a series of bases of different base strengths (i.e., tendencies to accept a proton). Hence, for a given solution, the term $a_{H^+}\gamma_B/\gamma_{BH^+}$ (called h_0) can be found experimentally if K_a is known. The negative of the logarithm of this term is defined as the Hammett acidity function H_0; therefore,

$$-\log K_a = pK_a = H_0 + \log \frac{C_{BH^+}}{C_B} \qquad (2\text{-}32)$$

In the limit of an infinitely dilute aqueous solution, a_{H^+} approaches C_{H^+} and the ratio of the activity coefficients γ_B/γ_{BH^+} approaches 1. Therefore, H_0 approaches the value of the pH as the solution approaches infinite dilution and the proton donor species become hydrated protons. In this limit,

$$pK_a = pH + \log \frac{C_{BH^+}}{C_B} \qquad (2\text{-}33)$$

The Hammett acidity function is just a more general definition than the pH of the solution's tendency to donate a proton to a neutral base. Values of H_0 for aqueous solutions of strong mineral acids are given in Fig. 2-12. There are significant differences between the familiar strong acids.[1]

Suppose that in a catalytic cycle taking place in an aqueous solution of a strong acid, the protonation of a neutral reactant molecule (base) is a kinetically significant step. One might then expect to find a relationship between the reaction rate and the Hammett acidity function. The expectation is borne out, for example, by the data of Fig. 2-13, which show that the logarithm of the rate of sucrose inversion is directly proportional to the acidity function $-H_0$ for a series of strong-acid solutions; the data show that the rate is proportional to the hydrogen ion concentration only in the limit as zero acid concentration is approached.

What explains this dependence of the rates of some acid-catalyzed reactions in aqueous solution on H_0? Consider the following mechanism as a hypothetical example:

$$R + H^+ \rightleftharpoons (RH^+)^\ddagger \rightarrow RH^+ \qquad (2\text{-}34)$$

$$RH^+ \rightleftharpoons H^+ + P \qquad (2\text{-}35)$$

According to the transition state theory,

$$r_{35} = \frac{kT}{h} C_{RH^+}^\ddagger \qquad (2\text{-}36)$$

and

$$K^\ddagger = \frac{a_{H^+} C_R}{C_{RH^+}^\ddagger} \frac{\gamma_R}{\gamma_{RH^+}^\ddagger} \qquad (2\text{-}37)$$

---

[1] Other acidity functions have been defined; for example, one such function is a measure of the tendency of a solution to donate a proton to a base with a charge of -1 [13].

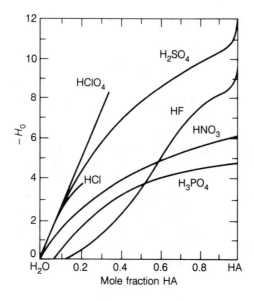

Figure 2-12
The Hammett acidity function H_0 for mineral acids [12].

Solving for $C_{RH^+}^\ddagger$ and substituting in Eq. 2-36 gives

$$r_{35} = \frac{kT}{hK^\ddagger} a_{H^+} C_R \frac{\gamma_R}{\gamma_{RH^+}^\ddagger} \tag{2-38}$$

Now, an important simplifying assumption is made on the basis of a supposed chemical similarity of B and R: $\gamma_R/\gamma_{RH^+}^\ddagger = \gamma_B/\gamma_{BH^+}$, where B refers to a Ham-

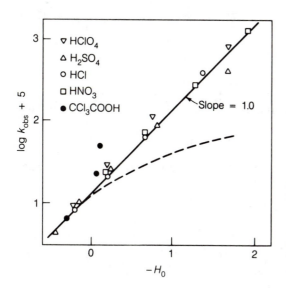

Figure 2-13
Dependence of the rate constants for sucrose inversion at 25°C on the Hammett acidity function [3]. Dashed line indicates the values expected if the rate were proportional to the concentration of H_3O^+. The rate equation is $r = k_{obs} C_{sucrose}$.

mett indicator base chosen to resemble R. Then,

$$r_{35} = \frac{kT}{hK^{\ddagger}} C_R h_0 \tag{2-39}$$

This result accounts for a linear dependence of the logarithm of the observed rate constant on the Hammett acidity function H_0.

There are many examples of acid-catalyzed reactions in concentrated strong-acid solutions for which the rate is a simple function of H_0, just as there are many examples for which the rate in a dilute solution is a simple function of the hydrogen ion concentration. Unfortunately, it is difficult to predict when either of these dependencies will prevail, and often neither one describes the data. Some of the complications arise from the various roles of water in the catalytic cycles; consideration of the hydration of various species leads to a more general representation [14], which is beyond the scope of this discussion and which still fails to unify a highly complex subject.

2.2.5 Catalysis by Bases

Thus far in this treatment of acid–base catalysis, the reactant has been the base, accepting a proton from the catalyst, HA, or H_3O^+, or other proton donors. The developments are similar when the roles are reversed and the catalyst is the base. Many reactions are catalyzed efficiently in either acidic or basic solution, as exemplified in Fig. 2-9 for the oximation of acetone. Consider one of the best understood of these reactions, the iodination of acetone. In acid solution, the reaction is believed to proceed via the catalytic cycle shown in Fig. 2-14. The sequence of elementary steps is written as

$$CH_3-\overset{\overset{\displaystyle O}{\|}}{C}-CH_3 + HA \underset{}{\overset{K_{40}}{\rightleftharpoons}} CH_3-\overset{\overset{\displaystyle \overset{H}{O^+}}{\|}}{C}-CH_3 + A^- \tag{2-40}$$

$$A^- + CH_3-\overset{\overset{\displaystyle \overset{H}{O^+}}{\|}}{C}-CH_3 \underset{}{\overset{k_{41}}{\rightleftharpoons}} CH_3-\overset{\overset{\displaystyle \overset{H}{O}}{|}}{C}=CH_2 + HA \tag{2-41}$$

The second step is rate determining, and the steps following it sum up to give

$$CH_3-\overset{\overset{\displaystyle \overset{H}{O}}{|}}{C}=CH_2 + I_2 \longrightarrow CH_3-\overset{\overset{\displaystyle O}{\|}}{C}-CH_2I + H^+I^- \tag{2-42}$$

and the overall reaction stoichiometry is

$$I_2 + CH_3-\overset{\overset{\displaystyle O}{\|}}{C}-CH_3 \longrightarrow CH_3-\overset{\overset{\displaystyle O}{\|}}{C}-CH_2I + H^+ + I^- \tag{2-43}$$

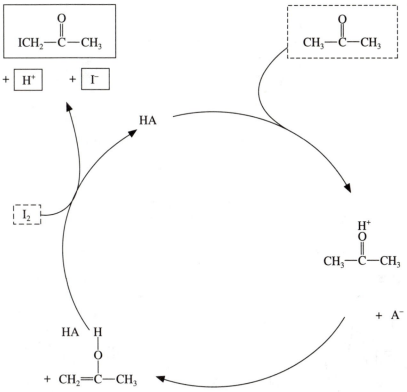

Figure 2-14
Catalytic cycle for general-acid-catalyzed iodination of acetone in an aqueous solution [15]. Arrows at left represent a sum of elementary steps rather than a single elementary step.

The kinetics is easily formulated as

$$r = k_{41}C_{A}-C_{\text{protonated acetone}} = k_{41}K_{40}C_{\text{acetone}}C_{\text{HA}} \tag{2-44}$$

The reaction is first order in one of the reactants (acetone) and zero order in the other (iodine). The appropriateness of the mechanism is confirmed by the result that other halogens react at the same rate as iodine. (What other measurable reactions would you expect to take place at this same rate?)

Contrast this mechanism with that prevailing in alkaline solution (Fig. 2-15). The mechanism of this general-base-catalyzed reaction can be represented as

$$\underset{\text{CH}_3-\overset{\text{O}}{\overset{\|}{\text{C}}}-\text{CH}_3 + \text{B}}{} \xrightarrow[\text{slow}]{k_{45}} \underset{\text{CH}_3-\overset{\text{O}}{\overset{\|}{\text{C}}}-\text{CH}_2^- + \text{BH}^+}{} \tag{2-45}$$

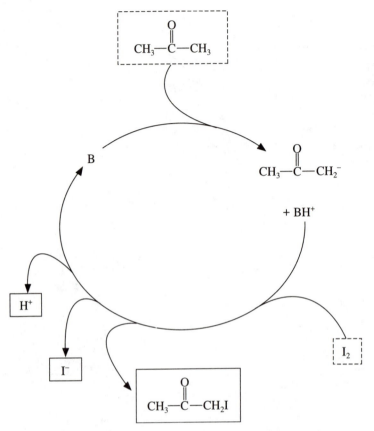

Figure 2-15
Catalytic cycle for general-base-catalyzed iodination of acetone [16].

$$CH_3-\overset{\overset{\displaystyle O}{\|}}{C}-CH_2^- + I_2 \rightarrow CH_3-\overset{\overset{\displaystyle O}{\|}}{C}-CH_2I + I^- \qquad (2\text{-}46)$$

$$BH^+ \rightarrow B + H^+ \qquad (2\text{-}47)$$

and again the overall stoichiometry is

$$CH_3-\overset{\overset{\displaystyle O}{\|}}{C}-CH_3 + I_2 \rightarrow CH_3-\overset{\overset{\displaystyle O}{\|}}{C}-CH_2I + H^+ + I^- \qquad (2\text{-}42)$$

The kinetics is simply

$$r = k_{45}C_{\text{acetone}}C_B \qquad (2\text{-}48)$$

The kinetics fails, however, to distinguish between (general base) catalysis by B and (specific base) catalysis by OH$^-$, since the following postulated sequence of steps is also consistent with the kinetics:

$$CH_3\!-\!\overset{\overset{\displaystyle O}{\|}}{C}\!-\!CH_3 + HB \;\rightleftharpoons\; CH_3\!-\!\overset{\overset{\displaystyle \overset{HB}{\cdot\!\cdot}}{\underset{}{\|}}{\underset{\displaystyle O}{}}}{C}\!-\!CH_3 \qquad\qquad (2\text{-}49)$$

$$OH^- + CH_3\!-\!\overset{\overset{\displaystyle \overset{HB}{\cdot\!\cdot}}{\underset{}{\|}}{\underset{\displaystyle O}{}}}{C}\!-\!CH_3 \;\rightarrow\; CH_3\!-\!\overset{\overset{\displaystyle O^-}{|}}{C}\!=\!CH_2 + HB + H_2O \qquad (2\text{-}50)$$

This ambiguity in the kinetics is typical for acid–base-catalyzed reactions. One can attempt to resolve the ambiguity by performing experiments in addition to kinetics. The most useful provide direct (e. g., spectroscopic) identification of reaction intermediates.

Other experiments may provide indirect evidence of the identities of reaction intermediates. For example, as the sulfuric acid concentration is varied from 85% to 94% in water, the solution becomes less active in catalyzing the racemization of the ketone D-α-phenylisocaprophenone [17]. In these solutions the ketone is almost completely converted into the protonated form. The decrease in reaction rate with increased acid concentration therefore suggests that the slow step requires a base to abstract a proton from the protonated ketone. This result suggests the following mechanism for acid-catalyzed ketone enolization in these solutions:

$$CH_3\!-\!\overset{\overset{\displaystyle O}{\|}}{C}\!-\!CH_3 + HA \;\rightleftharpoons\; CH_3\!-\!\overset{\overset{\displaystyle \overset{H}{\underset{}{O^+}}}{\|}}{C}\!-\!CH_3 + A^- \qquad (2\text{-}51)$$

$$CH_3\!-\!\overset{\overset{\displaystyle \overset{H}{\underset{}{O^+}}}{\|}}{C}\!-\!CH_3 + A^- \;\rightarrow\; CH_3\!-\!\overset{\overset{\displaystyle \overset{H}{\underset{}{O}}}{|}}{C}\!=\!CH_2 + HA \qquad (2\text{-}52)$$

2.2.6 Stepwise and Concerted Reactions

In all but one of the examples considered thus far, the reactions have been represented as stepwise processes, with the proton transfer usually being fast. The exception (Fig. 2-5) was the HBr-catalyzed dehydration of t-butyl alcohol. In this reaction, the various bond-breaking and bond-forming steps may take place almost simultaneously, and when this happens, the process is referred to as a **concerted** or *synchronous* **reaction**. Concerted reactions are common in acid–base catalysis in the absence of water or other good proton acceptors. An illustration of a plausible concerted reaction step, shown in Fig. 2-16, explains a rate equation of the form

$$r = kC_{\text{acetone}}C_{\text{HA}}C_{\text{X}^-} \qquad\qquad (2\text{-}53)$$

for the acetone halogenation reaction considered above. The intermediate shown in Fig. 2-16 is a hydrogen-bonded complex; such species are common, and in the presence of water, they may be hydrated, possibly incorporating water in a structure such as

$$
\begin{array}{c}
H\diagdown\ \ \ \ .HA \\
\ddot{O} \\
| \\
H \\
\ddot{} \\
H\ \ \ddot{O}\ \ H \\
|\ \ \ \|\ \ \ | \\
H-C-C-C-H\cdot\cdot A^- \\
|\ \ \ \ \ \ \ | \\
H\ \ \ \ \ \ \ H
\end{array}
$$

But the kinetics term of Eq. 2-53 can be equally well explained by a stepwise rather than a concerted reaction, as follows:

$$
\begin{array}{ccc}
& HA & \\
& \ddot{} & \\
O & \ddot{O} & OH \\
\| & \| & | \\
CH_3-C-CH_3 + HA \rightleftharpoons CH_3-C-CH_3 \xrightarrow{A^-} HA + A^- + CH_3-C=CH_2 \\
\end{array}
$$

$$(2\text{-}54)$$

where the second step is rate determining.

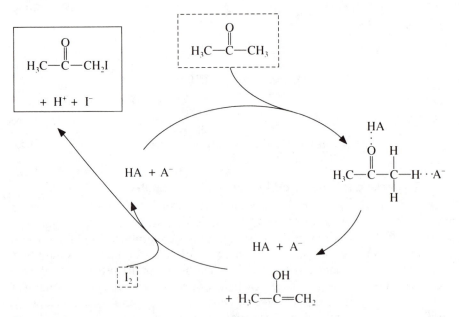

Figure 2-16
A concerted mechanism for the acetone halogenation reaction [18]. Arrows at left represent a sum of elementary steps rather than a single elementary step.

The differences between concerted and stepwise reactions are usually subtle, and distinguishing them experimentally is difficult. The differences amount to just where the protons, anions, and solvent molecules are located at particular stages of completion of the reaction. It is an oversimplification to refer to the stepwise and concerted reactions as the only two possibilities. These represent limiting cases, with a whole spectrum between them. It is often a good approximation that any one mechanism is uniquely important, but a spectrum of similar mechanisms may prevail, and the dominant mechanism may change with changing reaction conditions.

Concerted proton transfer processes are facile in pairs of molecules that can cycle between two tautomeric states. For example,

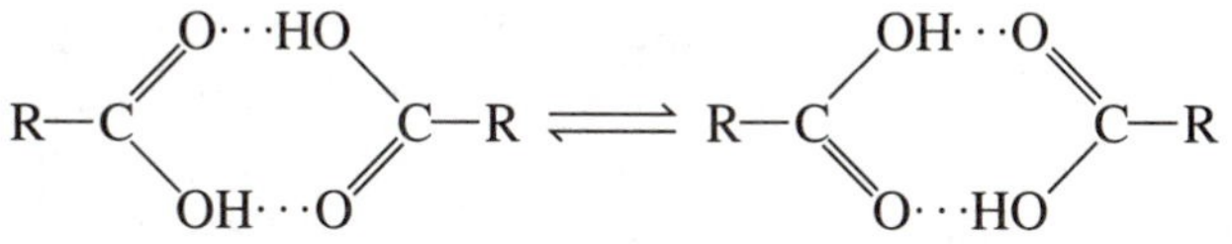

$$R-C \overset{O\cdots HO}{\underset{OH\cdots O}{\diagdown}} C-R \rightleftharpoons R-C \overset{OH\cdots O}{\underset{O\cdots HO}{\diagdown}} C-R \qquad (2\text{-}55)$$

Similar proton switches can take place with paired hydrogen-bonded molecules in a catalytic cycle. The mechanism is often referred to as a "push–pull" mechanism, indicating that one proton transfer almost simultaneously compensates the other.

For example, the following step has been suggested [19] to take place in the mutarotation of 2,3,4,6-tetramethyl-D-glucose catalyzed by 2-hydroxypyridine (where, for clarity, the substituent groups are omitted from the glucose):

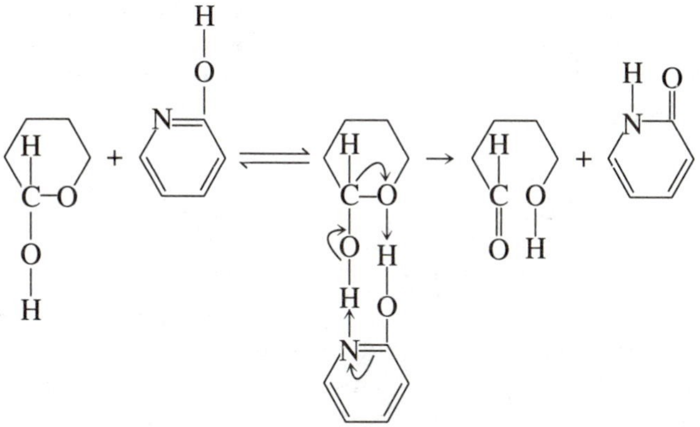

$$(2\text{-}56)$$

This "tautomeric catalysis" occurs in benzene solution [20]. A bifunctional catalyst like 2-hydroxypyridine is effective when its geometry allows it to form a doubly hydrogen-bonded intermediate with the reactant without the formation of high-energy ionic intermediates. This requirement implies that the pair of molecules can exist in two tautomeric forms of comparable energy. In tautomeric catalysis, the relative acid and base strengths of the molecules may be of only secondary importance in determining the catalytic activity; attempts to predict an activity from acid and base strengths may lead to estimates that are many orders of magnitude too small.

Tautomeric catalysis is a special case of acid–base catalysis that becomes important only when a proper geometric match exists between the catalyst and reactant, when the tautomeric forms are energetically nearly the same, and when an especially low activation energy is associated with the concerted proton transfers.

2.2.7 Catalysis by Metal Ions

Catalysis by hydrogen ions is the most common type of acid catalysis in solutions, but not the only type. Metal ions other than hydrogen may function similarly. Examples are manganese, copper, and zinc ions in aqueous solutions. Metal ions act as catalysts by bonding to organic reactants, introducing charge into them, and often polarizing them and inducing formation of carbenium ions. A metal ion (a Lewis acid) may be superior to a Brønsted acid (proton donor) in several respects. It may introduce a multiple positive charge into a reactant, whereas a proton can introduce only a single positive charge; it can exist in a neutral solution, not just acidic solutions; and it can coordinate to several donor atoms simultaneously, whereas a proton usually coordinates to only one.

A metal ion may act as a catalyst without changing its oxidation state. In contrast, there are examples encountered later in this chapter of cations playing catalytic roles that require them to change oxidation states (Section 2.4).

Cations catalyze numerous reactions, including ester and amide hydrolysis reactions, decarboxylation reactions (forming CO_2), some of which are important in biology, and phosphate cleavage reactions, which are also important in biology. Cations in enzymes play roles similar to those in solution, as described in Chapter 3.

2.2.8 Hydrocarbon Conversion

CARBENIUM ION REACTIONS [21, 22]

The examples of acid–base catalysis mentioned to this point have been restricted to reactant–catalyst combinations that include a good base such as a compound containing an N, O, or S atom. Some very weak bases, hydrocarbons, are important reactants in industrial chemistry, since hydrocarbons are the major constituents of petroleum, natural gas, and coal-derived liquids. The classes of compounds constituting one kind of crude petroleum are shown in Fig. 2-17.

Aromatic compounds are protonated by strong acids. Olefins are better proton acceptors; these are not present in crude oil or coal-derived liquids, but they are produced in large quantities in petroleum refining processes, some of which are considered later. Olefins are protonated in solutions of mineral acids to give carbenium ions:

$$R-\underset{\underset{H}{|}}{C}=\underset{\underset{H}{|}}{C}-R' + H^+ \rightarrow R-\underset{\underset{H}{|}}{C}-\overset{\overset{H}{|}}{\underset{\underset{H}{|}}{C}}-R' \tag{2-57}$$

Carbenium ions are usually solvated or present in ion pairs; but for simplicity, the carbenium ion structure alone is usually written.

Many strongly acidic solutions are capable of protonating olefins and aromatic compounds; but since the identity of the proton donor species is often unknown, it is represented by the shorthand symbol H^+. This shorthand does not necessarily imply that the proton is donated by a dissociated acid.

The essence of the catalytic chemistry of hydrocarbon conversions catalyzed by strong acids is carbenium ion chemistry. There is much more to the chemistry than proton transfer; the reactions of carbenium ions involve other ion transfers, for example, hydride ion transfer.

Much has been learned about the reactions of carbenium ions as a result of the discovery of extremely strong acids—called *superacids*—formed from combinations of Lewis and Brønsted acids. The combinations include the following:

Brønsted Acid	Lewis Acid
HF	BF_3
HF	SbF_5
FSO_3H	SbF_5

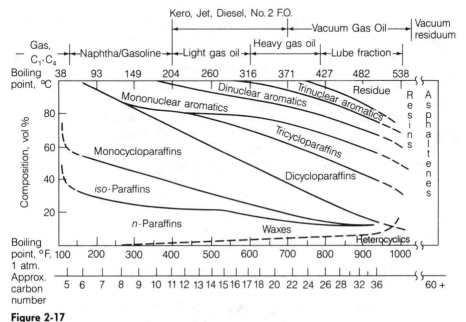

Figure 2-17
Relative quantities and boiling range of the major hydrocarbon classes in the crude oil from the Ponca City Field, Ponca City, Oklahoma [23].

Superacid solutions have Hammett acidity functions of even -20 and less [23]; the value for pure H_2SO_4 is only -12 (Fig. 2-12).

The great proton donor strengths of superacids result from reactions that stabilize protonated forms of the original Brønsted acids. For example, in the presence of SbF_5, $[FSO_3H_2]^+$ is formed as part of an ion pair:

$$2FSO_3H + SbF_5 \rightleftharpoons \begin{bmatrix} & F & \\ F & | & F \\ & Sb & \\ F & | & F \\ & O \quad O & \\ & S & \\ & O \quad F & \end{bmatrix}^- \quad [FSO_3H_2]^+ \qquad (2\text{-}58)$$

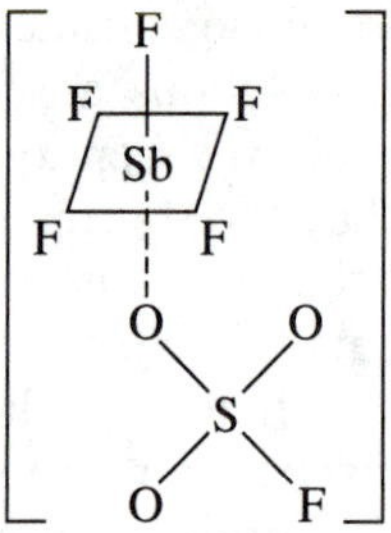

When HF is used with SbF_5, the superacidic proton donor is H_2F^+; these species can form because of the great stability of the anions $SbF_5(SO_3F)^-$ and SbF_6^-.

Carbenium ions are the key intermediates in hydrocarbon conversions taking place in the presence of strong acids, and the superacids allow quantitative investigation of carbenium ion reactivities. The superacids are such strong proton donors that the reactions forming carbenium ions are driven almost completely to the right:

$$R\text{—}CH\text{=}CH\text{—}R' + H^+ \rightarrow R\text{—}CH_2\text{—}\overset{+}{C}H\text{—}R' \qquad (2\text{-}59)$$

Carbenium ions react readily with olefins, but the superacids eliminate this complication. Investigations of carbenium ions in superacids with proton and ^{13}C NMR spectroscopy have provided a wealth of quantitative information about reactivity that helps explain catalytic conversions of hydrocarbons.

According to the IUPAC system of nomenclature, the parent ion, CH_3^+, is *carbenium ion*, also known as methyl cation. (Earlier, this had been referred to as a *carbonium ion*, and many people still cling to this term; the correct usage of the term "carbonium ion" is stated below.) When a CH_2 substituent is added to carbenium ion, the new ion formed is methylcarbenium ion, $CH_3CH_2^+$, or ethyl cation; trimethylcarbenium ion,

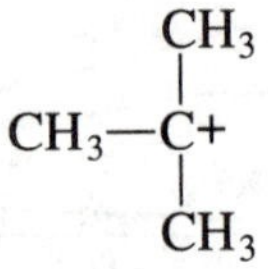

$$\begin{array}{c} CH_3 \\ | \\ CH_3\text{—}C\text{+} \\ | \\ CH_3 \end{array}$$

also called *t*-butyl cation, is a frequently encountered example in hydrocarbon conversion catalysis.

Tertiary carbenium ions are much more stable than secondary carbenium ions, which are much more stable than primary carbenium ions:

$$
CR_3\overset{+}{\underset{\underset{CR_3}{|}}{C}}CR_3 \;>\; CR_3\overset{+}{\underset{\underset{H}{|}}{C}}CR_3 \;>\; CR_3\overset{H}{\underset{\underset{H}{|}}{\overset{|}{C}}}\overset{+}{C}H_2 \;>\; \overset{+}{C}H_3
$$

Tertiary	more stable than	secondary	more stable than	primary	more stable than

Calorimetric measurements have shown that the difference in stability between tertiary and secondary carbenium ions having the same carbon skeleton is about 13 kcal/mol [24]. Because the entropy effect is likely to be negligible, the free-energy difference is nearly the same as the enthalpy difference. The equilibrium constant giving the ratio of tertiary to secondary carbenium ions having the same carbon skeleton is roughly 10^{10}. There are no significant differences in stabilizing effect of the different alkyl groups. The difference in stability between secondary and primary carbenium ions having the same carbon skeleton is greater than about 17 kcal/mol.

The important reactions of carbenium ions include isomerizations proceeding by 1,2-hydride and 1,2-alkyl shifts, symbolized by H~ and CH$_3$~, respectively:

$$
\text{C}-\underset{\underset{H}{|}}{\overset{\overset{C}{|}}{\overset{+}{C}}}-\overset{\overset{C}{|}}{C}-C \;\xrightleftharpoons{\;H\sim\;}\; \text{C}-\underset{\underset{H}{|}}{\overset{\overset{C}{|}}{C}}-\overset{\overset{C}{|}}{\overset{+}{C}}-C \tag{2-60}
$$

$$
\text{C}-\underset{\underset{C}{|}}{\overset{\overset{C}{|}}{\overset{+}{C}}}-\overset{\overset{C}{|}}{C}-C \;\xrightleftharpoons{\;CH_3\sim\;}\; \text{C}-\underset{\underset{C}{|}}{\overset{\overset{C}{|}}{C}}-\overset{\overset{C}{|}}{\overset{+}{C}}-C \tag{2-61}
$$

These reactions are characterized by activation energies $\Delta G^{o\ddagger}$ of only several kcal/mol; the rate constants at $-120°C$ are large, about 10^7 to 10^8 s^{-1} [25]. (Why is the activation energy represented by $\Delta G^{o\ddagger}$ here?) Similar values characterize the rates of interconversion of secondary (or primary) carbenium ions via hydride or alkyl shifts.

These equilibrium and kinetics results allow predictions of the kinetics of some carbenium ion reactions. For example, consider the skeletal isomerization of an octyl cation:

$$
\text{C}-\underset{\underset{C}{|}}{\overset{\overset{C}{|}}{C}}-C-\overset{\overset{C}{|}}{\overset{+}{C}}-C \;\rightleftharpoons\; \text{C}-\overset{\overset{C}{|}}{\overset{+}{C}}-\overset{\overset{C}{|}}{C}-\overset{\overset{C}{|}}{C}-C \tag{2-62}
$$

This transformation takes place most readily via a hydride shift followed by a methyl shift:

$$
\underset{\substack{|\\C}}{\overset{\substack{C\\|}}{C-\underset{+}{C}-C-\underset{\substack{|\\C}}{\overset{\substack{C\\|}}{C}}-C}} \overset{H\sim}{\rightleftharpoons} \underset{\substack{|\\C}}{\overset{\substack{C\\|}}{C-\underset{+}{C}-C-\overset{\substack{C\\|}}{C}-C}} \overset{CH_3\sim}{\rightleftharpoons} \overset{\substack{C\;\;C\;\;C\\|\;\;|\;\;|}}{C-\underset{+}{C}-C-C-C} \qquad (2\text{-}63)
$$

The intermediate is a secondary carbenium ion; it is about 13 kcal/mol less stable than the reactant ion. Since the activation energy barrier for a hydride or methyl shift is only a few kcal/mol, the activation energy for the overall reaction is expected to be about 15 kcal/mol, as has been observed [25]. The energy diagram of Fig. 2-18 summarizes this information.

Similar predictions are valid for a wide range of isomerization reactions, but experiments have shown that there is a class of carbenium ion rearrangements, those leading to a change in the degree of branching, for which the simple hydride and alkyl shifts do not explain the observations. For example, the following isomerization, which converts a carbenium ion with a single branch to one with two branches,

$$
\overset{\substack{C\\|}}{C-\underset{+}{C}-C-C-C} \rightleftharpoons \overset{\substack{C\;\;C\\|\;\;|}}{C-\underset{+}{C}-C-C} \qquad (2\text{-}64)
$$

might at first be expected to proceed as follows:

$$
\overset{\substack{C\\|}}{C-\underset{+}{C}-C-C-C} \overset{H\sim}{\rightleftharpoons} \overset{\substack{C\\|}}{C-\underset{+}{C}-C-C-C} \overset{CH_3\sim}{\rightleftharpoons} \overset{\substack{C\;\;C\\|\;\;|}}{C-C-C-C^+} \overset{H\sim}{\rightleftharpoons} \overset{\substack{C\;\;C\\|\;\;|}}{C-\underset{+}{C}-C-C}
$$

$$(2\text{-}65)$$

The only imaginable sequence of hydride and methyl shifts accounting for the conversion requires the intermediacy of a high-energy primary carbenium ion. The expected activation energy (corresponding to the difference in stabilities of the tertiary and primary carbenium ions plus a few kcal/mol) is more than 30 kcal/mol. The observed activation energy, however, is only 17 or 18 kcal/mol [21]; there must be a more efficient mechanism than that postulated in Eq. 2-65.

The accepted mechanism involves an intermediate incorporating a protonated cyclopropane ring, a delocalized electron-deficient structure that is much lower in energy than a primary carbenium ion:

$$
\overset{\substack{C\\|}}{C-\underset{+}{C}-C-C-C} \rightleftharpoons C-\overset{\substack{C\;C\;\;H^+}}{\underset{\diagup\!\diagdown}{C}}-C \rightleftharpoons \overset{\substack{C\;\;C\\|\;\;|}}{C-\underset{+}{C}-C-C} \qquad (2\text{-}66)
$$

Carbenium ion rearrangements leading to changes in the degree of branching are believed generally to proceed via such intermediates.

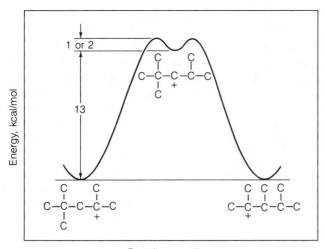

Figure 2-18
Energetics and reaction pathway for isomerization of a carbenium ion involving hydride and methyl shifts. The activation energy is about 15 kcal/mol.

Another important reaction of carbenium ions is β-scission of carbon–carbon bonds; this is the essential step in catalytic cracking, discussed in Chapter 5. For example,

$$\underset{\substack{|\\C}}{\overset{\substack{C\qquad C\\|\qquad|}}{C-C-C-C-C}} \rightarrow \overset{\substack{C\\|}}{C-C^+} + \overset{\substack{C\\|}}{C=C-C} \tag{2-67}$$

The reaction generates an olefin and another carbenium ion. The rate depends on the relative stabilities of the reactant and product carbenium ions. When both are tertiary, as in this example, then the reaction is fast, faster than the branching rearrangements described above but slower than the nonbranching rearrangements. A rate constant of $5 \times 10^{-4}\,\mathrm{s}^{-1}$ was measured for reaction 2-67 at $-73°C$ [27, 28].

The β-scission proceeds slowly when less stable carbenium ions are formed. For example, the reaction of the following dibranched carbenium ion might be envisioned to proceed by either of the following pathways:

$$\tag{2-68}$$

The upper pathway is slow because the secondary carbenium ion produced is about 13 kcal/mol less stable than the tertiary ion. The lower pathway is slow because the maximum (equilibrium) concentration of the secondary carbenium ion intermediate is about 10^{10} times less than that of the reactant carbenium ion. A consequence of these relative rates in the conversion of trimethylpentyl (or larger tribranched) cations in superacids is that there will be a fast equilibration with the other tribranched cations and β-scission of only tertiary cations. Rates of branching reactions are negligible relative to the cracking. The pattern is illustrated in Fig. 2-19.

Carbenium ions formed from the five- and six-membered ring hydrocarbons found in petroleum undergo β-scission only extremely slowly, because the lack of free rotation about the C—C bond prevents overlap of the p_z orbital on the α-carbon with that on the neighboring carbon, which would facilitate formation of the C=C bond.

Another important reaction of carbenium ions, which are strong Lewis acids, is hydride abstraction, another of the ion transfer reactions. For example,

$$
\underset{\substack{|\\ CH_3}}{\overset{\substack{CH_3\\ |}}{CH_3—C^+}} + \underset{\substack{|\\ CH_3}}{\overset{\substack{CH_3\\ |}}{H—C—CH_3}} \rightarrow \underset{\substack{|\\ CH_3}}{\overset{\substack{CH_3\\ |}}{CH_3—C—H}} + \underset{\substack{|\\ CH_3}}{\overset{\substack{CH_3\\ |}}{^+C—CH_3}} \tag{2-69}
$$

This bimolecular reaction is extremely fast when both reactant and product cations are tertiary. The rate constant for reaction 2-69 at $-40°C$ is 10^4 L/(mol·s) with an activation energy of 3.5 kcal/mol [22]. Hydride transfer reac-

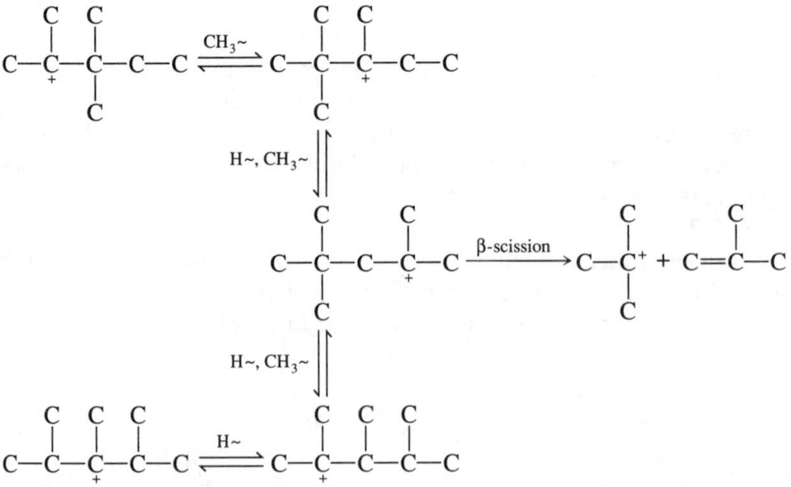

Figure 2-19
Reactions of tertiary octyl cations in a superacid solution. The hydride and methyl shifts are rapid in comparison with the β-scission; this β-scission step is the significant one, since it involves tertiary cations both as the reactant and product ions [21].

tions involving formation of less stable carbenium ions are much slower. For example, formation of a secondary carbenium ion by reaction of a paraffin with a tertiary carbenium ion is characterized by an activation energy of about 14 kcal/mol, slightly more than the free-energy change of the reaction.

Another bimolecular reaction of carbenium ions is alkylation, the reverse of β-scission. The reactants are the carbenium ion (a Lewis acid) and an olefin (a Lewis base); the product is a carbenium ion. For example,

$$
\begin{array}{c}
\overset{\displaystyle C}{\underset{\displaystyle C}{|}}\overset{\displaystyle C}{|}\overset{\displaystyle C}{|}\overset{\displaystyle C}{|} \\
C{-}\overset{+}{C}\; + \; C{=}C{-}C \;\rightarrow\; C{-}C{-}C{-}\overset{+}{C}{-}C
\end{array} \tag{2-70}
$$

In alkylation processes carried out on a commercial scale, typically with iso-butane and propylene as the reactants, C_7 (gasoline range) products are formed by this kind of carbon–carbon bond formation. Isomerizations take place rapidly, and additional carbon–carbon bond formations lead to higher-molecular-weight products, including polymers. (Commercial alkylation processes, described below, are carried out at relatively low temperatures, whereas cracking processes require high temperatures; can you figure out why?)

EXAMPLE 2-2 Cracking of Carbenium Ions Derived From Octane

Problem

Predict the important pathways involved in the cracking of carbenium ions formed from *n*-octane in a superacid solution.

Solution

Abstraction of hydride ions from *n*-octane gives secondary octyl cations. These are cracked only very slowly. Instead of cracking, they undergo rearrangements to give tertiary carbenium ions, which include isomers that are cracked slowly and one isomer that is cracked rapidly,

$$
\begin{array}{c}
\overset{\displaystyle C}{|}\overset{\displaystyle C}{|} \\
C{-}C{-}C{-}\overset{+}{C}{-}C \\
\underset{\displaystyle C}{|}
\end{array}
$$

All the carbenium ions abstract hydride ions from paraffins to form product paraffins, which are isomers of octane. This set of reactions is summarized in Fig. 2-20. ■

EXAMPLE 2-3 Mechanism and Kinetics of Isomerization of *n*-Butane and *n*-Pentane

Problem

The rate of skeletal isomerization of *n*-pentane in a superacid solution has been found to be much greater than the rate of skeletal isomerization of *n*-butane. Explain; predict approximately the relative rates. Also explain

n-Octane $\xrightleftharpoons[\text{fast}]{\text{hydride abstraction}}$ Secondary n-octyl cations $\xrightarrow[\text{very slow}]{\beta-\text{scission}}$

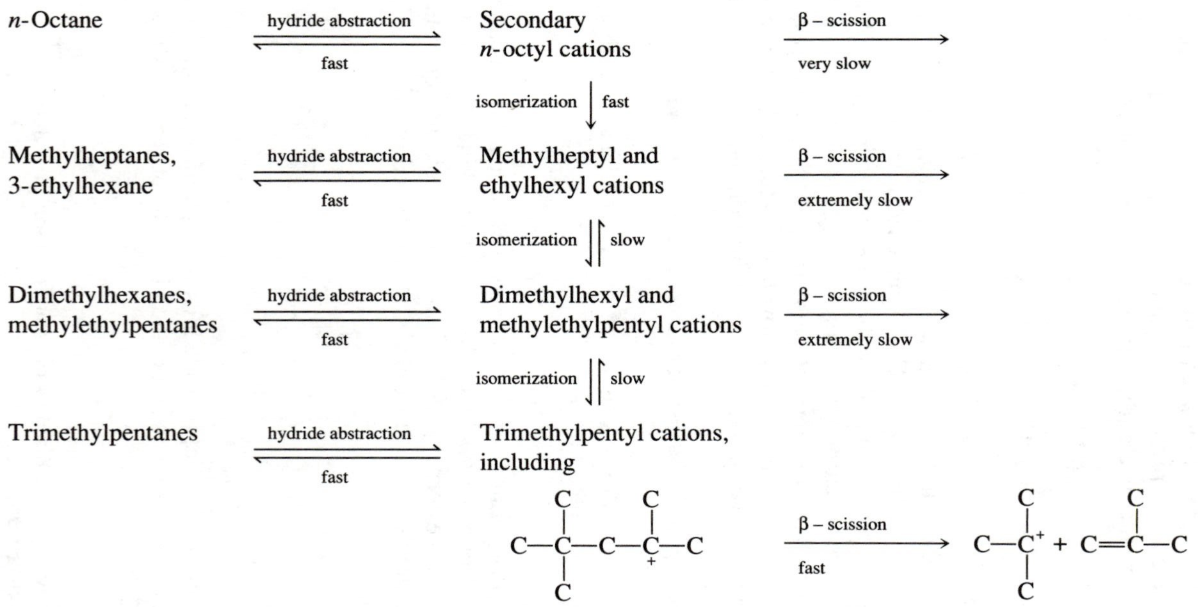

Secondary n-octyl cations $\xrightarrow[\text{isomerization}]{\text{fast}}$ Methylheptyl and ethylhexyl cations

Methylheptanes, 3-ethylhexane $\xrightleftharpoons[\text{fast}]{\text{hydride abstraction}}$ Methylheptyl and ethylhexyl cations $\xrightarrow[\text{extremely slow}]{\beta-\text{scission}}$

Methylheptyl and ethylhexyl cations $\underset{\text{isomerization}}{\overset{\text{slow}}{\rightleftharpoons}}$ Dimethylhexyl and methylethylpentyl cations

Dimethylhexanes, methylethylpentanes $\xrightleftharpoons[\text{fast}]{\text{hydride abstraction}}$ Dimethylhexyl and methylethylpentyl cations $\xrightarrow[\text{extremely slow}]{\beta-\text{scission}}$

Dimethylhexyl and methylethylpentyl cations $\underset{\text{isomerization}}{\overset{\text{slow}}{\rightleftharpoons}}$ Trimethylpentyl cations, including

Trimethylpentanes $\xrightleftharpoons[\text{fast}]{\text{hydride abstraction}}$ Trimethylpentyl cations $\xrightarrow[\text{fast}]{\beta-\text{scission}}$

Figure 2-20
Predominant reactions of C_8 hydrocarbons in a superacid solution. The pathway for cracking of n-octane proceeds from the upper left to the n-octyl cations, then to the trimethylpentyl cations, including that which cracks to give the tertiary butyl cation [21].

why the isomerization leading to scrambling of the carbon atoms, $CH_3CH_2CH_2{}^{13}CH_3 \rightarrow CH_3CH_2{}^{13}CH_2CH_3$, is much faster than the skeletal isomerization of n-butane; predict the relative rates.

Solution

The n-pentane isomerization is expected to proceed via the protonated cyclopropane ring intermediate. The secondary carbenium ion can be formed from pentane by hydride abstraction:

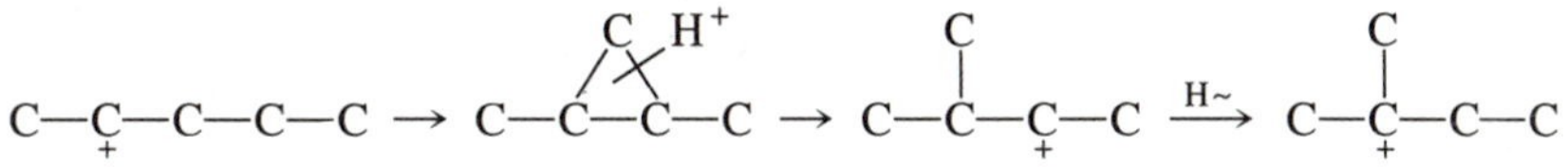

$$\text{C—}\underset{+}{\text{C}}\text{—C—C—C} \rightarrow \text{C—C}\overset{\overset{\text{C}}{\diagdown}\text{H}^+}{\diagup}\text{C—C} \rightarrow \text{C—}\overset{\text{C}}{\underset{+}{\text{C}}}\text{—}\underset{+}{\text{C}}\text{—C} \xrightarrow{\text{H}\sim} \text{C—}\overset{\text{C}}{\underset{+}{\text{C}}}\text{—C—C}$$

$$(2\text{-}71)$$

Hydride abstraction by the tertiary cation then generates the isopentane. The value of $\Delta G^{o\ddagger}$ for the reverse reaction has been found to be about 17 kcal/mol [21] as shown in Fig. 2-21.

The isomerization of the secondary butyl cation formed from $CH_3CH_2CH_2{}^{13}CH_3$ is also expected to proceed via a cyclic intermediate (where ^{13}C is represented as C^*):

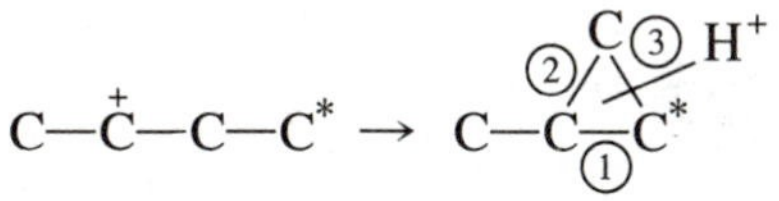

$$\text{C—}\underset{+}{\text{C}}\text{—C—C}^* \rightarrow \text{C—C}\overset{\overset{\text{C}③}{②\diagdown}\text{H}^+}{\underset{①}{\diagup}}\text{C}^*$$

$$(2\text{-}72)$$

If bond ① is broken, the reaction is reversed; if bond ② is broken, the secondary carbenium ion

$$\text{C—}\underset{+}{\text{C}}\text{—C}^*\text{—C}$$

is formed; and if bond ③ is broken, the primary carbenium ion

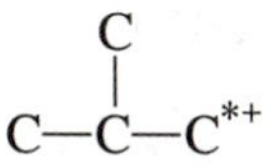

$$\text{C—}\overset{\text{C}}{\text{C}}\text{—C}^{*+}$$

is formed, which then undergoes a rapid hydride shift to give the tertiary butyl cation and ultimately isobutane. Breaking of bond ② gives the scrambled product and is characterized by nearly the same $\Delta G^{o\ddagger}$ value as the pentane isomerization. But because the breaking of bond ③ leads to a primary carbenium ion, the activation energy for the skeletal isomerization is greater. The prediction is that it would be about 18 or 19 kcal/mol, and it has been observed to be about 18 kcal/mol [25]. This reaction is roughly 10^{10} times slower than the others at room temperature. ∎

There is another class of reactions involving hydrocarbon cations (carbocations) in superacids. The proton donor strengths of these solutions are so

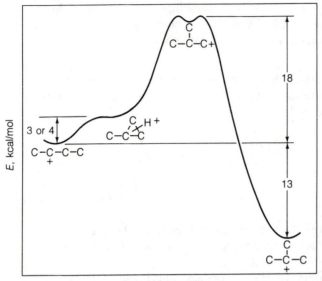

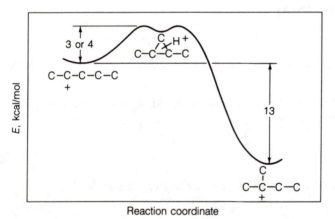

Figure 2-21
Energy diagrams for the branching isomerization reactions of butyl and pentyl cations.

great that even such weak bases as paraffins can be protonated [29]:

$$H^+ + CH_4 \rightleftharpoons CH_5^+ \tag{2-73}$$

The product of this reaction, called *carbonium ion*, often depicted as has a two-electron three-center bond:

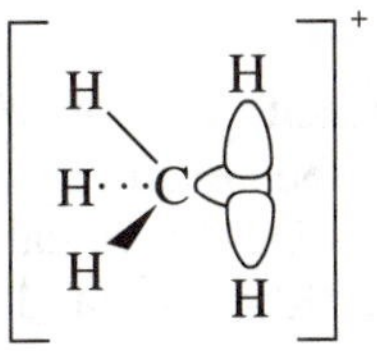

$$\left[\begin{array}{c} H \\ H\cdots C \\ H \end{array} \begin{array}{c} H \\ \\ H \end{array} \right]^{+}$$

Carbonium ions related to this, the parent ion, have been detected in superacid solutions by means of NMR and Raman spectroscopy and X-ray photoelectron spectroscopy (XPS, also known as ESCA, electron spectroscopy for chemical analysis).

There are other members of the family of reactions involving attack of H^+ or of R^+ on the C—H or C—C bonds of paraffins; the reactions result in the formation of new C—H or C—C bonds and the expulsion of H^+ or R^+. Examples are the following [21]:

$$R^+ + HR' \rightleftharpoons RH + R'^+ \tag{2-74}$$

$$H^+ + HR \rightleftharpoons H_2 + R^+ \tag{2-75}$$

$$H^+ + R—H \rightleftharpoons HR + H^+ \tag{2-76}$$

$$H^+ + R—R' \rightleftharpoons H—R + R'^+ \tag{2-77}$$

Equation 2-74 is the hydride abstraction discussed above. The reverse of reaction 2-75 is the reduction of a carbenium ion with molecular hydrogen, which therefore can suppress catalytic reactions that proceed via carbenium ion intermediates.

The rates of many of these reactions in superacid solutions of SbF_5 in HF (containing H_2F^+) have been determined. For example, for the reaction

$$H^+ + H—\overset{\displaystyle C}{\underset{\displaystyle C}{\overset{|}{\underset{|}{C}}}}—C \underset{k_r}{\overset{k_f}{\rightleftharpoons}} H_2 + {}^+\overset{\displaystyle C}{\underset{\displaystyle C}{\overset{|}{\underset{|}{C}}}}—C \tag{2-78}$$

at 20°C, the values of k_f and k_r are 2×10^{-3} and 8×10^{-3} L/(mol·s), respectively [22]. These reactions are orders of magnitude slower than the hydride transfer between tertiary cations and isobutane.

This kind of chemistry explains the conversion of methane, ethane, and propane in superacid solutions [21, 30]. For example, methane can react by the following sequence:

$$H^+ + CH_4 \rightleftharpoons CH_5^+ \tag{2-79}$$

$$CH_5^+ \rightleftharpoons CH_3^+ + H_2 \tag{2-80}$$

$$CH_3^+ + CH_4 \rightarrow CH_3CH_3 + H^+ \tag{2-81}$$

$$CH_3^+ + C_2H_6 \rightarrow C_3H_8 + H^+ \tag{2-82}$$

or

$$C_2H_5^+ + CH_4 \rightarrow C_3H_8 + H^+ \tag{2-83}$$

and so forth. When C_4H_{10} is formed, it is converted into tertiary butyl cations; these are so stable that their further reactions are negligible.

The superacids catalyze potentially valuable reactions, but the superacid solutions are so corrosive and difficult to handle and to dispose of in environmentally acceptable ways that they are scarcely used. However, some solid acids are superacids, and they are practical catalysts used on a large scale, as described in Chapters 5 and 6.

CATALYTIC REACTIONS INVOLVING CARBENIUM ION INTERMEDIATES

Consider some of the reactions of olefins in strongly acidic solutions, such as concentrated H_2SO_4. If but-1-ene is added to the solution, it becomes protonated, forming a secondary carbenium ion:

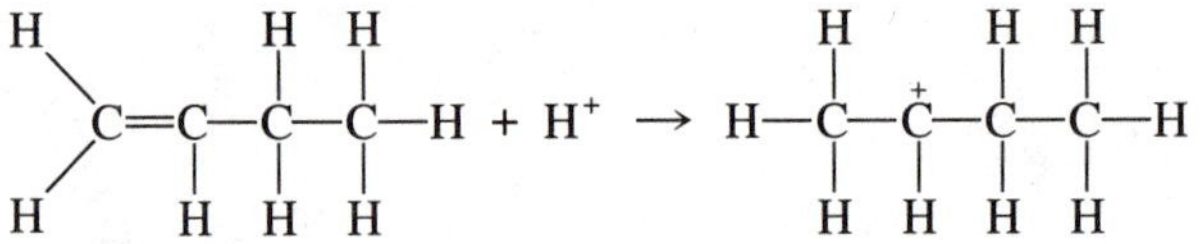

$$\tag{2-84}$$

The much less stable primary carbenium ion is present in negligibly low concentrations. The secondary carbenium ion can give back a proton to an anion, losing the proton from either of the carbon atoms neighboring the charged carbon atom. The hydrocarbon products formed as a result of this proton donation are but-1-ene (again) and the two isomers of but-2-ene (*cis-* and *trans-*but-2-ene). It follows that the catalytic isomerization of butenes takes place in the acidic solution, and the essence of the reaction mechanism is the following:

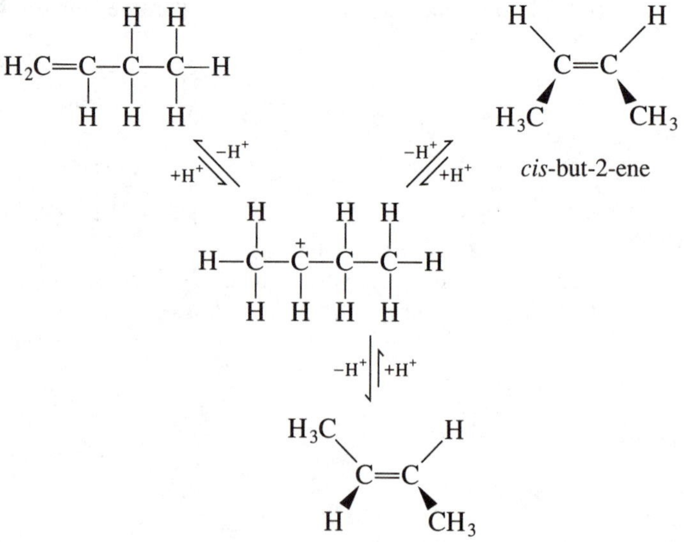

$$\tag{2-85}$$

trans-but-2-ene

In solutions that are very strongly acidic, the lifetime of the carbenium ion is great enough that a methyl migration reaction can occur to a significant degree, giving the fourth butene isomer, isobutylene:

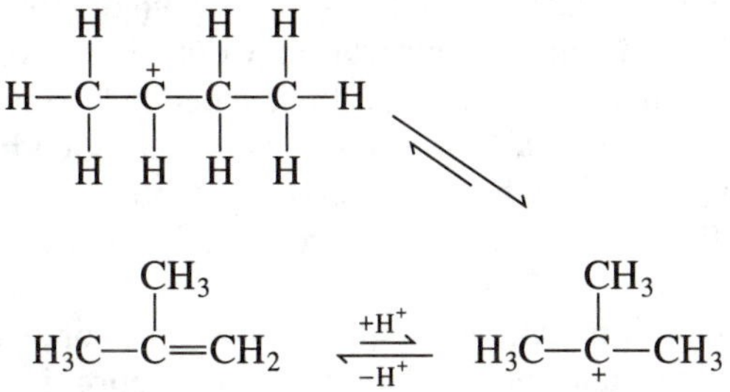

$$
\begin{array}{c}
\overset{\displaystyle H}{\underset{\displaystyle H}{|}}\quad \overset{\displaystyle H}{\underset{\displaystyle H}{|}}\ \overset{\displaystyle H}{\underset{\displaystyle H}{|}} \\
H{-}\overset{|}{C}{-}\overset{+}{\underset{|}{C}}{-}\overset{|}{C}{-}\overset{|}{C}{-}H
\end{array}
\tag{2-86}
$$

$$
\underset{H_3C-\overset{\displaystyle CH_3}{\overset{|}{C}}{=}CH_2}{}\quad \underset{-H^+}{\overset{+H^+}{\rightleftharpoons}}\quad \underset{H_3C-\overset{\displaystyle CH_3}{\overset{|}{\underset{+}{C}}}{-}CH_3}{}
$$

This skeletal isomerization is much slower than the double-bond migration. (What reaction intermediates occur in this isomerization?)

Other reactions take place simultaneously, giving high-molecular-weight hydrocarbon products. To understand these, first consider oligomerization, that is, polymerization that proceeds to a small degree, giving products consisting of only a few monomer units. Oligomerization of olefins like propylene is commercially applied for manufacture of products like nonenes.

Propylene is considered as an example because there is no complicating olefin isomerization that can occur. The first reaction again is a protonation to form a secondary carbenium ion:

$$
C{=}C{-}C + H^+ \rightarrow C{-}\overset{+}{C}{-}C
\tag{2-87}
$$

This ion is a strong Lewis acid and has an affinity for the bases in the reactant solution, which include propylene, with its electron-rich double bond. The kinetically significant reaction step initially leads to formation of a carbon–carbon bond:

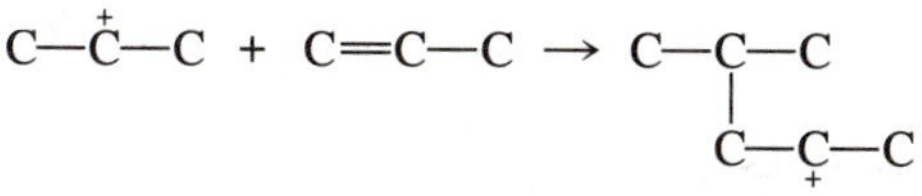

$$
C{-}\overset{+}{C}{-}C + C{=}C{-}C \rightarrow
\begin{array}{c}
C{-}\overset{|}{C}{-}C \\
\overset{|}{C}{-}\overset{|}{\underset{+}{C}}{-}C
\end{array}
\tag{2-88}
$$

This secondary carbenium ion can donate a proton to a base and give an olefinic product,

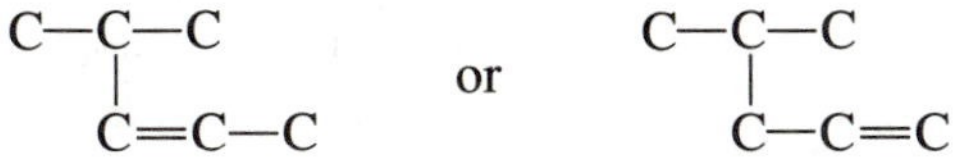

$$
\begin{array}{c}
C{-}\overset{|}{C}{-}C \\
\overset{|}{C}{=}C{-}C
\end{array}
\quad \text{or} \quad
\begin{array}{c}
C{-}\overset{|}{C}{-}C \\
\overset{|}{C}{-}C{=}C
\end{array}
$$

Alternatively, the secondary carbenium ion can react with an olefin such as propylene or the "propylene dimer" shown above:

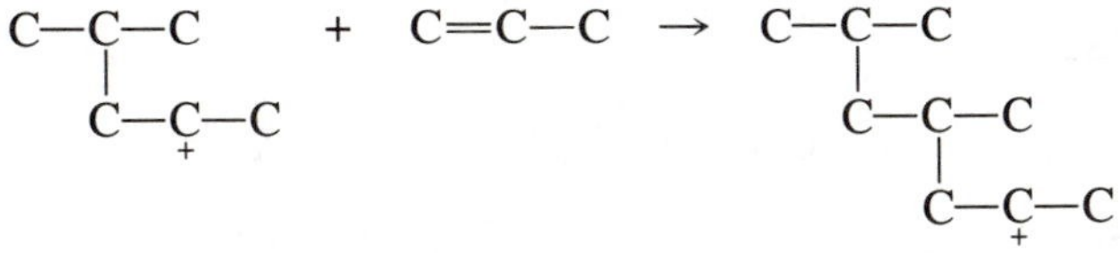

$$
\begin{array}{c}
C{-}\overset{|}{C}{-}C \\
\overset{|}{C}{-}\overset{|}{\underset{+}{C}}{-}C
\end{array}
+ C{=}C{-}C \rightarrow
\begin{array}{c}
C{-}\overset{|}{C}{-}C \\
\overset{|}{C}{-}\overset{|}{C}{-}C \\
\overset{|}{C}{-}\overset{|}{\underset{+}{C}}{-}C
\end{array}
\tag{2-89}
$$

This secondary carbenium ion can give up a proton to form one of the isomers of "propylene trimer," or it can react further with an olefin, and so forth.

When the starting olefin is propylene, the carbenium ions are virtually all secondary, and the main reaction products may be oligomers. When the starting olefin is ethylene, then the polymerization would have to proceed through primary carbenium ions; consequently, it is exceedingly slow. When the starting olefin has four or more carbon atoms, then tertiary carbenium ions form, the reaction is faster, and high-molecular-weight polymers form. A catalytic cycle illustrating the beginning of this process, isobutylene dimerization, is shown in Fig. 2-22.

When these reactions take place in aqueous solution, side reactions such as olefin hydration can take place. This type of reaction is the reverse of the alcohol dehydration reaction considered earlier (Fig. 2-5). The alcohol can now be dehydrated not only to give back propylene but also to give diisopropyl ether; other related reactions also take place.

Another important set of acid-catalyzed hydrocarbon conversion reac-

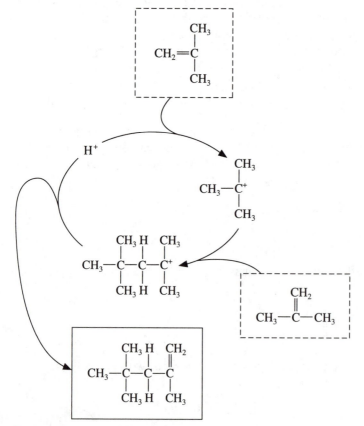

Figure 2-22
Catalytic cycle for the dimerization of isobutylene.

tions involves aromatic compounds; a familiar example is the Friedel–Crafts alkylation of benzene. These reactions, like the olefin polymerization, involve carbenium ion intermediates and the formation of C—C bonds. Reactants mentioned frequently in organic chemistry textbooks include benzene with alkylating agents such as alcohols, alkyl halides, olefins, acetylenes, and esters. The catalysts are Brønsted acids such as HF and H_2SO_4; the Lewis acid $AlCl_3$ is also used, and small amounts of compounds like water generate strong Brønsted acids by reacting with it. The Lewis acid is often used in a nearly anhydrous form and in combination with a proton donor like HCl (a cocatalyst) forming another proton donor which is stronger than HCl, since the $AlCl_4^-$ ion is quite stable. The $AlCl_3$-HCl combination is a superacid.

In some reactions, the $AlCl_3$ may alone be a catalyst, forming a complex with the reactant, as in the following example:

$$ROH + AlCl_3 \rightarrow ROH \cdot AlCl_3 \qquad (2\text{-}90)$$

The complexation increases the reactivity of the alcohol. Sometimes the $AlCl_3$ is consumed in related reactions:

$$ROH + AlCl_3 \rightarrow ROH \cdot AlCl_3 \rightarrow ROAlCl_2 + HCl \rightarrow RCl + AlOCl \qquad (2\text{-}91)$$

Other Lewis acids, including BF_3, $SbCl_5$, SbF_5, $FeCl_3$, and $ZnCl_2$, are also active aromatic alkylation catalysts. The most important one in industrial practice is $AlCl_3$; it requires a solvent such as nitrobenzene or carbon disulfide.

The usual industrial alkylating agents in aromatic alkylation are olefins, since they are readily available from petroleum refinery streams. The common catalysts are concentrated H_2SO_4, HF, and $AlCl_3$ in combination with a cocatalyst such as HCl.

A catalytic cycle for benzene alkylation with propylene is suggested in Fig. 2-23. The reaction involves electrophilic aromatic substitution by the s-propyl cation; an intermediate structure like the following is envisioned:

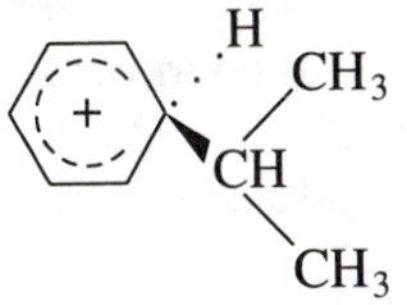

It is evident that olefins that can be protonated to give tertiary carbenium ions are more reactive than propylene; ethylene, on the other hand, can give only the primary ethyl cation and is quite unreactive. Aromatics like phenol are more reactive than benzene because the electron-donating OH substituent activates the ring for substitution. Side reactions in aromatic alkylation include olefin isomerization and oligomerization, mentioned previously, and formation of di- and tri-alkylated aromatics. Other side reactions are mentioned in the following example.

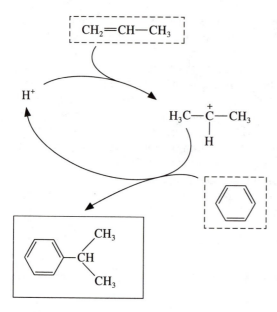

Figure 2-23
Catalytic cycle for benzene propylation in the presence of a strong proton donor.

EXAMPLE 2-4 Kinetics and Mechanism of Alkylbenzene Disproportionation Catalyzed by AlCl₃–HCl Solutions

Problem

In solution with nitromethane containing $AlCl_3$ and HCl, ethylbenzene, isopropylbenzene, and *t*-butylbenzene each reacts to give benzene and the corresponding dialkylbenzene:

$$2 \quad \underset{}{\bigcirc}{}^{R} \rightarrow \bigcirc + \underset{}{\bigcirc}{}^{R}{-}R \qquad (2\text{-}92)$$

Rearranged alkyl groups were not observed; the reaction of each alkylbenzene was found to be second order in alkylbenzene, first order in AlCl₃, and first order in HCl [31]. At a temperature of about 70°C, the relative reactivities were roughly as follows:

Ethylbenzene	Isopropylbenzene	*t*-Butylbenzene
1	10^2	10^5

Explain these results in terms of reaction mechanisms and identify the rate-determining step.

Solution

The results point to carbenium ion intermediates, the most stable of them (the *t*-butyl cation) being formed from the most reactive of the alkylbenzenes:

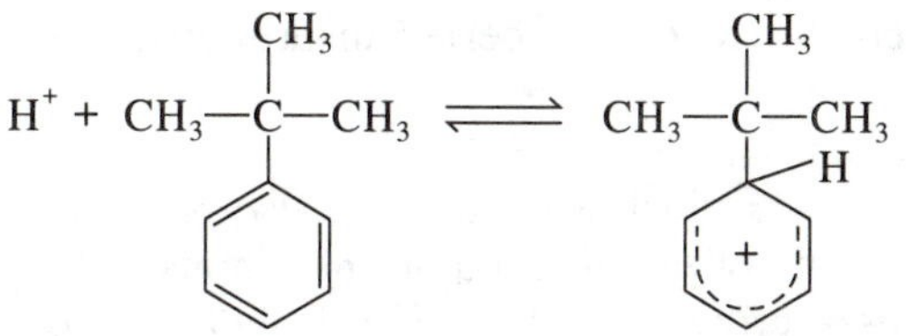

$$H^+ + CH_3-\underset{\underset{\text{Ph}}{|}}{\overset{\overset{\displaystyle CH_3}{|}}{C}}-CH_3 \;\rightleftharpoons\; CH_3-\underset{\underset{\text{(protonated ring, +)}}{|}}{\overset{\overset{\displaystyle CH_3}{|}}{C}}\!\!\begin{array}{c}-CH_3\\ -H\end{array} \qquad (2\text{-}93)$$

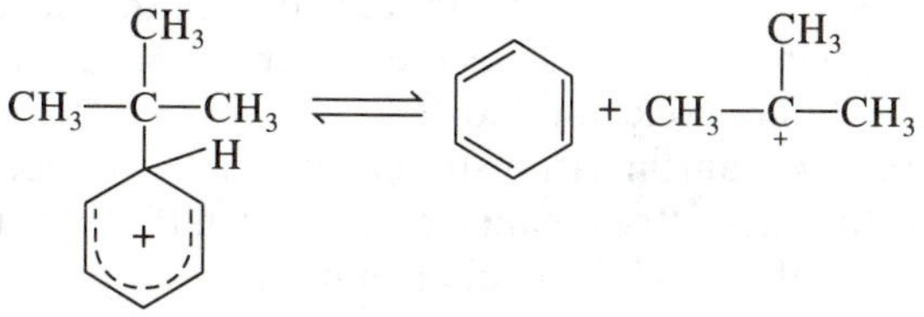

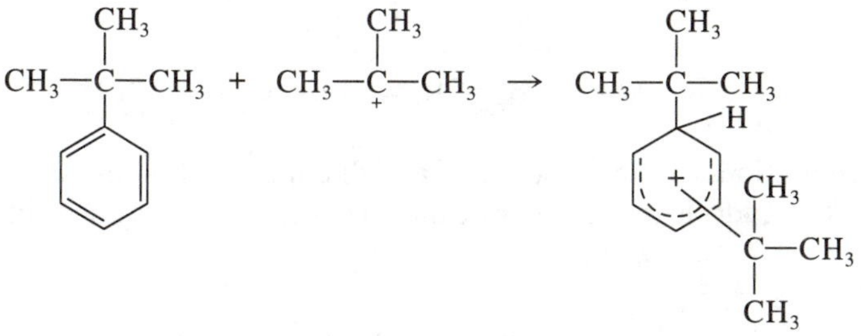

$$CH_3-\underset{\underset{\text{(protonated ring, +)}}{|}}{\overset{\overset{\displaystyle CH_3}{|}}{C}}\!\!\begin{array}{c}-CH_3\\ -H\end{array} \;\rightleftharpoons\; \bigcirc + CH_3-\overset{\overset{\displaystyle CH_3}{|}}{\underset{+}{C}}-CH_3 \qquad (2\text{-}94)$$

$$CH_3-\underset{\underset{\text{Ph}}{|}}{\overset{\overset{\displaystyle CH_3}{|}}{C}}-CH_3 \;+\; CH_3-\overset{\overset{\displaystyle CH_3}{|}}{\underset{+}{C}}-CH_3 \;\rightarrow\; CH_3-\overset{\overset{\displaystyle CH_3}{|}}{C}\!\!\begin{array}{c}-CH_3\\ -H\end{array} \qquad (2\text{-}95)$$

(with ring bearing $+$ and a $-\overset{\overset{\displaystyle CH_3}{|}}{\underset{\underset{\displaystyle CH_3}{|}}{C}}-CH_3$ substituent)

$$\underset{\text{(ring }+\text{, substituents)}}{H_3C-\overset{\overset{\displaystyle CH_3}{|}}{C}\!\!\begin{array}{c}-CH_3\\ H\end{array}} \quad \xrightarrow{} \;H^+ \;+\; \underset{\text{(product)}}{H_3C-\overset{\overset{\displaystyle CH_3}{|}}{C}-CH_3} \qquad (2\text{-}96)$$

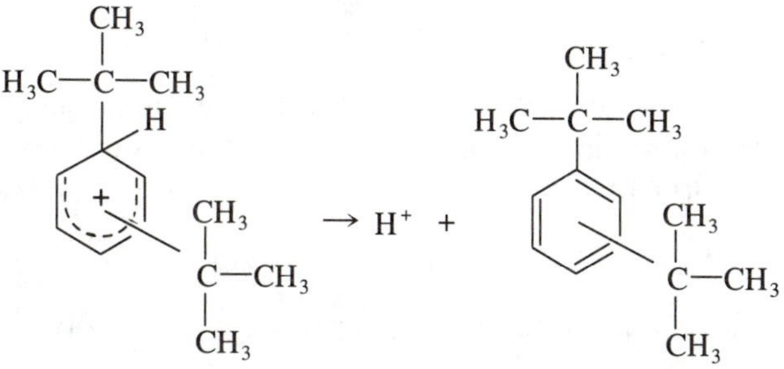

If step 2-95 is rate determining, the form of the kinetics is accounted for (and there would be inhibition by benzene). Similar steps account for the conversion of the other alkylbenzenes, their lower reactivities corresponding to the lower stabilities of secondary and primary carbenium ions, respectively. (How does the rate constant depend on the structure of the carbenium ion?) The lack of rearrangement of the alkyl groups is accounted for by the fact that the necessary carbenium ion rearrangements would require conversion of carbenium ion intermediates into less-stable carbenium ion intermediates.

Concerted reactions may play a role in these examples; the likelihood is greatest in the reaction of ethylbenzene, because the formation of the unstable ethyl cation would be avoided. (What are plausible structures of intermediates in the concerted disproportionation of ethylbenzene?) ∎

In alkylation, it is not just aromatics but also paraffins that can be used. The alkylation reactions of greatest importance industrially involve the reactions of light olefins such as propylene with isobutane (both products of petroleum refining) to give valuable gasoline components. Again, the reactions involve protonation of olefins to give carbenium ion intermediates and then formation of C—C bonds. Strong acids are required; those used commercially are concentrated HF and concentrated H_2SO_4.

The reaction mechanism is illustrated with the isobutane–ethylene alkylation, because the chemistry is simpler than that of the isobutane–propylene alkylation. As usual, the catalytic cycle is initiated as a proton is donated by an acid to a relatively strong base; in this case, the strongest available base is ethylene, and the resulting ionic species can only be a primary carbenium ion:

$$H^+ + CH_2{=}CH_2 \rightleftharpoons \overset{+}{H_2}C{-}CH_3 \qquad (2\text{-}97)$$

This carbenium ion, in the presence of a high concentration of isobutane, rapidly abstracts a hydride ion to generate the relatively stable tertiary butyl cation:

$$\underset{\underset{CH_3}{|}}{\overset{\overset{CH_3}{|}}{CH_3{-}C{-}H}} + \overset{+}{H_2}C{-}CH_3 \rightleftharpoons \underset{\underset{CH_3}{|}}{\overset{\overset{CH_3}{|}}{CH_3{-}\overset{+}{C}}} + CH_3{-}CH_3 \qquad (2\text{-}98)$$

The tertiary carbenium ion can now react with the olefin according to the pattern illustrated for olefin oligomerization. This step is slow, since a primary carbenium ion is formed:

$$\underset{\underset{CH_3}{|}}{\overset{\overset{CH_3}{|}}{CH_3{-}\overset{+}{C}}} + H_2C{=}CH_2 \rightleftharpoons \underset{\underset{CH_3}{|}}{\overset{\overset{CH_3}{|}}{CH_3{-}C{-}CH_2{-}\overset{+}{C}H_2}} \qquad (2\text{-}99)$$

This primary carbenium ion can now react in the same way that the ethyl cation reacts, abstracting a hydride ion from isobutane:

$$\underset{\underset{CH_3}{|}}{\overset{\overset{CH_3}{|}}{CH_3{-}C{-}CH_2{-}\overset{+}{C}H_2}} + \underset{\underset{CH_3}{|}}{\overset{\overset{CH_3}{|}}{H{-}C{-}CH_3}} \rightleftharpoons \underset{\underset{CH_3}{|}}{\overset{\overset{CH_3}{|}}{CH_3{-}C{-}CH_2{-}CH_3}} + \underset{\underset{CH_3}{|}}{\overset{\overset{CH_3}{|}}{\overset{+}{C}{-}CH_3}}$$

$$\text{alkylate}$$

$$(2\text{-}100)$$

The hydrocarbon product (2,2-dimethylbutane) is called *alkylate*. The tertiary carbenium ion consumed in reaction 2-99 is regenerated and can react further with ethylene. This carbenium ion is the predominant *chain carrier* in the chain reaction mechanism.

Although reaction 2-100 takes place, it is relatively unimportant in the formation of product alkylate, because one of the reactants, the primary carbenium ion, rapidly rearranges to give more stable secondary and tertiary carbenium ions. The following reactions are also important in the formation of alkylate:

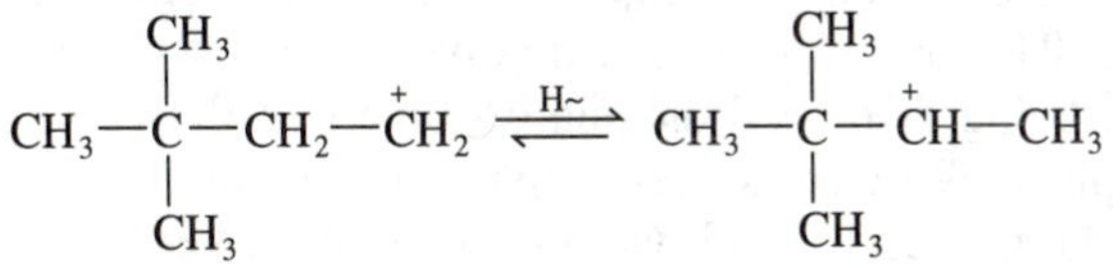

$$
\underset{\substack{|\\ CH_3}}{\overset{\substack{CH_3\\|}}{CH_3-C-CH_2-\overset{+}{C}H_2}} \underset{\xrightarrow{\;H\sim\;}}{\rightleftharpoons} \underset{\substack{|\\ CH_3}}{\overset{\substack{CH_3\\|}}{CH_3-C-\overset{+}{C}H-CH_3}} \qquad (2\text{-}101)
$$

$$
\underset{\substack{|\\ CH_3}}{\overset{\substack{CH_3\\|}}{CH_3-C-\overset{+}{C}H-CH_3}} + \underset{\substack{|\\ CH_3}}{\overset{\substack{CH_3\\|}}{H-C-CH_3}} \rightarrow \underset{\substack{|\\ CH_3\\ \text{alkylate}}}{\overset{\substack{CH_3\\|}}{CH_3-C-CH_2-CH_3}} + \underset{\substack{|\\ CH_3\\ \text{chain carrier}}}{\overset{\substack{CH_3\\|}}{\overset{+}{C}-CH_3}}
$$

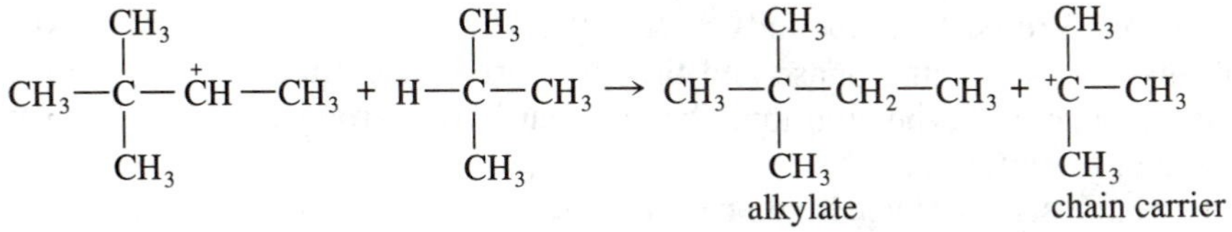

$$(2\text{-}102)$$

(The product is again 2,2-dimethylbutane.)

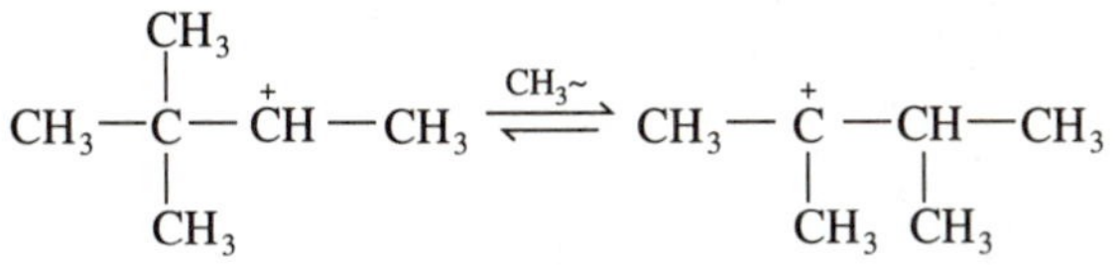

$$
\underset{\substack{|\\ CH_3}}{\overset{\substack{CH_3\\|}}{CH_3-C-\overset{+}{C}H-CH_3}} \underset{\xrightarrow{\;CH_3\sim\;}}{\rightleftharpoons} \underset{\substack{|\;\;|\\ CH_3\;CH_3}}{CH_3-\overset{+}{C}-CH-CH_3} \qquad (2\text{-}103)
$$

$$
\underset{\substack{|\;\;|\\ CH_3\;CH_3}}{CH_3-\overset{+}{C}-CH-CH_3} + \underset{\substack{|\\ CH_3}}{\overset{\substack{CH_3\\|}}{H-C-CH_3}} \rightarrow \underset{\substack{|\;\;|\\ CH_3\;CH_3\\ \text{principal}\\ \text{alkylate}}}{CH_3-CH-CH-CH_3} + \underset{\substack{|\\ CH_3\\ \text{chain carrier}}}{\overset{\substack{CH_3\\|}}{\overset{+}{C}-CH_3}}
$$

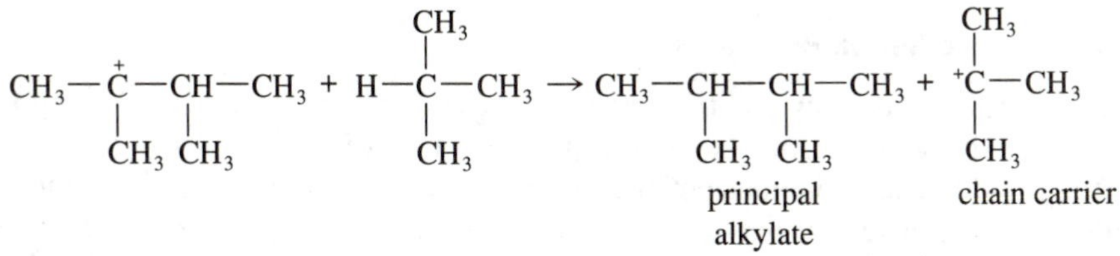

$$(2\text{-}104)$$

The product here, 2,3-dimethylbutane, is expected to be the principal one, since it is formed from a relatively stable tertiary carbenium ion.

Alkylation with propylene proceeds much more rapidly than alkylation with ethylene, because the secondary propyl cation is more easily formed than the primary ethyl cation. Alkylation with butenes proceeds still more rapidly. The product distributions in these latter reactions are complex, since many different carbenium ion intermediates can be formed, which leads to many products.

The process of isoparaffin–olefin alkylation is carried out in refrigerated reactors in petroleum refineries.[2] In one respect it is more complicated than all the reactions considered thus far, since it involves not one but two liquid phases. One is an aqueous phase, containing almost all the H_2SO_4 catalyst but only about 100 ppm of the nearly insoluble hydrocarbon; and the other is an organic (or "oil") phase, containing virtually all the reactants and products but only about 100 ppm of the nearly insoluble acid catalyst. Since the catalyst and reactants are so inefficiently brought in contact with each other in these phases, the rate of alkylation would be much too low to be economical, unless the process were somehow designed to maximize contacting of the reactants and catalyst. The practical design involves a vigorously stirred reactor, one liquid phase being dispersed as small droplets in the other phase, called the continuous phase. When this operation is carried out efficiently, high interfacial areas are obtained, and the alkylation takes place very near the liquid–liquid interface. Rates of reaction can be increased by adding surfactants to the reaction mixture [32]. Compounds such as $(CH_3)_4N^+Cl^-$, like soaps, have some affinity for each liquid phase and therefore are concentrated at the interface and stabilize it. Carbenium ions, being polar hydrocarbons, also are concentrated at the interface.

Catalysis in a single fluid phase is called *homogeneous catalysis*; catalysis occurring in the presence of more than one phase is called *heterogeneous catalysis*. Most practically important examples of heterogeneous catalysis involve solid catalysts and gas phase reactants. In these cases, considered in the following chapters, the reactions again take place at interfaces, now solid surfaces.

2.3
CATALYSIS BY ELECTRON TRANSFER

2.3.1 Simple Redox Reactions

The preceding sections illustrate a rich catalytic chemistry in which many of the elementary steps are transfers of protons or hydride ions. These steps are often rapid because the mass of the hydrogen atom is so low. Another kind of elementary transfer process is also rapid: the transfer of single electrons. These are the predominant processes in redox reactions, that is, reactions that involve changes in reactant oxidation states.

An example of a redox reaction taking place in aqueous solution is

$$V^{3+} + Cu^{2+} \rightleftharpoons V^{4+} + Cu^+ \tag{2-105}$$

---

[2] The alkylation process was invented by V. Ipatieff and H. Pines and coworkers in the U.S.A. in the early 1940s, and it was important in determining the outcome of the second world war: High-octane alkylate aviation fuel supplied from the U.S.A. gave British fighter plane pilots an advantage in acceleration over their German opponents that was decisive in the Battle of Britain.

This is a transfer of one electron and an example of one of the simplest kinds of redox reactions. It can result from the collision of the two cations, with the transfer of an outer-sphere electron; but collisions of ions of the same charge are not very likely, and water of hydration may hinder the process. Alternatively, water may serve as an intermediary, receiving an electron to form a hydrated electron and then donating an electron to another ion. Recall that water can similarly act as an intermediary in acid–base catalysis as H_3O^+ is formed.

A more complex redox reaction is, for example,

$$Tl^+ + 2Ce^{4+} \rightleftharpoons Tl^{3+} + 2Ce^{3+} \tag{2-106}$$

which involves a change of two units in the oxidation state of thallium ion. One might at first expect the elementary reaction step to involve two Ce^{4+} ions and one Tl^+ ion and therefore to proceed relatively slowly because of the infrequency of three-body collisions of species all having the same charge. A more efficient way to carry out reaction 2-106 is to take advantage of an intermediary, a redox pair such as Cu^{2+}/Cu^+ or Ag^{2+}/Ag^+, which can be involved in rapid one-electron transfers:

$$Ce^{4+} + Ag^+ \rightleftharpoons Ce^{3+} + Ag^{2+} \tag{2-107}$$

$$Ag^{2+} + Tl^+ \rightarrow Tl^{2+} + Ag^+ \tag{2-108}$$

$$Tl^{2+} + Ce^{4+} \rightarrow Tl^{3+} + Ce^{3+} \tag{2-109}$$

The reaction proceeds via the unstable Tl^{2+} intermediate.

Kinetics data [33] show that step 2-108 is rate determining:

$$r = kC_{Ag^{2+}}C_{Tl^+} \tag{2-110}$$

The silver ions play a catalytic role in the redox reaction, which is summarized in the cycle of Fig. 2-24.

The intervention of the Cu^+/Cu^{2+} redox pair can even accelerate redox reactions that otherwise might take place via a simple one-electron transfer, such as

$$Fe^{2+} + V^{4+} \rightarrow Fe^{3+} + V^{3+} \tag{2-111}$$

The reason for the catalytic activity in this example is that the one-electron transfers involving copper are more rapid than the one-electron transfer involving iron and vanadium (reaction 2-111), which is hindered by the solvation shells of the ions.

Electron transfer processes are also important in organic chemistry, for

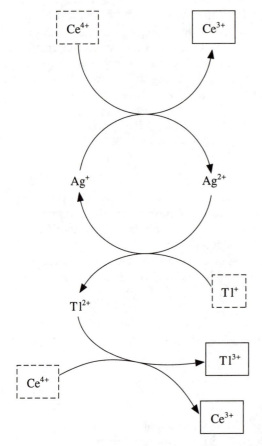

Figure 2-24
Catalytic cycle for the redox reaction
$Tl^+ + 2Ce^{4+} \rightarrow Tl^{3+} + 2Ce^{3+}$.

example, in the quinone–hydroquinone conversion:

$$
\underset{\substack{p\text{-benzoquinone} \\ (\text{quinone})}}{\text{(quinone)}} \quad \xrightleftharpoons[\substack{\text{oxidation} \\ (\text{e.g., } Fe^{3+})}]{\substack{\text{reduction} \\ (\text{e.g., } SO_3{}^{2-})}} \quad \underset{\text{hydroquinone}}{\text{(hydroquinone)}}
\tag{2-112}
$$

The organic redox reactions often involve free-radical intermediates, and many occur in catalytic oxidation, which is considered next.

2.3.2 Oxidation Involving Free-Radical Intermediates

Catalytic oxidation involves redox cycles. The reactions are characterized by complex reactivity patterns and complex product distributions, since O_2 has a high reactivity with organic reactants, with metal centers, and with many ancillary ligands. Reactions leading to the formation of organic free radicals are common, and often organic peroxides may form. The metal typically plays a role in the initiation process, helping to generate free radicals and start a chain reaction; often free radicals are generated by metal-catalyzed decomposition of organic hydroperoxides. Metal complexes differ greatly in their abilities to catalyze oxidation reactions *selectively*, and in recent years much research has been done with the objective of finding new routes for selective oxidation of petroleum-derived hydrocarbons, especially olefins and, to a smaller extent, paraffins, under mild conditions [34, 35].

Many oxidations of organic reactants in the liquid phase take place almost spontaneously at low temperatures and low O_2 concentrations and are called autoxidations. Most proceed via chain reactions described by the following general scheme:

Initiation

$$In_2 \rightarrow 2In^{\cdot} \tag{2-113}$$

$$In^{\cdot} + RH \rightarrow InH + R^{\cdot} \tag{2-114}$$

In_2 represents the initiator, which may be an impurity in the reactant, such as a hydroperoxide, and RH is the hydrocarbon reactant.

Propagation

$$R^{\cdot} + O_2 \rightarrow RO_2^{\cdot} \tag{2-115}$$

$$RO_2^{\cdot} + RH \rightarrow RO_2H + R^{\cdot} \tag{2-116}$$

This chain reaction cycle, which may occur many times for each occurrence of an initiation step, generates the hydroperoxide product, consuming the hydrocarbon and oxygen. Since hydroperoxide may act as an initiator, the process may be autocatalytic.

Termination
The termination steps are those that cause a net destruction of free radicals, for example,

$$R^{\cdot} + RO_2^{\cdot} \rightarrow RO_2R \tag{2-117}$$

$$2RO_2^{\cdot} \rightarrow RO_4R \rightarrow O_2 + \text{nonradical products} \tag{2-118}$$

The hydroperoxide product may be obtained in high yield, but it may also be an intermediate, undergoing further conversion. A common pathway for liquid phase autoxidations involves metal-catalyzed decomposition of alkyl hydroperoxides. The following cycle may occur in the presence of trace amounts of iron, manganese, cobalt, or copper naphthenates [35]:

$$RO_2H + M^{(n-1)+} \rightarrow RO^{\cdot} + M^{n+} + OH^- \qquad (2\text{-}119)$$

$$RO_2H + M^{n+} \rightarrow RO_2^{\cdot} + M^{(n-1)+} + H^+ \qquad (2\text{-}120)$$

The following chain can then be set up, not involving the metal:

$$2RO_2^{\cdot} \rightarrow 2RO^{\cdot} + O_2 \qquad (2\text{-}121)$$

$$RO^{\cdot} + RO_2H \rightarrow RO_2^{\cdot} + ROH \qquad (2\text{-}122)$$

If step 2-122 occurs twice for each occurrence of step 2-121, the overall reaction is

$$2RO_2H \rightarrow 2ROH + O_2 \qquad (2\text{-}123)$$

The role of the metal in this scheme is that of a redox initiator rather than a catalyst in the traditional sense. The only appropriate metals are those that undergo electron transfer giving changes of one unit in the oxidation state. The rate of the overall process depends on the redox potential of the $M^{n+}/M^{(n-1)+}$ couple.

In a number of practical applications in which transition metals accelerate autoxidations, the reactions are carried out in polar solvents, typically acetic acid, in the presence of relatively large amounts of metal catalyst, usually metal carboxylates and especially the acetate. Often the rate is high, with the conversion being carried on through oxidation of a hydroperoxide intermediate. An industrially important example is the autoxidation of cyclohexane in acetic acid in the presence of cobalt acetate or cobalt naphthenate; the desired product, formed in high yield, is adipic acid. This example is considered again below.

In such processes the metal may do more than just react with intermediate hydroperoxides (as in the case of Eqs. 2-119 and 2-120); it may play another role by reacting directly with the hydrocarbon reactant and/or with secondary autoxidation products. Two likely routes for the formation of free radicals resulting from the direct interaction of metal oxidants with hydrocarbons are electron transfer and electrophilic substitution. These possibilities are illustrated below for reaction of a metal triacetate with a hydrocarbon [35]:

Electron Transfer

$$RH + M(OAc)_3 \rightarrow [RH]^{\cdot +} + M(OAc)_2 + AcO^- \qquad (2\text{-}124)$$

$$[RH]^{\cdot +} \rightarrow R^{\cdot} + H^+ \qquad (2\text{-}125)$$

Electrophilic Substitution

$$RH + M(OAc)_3 \rightarrow RM(OAc)_2 + HOAc \qquad (2\text{-}126)$$

$$RM(OAc)_2 \rightarrow R^{\cdot} + M(OAc)_2 \qquad (2\text{-}127)$$

Each process forms the radical R˙ with a one-electron reduction of the metal. Distinguishing between these two possibilities is difficult, because it requires detection of the intermediate species, that is, the radical cation [RH]˙⁺ or the complex RM(OAc)₂. The role of the metal here is not simply catalytic, and, again, the products of the redox steps may be free radicals that can become involved in the cyclic propagation steps of a chain reaction.

Consider the example of methylnaphthalene oxidation in the presence of manganic acetate. One mechanism has been suggested to involve electron transfer (Fig. 2-25). The steps include a one-electron transfer (Mn^{3+} to Mn^{2+}) and proton transfer, as is typical in autoxidation. The chemistry of this oxidation is complicated, and the above statement is far from complete. One of the complications involves inhibition of the electron transfer route of Fig. 2-25 by Mn^{2+}, possibly by virtue of its effect on the equilibrium of the first step in the sequence.

Inhibition of autoxidation reactions by transition metals in low oxidation states (e.g., Co^{2+} or Mn^{2+}) has been observed frequently [35]. As alkylperoxy

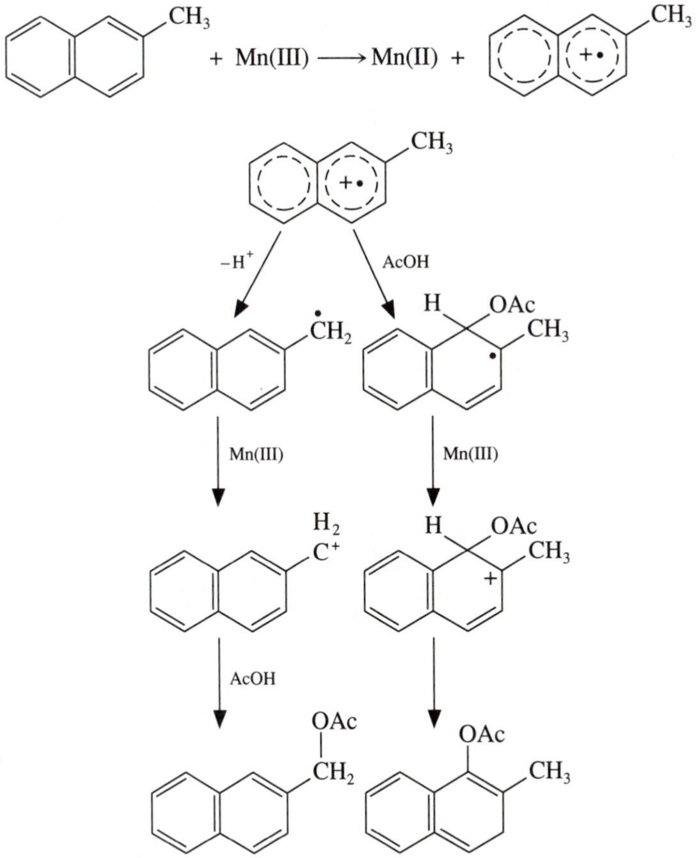

Figure 2-25
The suggested role of Mn^{3+} in the oxidation of 2-methylnaphthalene [35]. This scheme is only part of a more complex sequence.

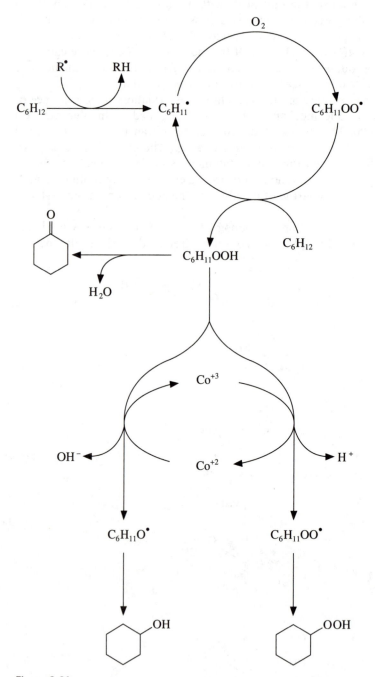

Figure 2-26
Catalytic cycle suggested for the free-radical oxidation of cyclohexane to give
nylon intermediates [36].

radicals are relatively strong oxidizing agents, they can undergo redox reactions with the metal ions:

$$RO_2^{\cdot} + M^{(n-1)+} \rightarrow RO_2M^{n+} \tag{2-128}$$

In the early stage of an autoxidation, which is characterized by an induction period before the free-radical concentration has reached a nearly time-invariant value, the metal ions can act as inhibitors by depleting the supply of chain-carrying free radicals; by reactions like that shown in Eq. 2-128, the metal ions provide new routes for chain termination.

A large-scale application of liquid phase autoxidation is the conversion of cyclohexane with air to give cyclohexanone and cyclohexanol, which can be oxidized further to give adipic acid. The catalyst is typically cobalt naphthenate used in low concentration, ca. 20 ppm. The conversion is kept to about 10%, since the selectivity for formation of the desired products decreases with increasing conversion, as the products are more reactive than cyclohexane [36].

The principal primary product is cyclohexyl hydroperoxide, as expected, and most of the cyclohexanone and cyclohexanol are formed from the hydroperoxide. A suggested catalytic cycle is shown in Fig. 2-26. The familiar free-radical chain with $R^{\cdot}$ (C_6H_{11}) and $ROO^{\cdot}$ ($C_6H_{11}OO^{\cdot}$) provides a route to the alkyl hydroperoxide; this is shown in the upper right of the figure. The role of Co^{3+} is probably that of an electron transfer agent undergoing a one-electron change, but the details of the lower part of the figure are obscure.

Most liquid phase transition-metal-catalyzed oxidations of organic compounds fall in the category of free-radical autoxidation reactions. Nonetheless, there are important examples of liquid phase oxidations that do not involve free radicals, as described in Section 2.4.2. Experiments to distinguish these possibilities are sometimes done with added free-radical scavengers. If these markedly slow the catalytic reaction, free radicals may be implicated (but the scavengers may also be ligands that inhibit reactions by competing with reactants for bonding sites on the metal). It is sometimes possible to observe free-radical intermediates by electron spin resonance spectroscopy, or to trap the radicals and to observe the products spectroscopically.

It is not appropriate to assume that an easy distinction between the two classes of reactions can always be made. Often there may be just too low a concentration of free radicals to allow detection, and even steps like oxidative addition (discussed below) may proceed via free radicals as, for example, in hydroformylation [37]. Free-radical intermediates may be more important in reactions other than oxidations than has been appreciated [38].

2.4
ORGANOMETALLIC CATALYSIS

2.4.1 Introduction

The examples of metals in catalysis presented thus far show that they act by acid–base polarization and by participating in one-electron transfer reactions.

But metals do far more than this in catalysis: they undergo cycles with two-electron reactions, and they bond to organic reactants in unique ways to activate them and to facilitate their conversion. The metals involved in this chemistry are the transition metals; many of the most useful are Group 8 metals. The recent renaissance in the chemistry of transition metal compounds has been motivated in large measure by the application of these compounds as soluble catalysts. A list of industrial catalytic reactions in this group is given in Table 2-1.

The subject of catalysis by transition metal compounds (or complexes) is unified by the chemistry of the complexes. The organic reactants are first bonded (coordinated) as ligands to the metal, and they are then converted in several kinds of reactions. The bonding, structure, and reactions of transition metal complexes are reviewed briefly here to provide a basis for understanding the catalysis.

BONDING AND STRUCTURE OF TRANSITION METAL COMPLEXES [37]

Transition metal complexes have metal atoms or ions bonded to atoms or groups of atoms called ligands. When there are metal–carbon bonds, the complexes are *organometallic*. The ligands surround the metal and form a polyhedron with the metal in the center. The most frequently observed geometries are illustrated in Fig. 2-27; they are the regular six surrounding (octahedral), two types of five surrounding (tetragonal pyramidal and trigonal bipyramidal), and two types of four surrounding (tetrahedral and square planar). There are often distortions from these idealized geometries.

The transition metals, listed in Table 2-2, are the elements that have partially filled d shells in some of their compounds. These metals exhibit multiple oxidation states and bond to variable numbers of ligands. The metal uses its partially filled d orbitals and the next higher s and p orbitals for the formation of the metal–ligand bonds in the complex. The Nd level of a transition metal cation is usually lower in energy that the $(N + 1)s$ level, which is lower in energy than the $(N + 1)p$ level (N is the principal quantum number). Consequently, it is conventional to treat the metal's valence shell electrons as though they were all in the Nd shell. The convention is usually appropriate for metals in relatively high oxidation states and for elements toward the right of the transition series, but it may not be valid for neutral complexes of early transition metals.

As the d levels are usually those having the highest energies, they are the ones to which electrons can most easily be added or from which electrons can most easily be removed. The d electrons are primarily associated with the metal, and because it is the d level to which electrons are added or from which they are removed, the number of these d electrons, d^n, is related to the oxidation state of the metal.

The formal oxidation state of a metal in a complex is defined as the charge remaining on the metal atom when the ligands are removed in their normal

Table 2-1
INDUSTRIAL REACTIONS CATALYZED BY
TRANSITION METAL COMPLEXES [36]

Carbonylations

$$CH_3CH{=}CH_2 + CO + H_2 \rightarrow C_3H_7CHO \text{ (plus other oxo products)}$$

$$RCH{=}CH_2 + CO + 2H_2 \rightarrow RCH_2CH_2CH_2OH \ (R > C_8H_{17})$$

$$CH_3OH + CO \rightarrow CH_3COOH$$

Monoolefin Reactions

$$CH_2{=}CH_2 + O_2 \rightarrow CH_3CHO$$

$$CH_3CH{=}CH_2 + ROOH \rightarrow CH_3CH\underset{O}{\overset{}{\diagup\!\!\diagdown}}CH_2 + ROH$$

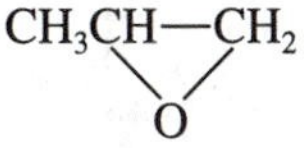

$$CH_2{=}CH_2 \rightarrow \text{polyethylene}$$

$$CH_2{=}CH_2 + CH_3CH{=}CH_2 + \text{diene} \rightarrow \text{EPDM rubber}$$

Diene Reactions

$$3CH_2{=}CHCH{=}CH_2 \rightarrow \text{cyclododecatriene}$$

$$C_4H_6 + CH_2{=}CH_2 \rightarrow \text{1,4-hexadiene}$$

$$C_4H_6 + 2HCN \rightarrow NC(CH_2)_4CN$$

$$C_4H_6 \rightarrow cis\text{-1,4-polybutadiene}$$

Oxidations

$$\text{cyclohexane} \xrightarrow{O_2} \text{cyclohexanol} + \text{cyclohexanone} \xrightarrow[HNO_3]{O_2 \text{ or}} \text{adipic acid}$$

$$H_3C\text{—}\langle\text{benzene ring}\rangle\text{—}CH_3 \xrightarrow{O_2} \text{terephthalic acid and esters}$$

$$n\text{-}C_4H_{10} \xrightarrow{O_2} CH_3COOH$$

$$CH_3CHO \xrightarrow{O_2} CH_3COOH$$

Other Reactions

$$CH_2{=}CHCHClCH_2Cl \rightleftharpoons ClCH_2CH{=}CHCH_2Cl$$

$$ClCH_2CH{=}CHCH_2Cl + 2NaCN \rightarrow NCCH_2CH{=}CHCH_2CN$$

$$ROOC\text{—}\langle\text{benzene ring}\rangle\text{—}COOR + HOCH_2CH_2OH \rightarrow \text{polyester}$$

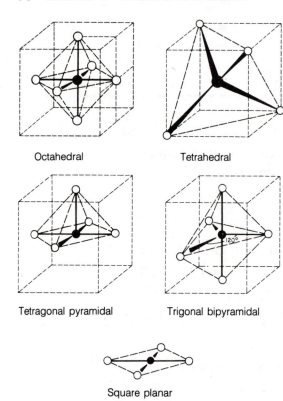

Octahedral

Tetrahedral

Tetragonal pyramidal

Trigonal bipyramidal

Square planar

Figure 2-27
Ligand arrangements in transition metal complexes. The ligand polyhedra are presumed to contain only a single type of ligand. Similar surroundings occur with more than one kind of ligand in the complex, but then symmetry is generally lower.

closed-shell configurations. It is important to realize that oxidation state is a formalism; it does not refer to the actual location of the electrons in a transition metal complex. The formalism provides a convenient and consistent basis for bookkeeping but not a basis for predicting properties of metal complexes.

The bookkeeping refers to the occupancy of the *d* orbitals. There is usually

Table 2-2
THE TRANSITION METALS AND THEIR NUMBERS OF *d*
ELECTRONS IN VARIOUS OXIDATION STATES [37]

Group Number		**4**	**5**	**6**	**7**	**8**	**9**	**10**	**11**
First row	3d	Ti	V	Cr	Mn	Fe	Co	Ni	Cu
Second row	4d	Zr	Nb	Mo	Tc	Ru	Rh	Pd	Ag
Third row	5d	Hf	Ta	W	Re	Os	Ir	Pt	Au
Oxidation State		d^n, Number of *d* Electrons							
Zero		4	5	6	7	8	9	10	—
I		3	4	5	6	7	8	9	10
II		2	3	4	5	6	7	8	9
III		1	2	3	4	5	6	7	8
IV		0	1	2	3	4	5	6	7

a maximum number of ligands allowed for each d^n, provided that the complex is *mononuclear* (i.e., has only one metal atom) and diamagnetic (n is an even number, with all electrons paired):

$$n + 2(CN)_{max} = 18 \qquad (2\text{-}129)$$

where n is the number of d electrons and CN is the metal coordination number (defined as the number of σ bonds formed between a metal and its ligands). Equation 2-129 is called the 18-electron rule.

The rule is almost always obeyed; scarcely any exceptions are encountered in the following discussion of catalytic reactions. Equation 2-129 refers to the *maximum* number of ligands. When there are fewer than the maximum number of ligands in a metal complex, the complex is then referred to as *coordinatively unsaturated*. Coordinatively unsaturated 14- and 16-electron complexes are reactive and important in catalysis.

To do electron counting in a transition metal complex, one needs d^n for the metal (Table 2-2) and the formal charge and coordination number of each ligand; the latter are summarized for some common ligands in Table 2-3. The choice of charge on a ligand is sometimes more a matter of convention than a representation of the chemistry. For example, H is considered as a hydride ligand, bearing a charge of -1 and being an electron pair donor; but for some metal complexes the assignment is clearly arbitrary [e.g., $HCo(CO)_4$, which is a strong acid]. It is also important to recognize that a ligand can have different coordination numbers. For example, H can be bonded to a single metal atom

Table 2-3
LIGANDS AND THEIR TYPICAL CHARGES AND
COORDINATION NUMBERS (ADAPTED FROM REF. 37)

Ligand	Charge[a]	Coordination Number[b]
X (Cl, Br, I)	-1	1 (2)
H	-1	1 (2, 3)
CH_3	-1	1 (2)
CO	0	1 (2, 3)
$R_2C{=}CR_2$	0 (-2)	1 (2)
$RC{\equiv}CR$	0 (-2)	1 (2)
η^3-Allyl[c]	-1	2
η^6-Benzene	0	3 (2, 1)
η^5-Cyclopentadienyl	-1	3
RCO	-1	1 (2)
R_3N	0	1
R_3P	0	1
O	-2	2
O_2	-2 (-1)	2 (1)

[a] Less common charges are stated in parentheses.

[b] Less common coordination numbers are stated in parentheses.

[c] The superscript 3 implies that all three carbon atoms of the allyl ligand interact with the metal.

(CN for hydrogen $= 1$) (it is a terminal ligand), or it can bridge two metal atoms (CN for hydrogen $= 2$). Hydride ligands frequently occur in transition metal complexes that are intermediates in catalytic reactions involving H_2. The metal–hydrogen bonds are usually strong (with bond energies of about 60 kcal/mol).

To understand the chemical bonding in transition metal complexes, a survey of ligands and their bonds with the metals is helpful. Ligands such as NH_3 and H_2O, which have only a filled orbital ("lone pair") for interaction with the metal, form classical coordination complexes with the metal. They are combined only by interaction of the ligand electrons with empty d, s, or p orbitals of the metal. These ligands are Lewis bases, and the metal is a Lewis acid. The bond formed is rotationally symmetric about the metal–ligand axis and is therefore designated as a σ bond. The ligands are unidentate, that is, they are bonded to the metal through single ligand atoms. The classical unidentate oxygen donors include H_2O, CH_3OH, and tetrahydrofuran. These are "hard" (small and weakly polarizable) and weakly basic and are only weakly bonded to transition metals in low oxidation states.

Ligands including Cl^-, Br^-, I^-, and OH^- have two or more filled orbitals which can interact with two empty metal orbitals, as illustrated in Fig. 2-28. One of the ligand orbitals (p_x) forms a σ bond, but the second (p_y), which must be oriented perpendicular to the metal–ligand axis, can only form a bond having no rotational symmetry; it is therefore called a π bond. For both the σ and π bonds, the electrons are donated by the ligand.

The halide ligands readily form bridges, which are easily broken in reactions with other ligands. The following structure, for example, is believed to be important in the Wacker reaction, discussed later:

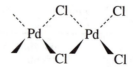

Metal   Ligand (Cl^-)

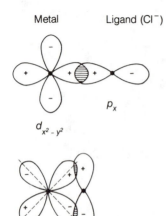

Figure 2-28
Double bond involving two filled ligand orbitals and two empty metal orbitals. The ligand may, for example, be Cl^-, Br^-, I^-, or OH^-.

The iodide ligand is important in catalysis. It is a large, polarizable ("soft") ligand, a strong nucleophile, and a weak proton acceptor; it forms strong bonds with transition metals in low oxidation states. It plays a role in the catalytic process for methanol carbonylation to give acetic acid, as discussed below.

There is a group of electron donor ligands containing P and, less important, As and Sb. The phosphines, especially triphenyl phosphine, are strong electron donors; they are the most common of these ligands encountered in catalysis. The tendency for the donor atoms to be coordinated with metals in high oxidation states decreases in the order P > As > Sb; the trend may be related to the increasing polarizability (or "softness") of the donor atoms in this series.

The tertiary phosphine ligands often incorporate bulky groups such as phenyl or *t*-butyl. Consequently, steric effects are important in many of the catalytic reactions, as is illustrated below for hydroformylation of olefins.

There are a number of bidentate and multidentate phosphine ligands, such as the following, known as DIPHOS:

$$\begin{array}{cc} CH_2 & -CH_2 \\ | & | \\ PPh_2 & PPh_2 \end{array}$$

This bidentate ligand can occupy two coordination sites on a metal atom.

Hydrocarbon ligands are important in organometallic catalysis. Among the simplest are alkyl ligands, which, like hydride, are regarded as negatively charged two-electron donors forming σ bonds with metals.

The acyl ligand bonded to a metal M,

is also regarded as a negatively charged two-electron donor.

Metal carbonyls are among the most common organotransition metal complexes, and CO is a common reactant in catalysis. CO bonds to a transition metal as an electron donor, but there is much more to the bonding than this. Figure 2-29 shows the interaction of a CO ligand and a metal. The C atom has a lone pair of electrons, which is the filled orbital, and an antibonding π^* orbital, which is the empty orbital. The combination of this filled bonding orbital and an empty d orbital on the metal constitutes the σ bond. But the symmetry of the d orbitals of the transition metal complements the symmetry of the orbitals of CO in such a way that there is overlap between the filled bonding orbitals of the metal and the empty antibonding orbitals of the CO (Fig. 2-29); hence, there is another kind of bonding (π bonding) between the ligand and metal. The CO ligand has the property of stabilizing lower metal oxidation states because the low-energy vacant orbitals of the CO have the correct symmetry to form π bonds with the metal by accepting electrons from filled metal d orbitals. The CO ligand is simultaneously an electron donor and an electron

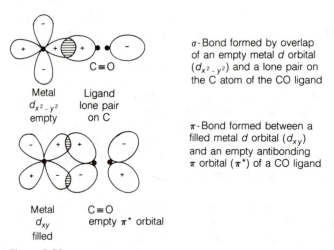

σ-Bond formed by overlap of an empty metal d orbital ($d_{x^2-y^2}$) and a lone pair on the C atom of the CO ligand

π-Bond formed between a filled metal d orbital (d_{xy}) and an empty antibonding π orbital (π*) of a CO ligand

Figure 2-29
Double bond involving a transition metal and a CO ligand. The metal, C, and O atoms are arranged linearly.

acceptor; the interaction is referred to as *backbonding* because electrons are transferred from CO to the metal and back to CO. Ligands capable of backbonding are referred to as π acids.

CO is a strong π acid; the degree of electron transfer back to the CO ligand is significant in many metal carbonyl complexes. One consequence of this backbonding is that it stabilizes the otherwise "nonbonding" d electrons, making them less susceptible to reaction. Another consequence is that it weakens the C—O bond and makes the ligand more reactive than uncoordinated CO, and this bond activation is important in catalysis.

There are numerous other π-acid ligands. Olefins, for example, undergo backbonding with transition metals, as depicted in Fig. 2-30. The σ donor bond results from the interaction of the filled π orbital of the olefin with an empty d orbital of the metal. The backbonding results from the interaction of a filled d orbital of the metal with the empty antibonding π* orbital of the olefin. This bonding weakens the C—C bond and activates the ligand. In an extreme case of strong bonding, the complex becomes a *metallacycle*, in which the metal forms two σ bonds to two carbon atoms.

Backbonding is one of the characteristics distinguishing transition metals from nontransition metals. Backbonding is reduced by positive charge at the metal atom. The bonding requires a close approach of the ligand to the metal so that orbital overlap can occur. The shortened metal–ligand bond distance provides an experimental criterion for the occurrence of backbonding. Another criterion is the weakening of the ligand bonds (such as C≡O and C=C), indicated by decreased force constants of the stretching vibrations shown by band shifts in the infrared and Raman spectra.

Many unsaturated hydrocarbons form complexes with transition metals. Aromatics, for example, are π-bonded in sandwich complexes:

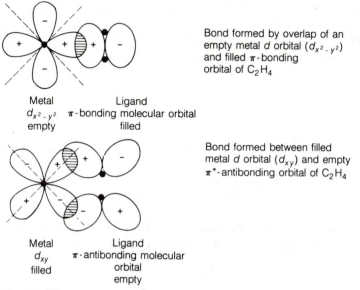

Bond formed by overlap of an empty metal d orbital ($d_{x^2-y^2}$) and filled π-bonding orbital of C_2H_4

Metal $d_{x^2-y^2}$ empty

Ligand π-bonding molecular orbital filled

Bond formed between filled metal d orbital (d_{xy}) and empty π^*-antibonding orbital of C_2H_4

Metal d_{xy} filled

Ligand π-antibonding molecular orbital empty

Figure 2-30
Double bond involving a transition metal and a $CH_2{=}CH_2$ ligand. The C–C axis is perpendicular to the metal–ligand bond.

M = Cr, Mo, or W

Cyclopentadienyl ligands form similar π-bonded complexes referred to as *metallocenes*:

Allyl groups are also π-bonded to metals; these are examples of *fluxional ligands* (i.e., those which exist in various structural forms). (What example of a fluxional ligand has already been encountered?) The following rearrangements have been found to be fast and reversible, occurring on the ^{1}H NMR time scale:

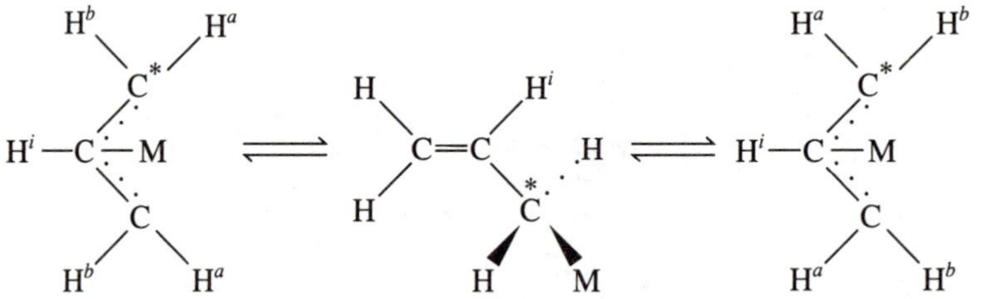

$$(2\text{-}130)$$

These complexes play a role as intermediates in catalytic conversion of olefins.

Dinitrogen forms weakly backbonded complexes with transition metals:

$$M\text{—}N\equiv N$$

These may be important in the reduction of N_2 to NH_3 catalyzed by enzymes containing metals (metalloenzymes).

Dioxygen also forms complexes with transition metals, with structures such as the following, containing bidentate and unidentate ligands:

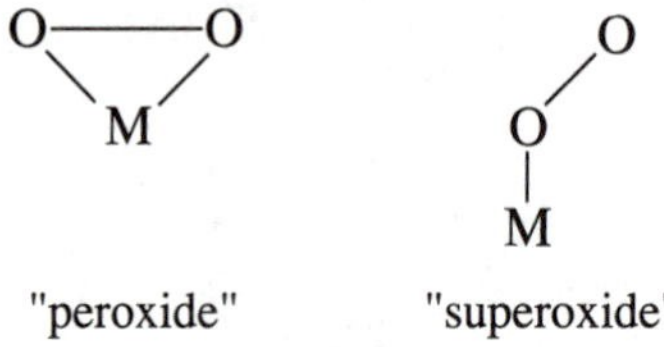

"peroxide"   "superoxide"

These play roles in catalytic oxidations, including those catalyzed by metalloenzymes.

EXAMPLE 2-5 Electron Counting in Transition Metal Complexes
Problem

(a) A set of isoelectronic octahedral metal complexes (those having the same bonding pattern) is $[V(CO)_6]^-$, $[Cr(CO)_6]$, and $[Mn(CO)_6]^+$. What are the charges on the metals? (b) What are the numbers of electrons in the following catalytic reaction intermediates: $[RhI_2(CO)_2]^-$, $[RhCl(PPh_3)_3]$, and $[H_2RhCl(PPh_3)_3]$? Which of these are coordinatively saturated? The first two incorporate Rh in the $+1$ oxidation state; the third incorporates Rh in the $+3$ oxidation state.

Solution

(a) Each of the isoelectronic octahedral complexes has 18 electrons; the carbonyl ligands supply two each, hence, 12 per complex. Each cation is therefore d^6; V is therefore V(-1); Cr is Cr(0); and Mn is Mn($+1$) (Table 2-2). The total charges check.

(b) From Table 2-2 it is evident that Rh(I) is d^8; two I^- ligands donate four electrons and two CO ligands four more, for a total of 16 in $[RhI_2(CO)_2]^-$; this complex is therefore coordinatively unsaturated. Similarly, $[RhCl(PPh_3)_3]$ is a 16-electron complex and coordinatively unsaturated. $[H_2RhCl(PPh_3)_3]$ is a coordinatively saturated 18-electron complex. ∎

REACTIONS OF TRANSITION METAL COMPLEXES [37]

A small group of elementary reactions accounts for all of the catalytic cycles involving reactants that are coordinated to transition metal complexes. One of these is *ligand dissociation*, which is simply the breaking of a metal–ligand bond:

$$ML_n \rightarrow ML_{n-1}\square + L \tag{2-131}$$

where the square (which is often omitted) represents a vacant coordination site.

In a reaction of a molecule A with a molecule B catalyzed by a transition metal complex, A (and/or B) often becomes coordinated to the metal, requiring a vacant coordination site, which can be formed by ligand dissociation. The coordinatively unsaturated complex ML_{n-1} may react with ligand A to give $ML_{n-1}A$, and dissociation of another ligand L followed by *association* of ligand B gives $ML_{n-2}AB$. The sequence of ligand dissociation–ligand association reactions is called *ligand exchange*.

Ligand dissociation is usually thermally initiated and often rapid; photodissociation of ligands (e.g., CO) may be even more rapid. In catalysis, it is important that the ligand–metal bond strength not be too great, or else it would prevent formation of vacant sites. Furthermore, the M—A and M—B bonds, which may be similar in strength to the M—L bonds, should not be too weak, because then A and B would not be bonded to the metal in sufficient concentrations to allow rapid reaction with each other. This reasoning suggests the idea of an optimum bond strength between the metal and a reactant. Similar considerations apply more generally in catalysis, as is illustrated in Chapter 6 in the discussion of Sabatier's principle.

Association of neutral ligands such as CO or C_2H_4 does not appreciably change the electron density on a metal, nor does it change the formal oxidation state of the metal. If, however, a molecule such as H_2 is to become bonded to the metal with dissociation of the H—H bond, the metal must donate two electrons when accepting the H atoms as ligands. Consequently, the metal is converted to a higher oxidation state. This type of reaction is known as *oxidative addition*, and the reverse is known as *reductive elimination*:

$$
\begin{array}{c}
\overset{\displaystyle .L}{\underset{\displaystyle L}{\diagup}} \\
L-\overset{..}{M^{n+}}-L \quad + \quad AB
\end{array}
\underset{\text{reductive elimination}}{\overset{\text{oxidative addition}}{\rightleftharpoons}}
\quad
\begin{array}{c}
B \quad .L \\
| \quad .. \\
L-M^{(n+2)+}-A \\
\diagup \quad | \\
L \quad L
\end{array}
\tag{2-132}
$$

Since two electrons must be given up by the metal ion, oxidative addition requires two vacant coordination sites and a transition metal, such as Rh, that has a tendency to occur in oxidation states which are separated by two units. Oxidative additions are the most important reactions in forming metal–carbon and metal–hydrogen bonds. Many compounds important in catalysis, including H_2, HI, and CH_3I, undergo oxidative addition reactions with metal complexes, and most catalytic cycles involving transition metal complexes involve oxidative addition and reductive elimination. Often oxidative additions are slow in comparison with ligand exchanges. Oxidative addition and reductive elimination are generic terms that do not imply particular mechanisms; the mechanistic issues are complex, and various intermediates may be involved, including free radicals.

For catalysis, one of the most important reactions that can occur in the coordination sphere of a transition metal is the *insertion* reaction, which is one of the steps in each of the cycles discussed below under the heading of organometallic catalysis. An insertion reaction is defined as one in which an atom or group of atoms is inserted between two atoms initially bound together. The insertion reaction is understood more specifically here to be a reaction taking place in the coordination sphere of a metal atom or ion, the result being the insertion of one ligand between the metal and another ligand. In a subsequent (and often rapid) step, an external ligand (or solvent molecule) usually fills the coordination site left open by the insertion.

The insertion reaction can be described in simplified terms on the basis of two different transition states. A three-center transition state explains the commonly observed CO insertion between a metal and an alkyl group:

$$
\begin{array}{c}
R_3 \\
C \\
| \\
-M-CO \\
|
\end{array}
\rightleftharpoons
\left[
\begin{array}{c}
R_3 \\
.C.. \\
-M\cdots\cdots CO \\
|
\end{array}
\right]^{\ddagger}
\rightleftharpoons
\begin{array}{c}
\square \quad CR_3 \\
-M-C \\
| \quad\quad\ \ O
\end{array}
\tag{2-133}
$$

Mechanistically, this reaction is described as an alkyl migration, and it is similar to the alkyl migrations in carbenium ions. The reactivity of transition metal alkyls toward insertion of the carbonyl ligand usually decreases with descent in a triad; for example, the rate decreases from Co to Rh to Ir complexes. These reactions can be induced by oxidation of the metal.

A four-center transition state explains reactions like the following one, which occurs in olefin polymerization:

$$
\begin{array}{c}
R_3 \\
C \\
| \quad CH_2 \\
-M-\ \| \\
| \quad CH_2
\end{array}
\rightleftharpoons
\left[
\begin{array}{c}
\ \ \ H \quad\ \ H \\
R_3 \quad \backslash\ /\ \\
C\cdots C \\
-M\cdots C \\
| \quad /\ \backslash \\
\ \ \ H \quad\ \ H
\end{array}
\right]^{\ddagger}
\rightleftharpoons
\begin{array}{c}
R_3 \\
C \\
| \\
HCH \\
\square \quad | \\
-M-\ \ C \\
| \quad /\ \backslash \\
\ \ \ H \quad\ H
\end{array}
\tag{2-134}
$$

The intramolecular migrations of hydride, alkyl, aryl, and acyl ligands to co-ordinated olefins and acetylenes occur in many catalytic cycles.

These insertion reactions are referred to as cis insertions because the reactants are bonded cis to each other in the coordination sphere of the metal. Often the insertions are slow in comparison with ligand exchanges, and they are rate determining in many catalytic cycles.

PATTERNS OF REACTIVITY OF THE TRANSITION METALS [37]

In predicting catalytic properties of transition metal complexes, it would be most helpful to know a pattern of reactivities of the metals. Unfortunately, only a few generalizations can be made.

The second- and third-row metals usually have stronger metal–hydrogen, metal–carbon, and metal–metal bonds than the first-row metals. The third-row complexes usually exhibit lower reactivity than the second- or first-row elements, and the most active catalysts are usually from the second or first row.

The metals on the left-hand side of the periodic table (Groups 4, 5, and 6) are more electropositive than those at the right and tend to form stronger bonds with "hard" donor atoms such as oxygen. These oxophilic elements have relatively few d electrons and therefore high coordination numbers. The π-acid ligands do not bond as strongly to Group 4 and 5 metals as to those at the right, and the migratory insertion reactions are often facile, with the equilibria lying to the side of the insertion products.

The Group 8, 9, and 10 metals are less electropositive than those at the left and are easily reduced to the zerovalent state. Soft ligands form strong bonds with these elements. Many of the complexes have d^8 or d^{10} configurations and readily undergo oxidative addition and reductive elimination reactions. The Group 8, 9, and 10 metals, especially in the second and third rows, often have low coordination numbers with relatively stable coordinatively unsaturated species. These metals often occur in complexes with strongly bonded π-acid ligands, such as CO and olefins.

2.4.2 Examples of Organometallic Catalysis

OLEFIN HYDROGENATION

Olefin Hydrogenation with Wilkinson's Catalyst

One of the best-understood catalytic cycles of all is the hydrogenation of an olefin in the presence of phosphine complexes of rhodium. The stoichiometry is

$$\diagdown \hspace{-0.3em} C \hspace{-0.3em} = \hspace{-0.3em} C \hspace{-0.3em} \diagup + H_2 \rightarrow \diagdown \hspace{-0.3em} HC \hspace{-0.3em} - \hspace{-0.3em} CH \hspace{-0.3em} \diagup \qquad (2\text{-}135)$$

This catalytic reaction takes place under conditions as mild as 1 atm and 25°C. Phosphine complexes of ruthenium and iridium are similar to the complexes of rhodium as hydrogenation catalysts [39]. Wilkinson and co-workers discovered the remarkable catalytic properties of the rhodium phosphine complex

RhCl(PPh$_3$)$_3$ for this reaction [40, 41] and of similar rhodium phosphine complexes [e.g., Rh(CO)Cl(PPh$_3$)$_2$] for the olefin hydroformylation reaction considered below. These discoveries played a central role in awakening interest in organometallic catalysis and opened the way to the discovery of some of the industrial processes listed in Table 2-1 and discussed in the following pages. Wilkinson and E. O. Fischer were awarded the Nobel prize in chemistry in 1973 in recognition of their pioneering work in organometallic chemistry.

The Wilkinson hydrogenation was investigated by Halpern and co-workers [42–44] (summarized in refs. 37 and 45), whose work led to identification of the reaction intermediates and determination of the kinetics of individual steps in the cycle. The quantitative kinetics is given with the cycle in Fig. 2-31. The statement of this cycle is exemplary for its depth and completeness. The research required to determine it included isolation and identification of individual rhodium complexes, measurements of equilibria of individual steps, determination of rates of individual steps under conditions of stoichiometric reaction (with certain reactants missing so that catalysis could not occur), and determination of rates of the overall catalytic reaction.

In the investigation of the catalytic hydrogenation, several rhodium complexes were directly observed and characterized, either being crystallized from the solution or identified in solution by ^{31}P NMR spectroscopy [42–44]. These complexes are RhCl(PPh$_3$)$_3$, RhCl(PPh$_3$)$_2$(olefin), Rh$_2$Cl$_2$(PPh$_3$)$_4$, RhH$_2$Cl(PPh$_3$), and Rh$_2$H$_2$Cl$_2$(PPh$_3$)$_4$. Examination of the catalytic cycle of Fig. 2-31 (including the quantitative kinetics) shows that these complexes are not involved in the kinetically significant steps of the cycle. In a sense, they represent dead ends; their presence indicates a draining away of Rh from the catalytic cycle. There is a generally important lesson about catalysis here: The species that can be observed during catalysis often are not directly involved in the cycle, and the identification of the species present in high concentrations can easily lead to a misinterpretation of the cycle. The exceptions are usually the cycles of least interest—those that involve slow catalytic reactions.

The Wilkinson hydrogenation cycle was resolved in a series of kinetics experiments, usually not involving the complete cycle [42–44]. In one set of experiments, the kinetics of hydrogenation of RhCl(PPh$_3$)$_3$ was determined spectrophotometrically; when less than an excess of triphenylphosphine was present, the reaction was so fast that stopped-flow techniques had to be used.

Analysis of the kinetics data led to the conclusion that the kinetically significant path for hydrogenation involved dissociation of a PPh$_3$ ligand from RhCl(PPh$_3$)$_3$, giving the coordinatively unsaturated (14-electron) complex RhCl(PPh$_3$)$_2$. (A solvent molecule may be weakly coordinated to the Rh in this complex.) The subsequent hydrogenation of RhCl(PPh$_3$)$_2$ (an oxidative addition) is extremely rapid. The resulting unsaturated complex undergoes a ligand association with PPh$_3$; this is a rapid (equilibrium) step, the product being RhClH$_2$(PPh$_3$)$_3$. This indirect route to this complex is much more efficient than the direct hydrogenation of RhCl(PPh$_3$)$_3$, the rate of which was measured separately; the reactivity of RhCl(PPh$_3$)$_2$ with H$_2$ is at least 10^4 times that of RhCl(PPh$_3$)$_3$. The three-coordinate complex RhCl(PPh$_3$)$_2$ has a strong tendency

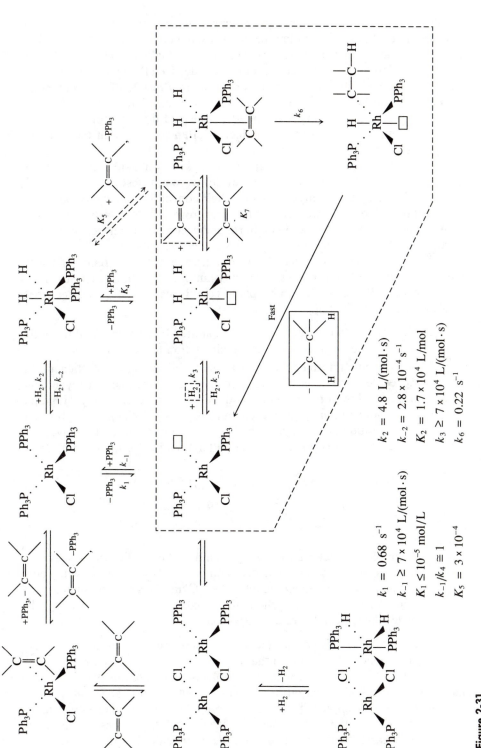

Figure 2-31
Catalytic cycle for the Wilkinson hydrogenation, as determined by Halpern et al. and given in ref. 37.

$k_2 = 4.8$ L/(mol·s)
$k_{-2} = 2.8 \times 10^{-4}$ s^{-1}
$K_2 = 1.7 \times 10^4$ L/mol
$k_3 \geq 7 \times 10^4$ L/(mol·s)
$k_6 = 0.22$ s^{-1}

$k_1 = 0.68$ s^{-1}
$k_{-1} \geq 7 \times 10^4$ L/(mol·s)
$K_1 \leq 10^{-5}$ mol/L
$k_{-1}/k_4 \cong 1$
$K_5 = 3 \times 10^{-4}$

to dimerize, giving the species with bridging Cl ligands shown in Fig. 2-31; the equilibrium constant for the dimerization is extremely large. But H_2 intercepts the highly reactive intermediate during catalysis virtually as fast as it is formed and thereby prevents its diversion into the "dead end" dimer. In other experiments, the kinetics of the reaction between $RhClH_2(PPh_3)_3$ and cyclohexene (present in large excess) was established, and kinetics of the overall cycle (illustrated in Example 2-6 below) established the rate constant for the virtually irreversible insertion step, which is often so small that the step may be approximated as rate determining.

It should be emphasized that all the species shown inside the dashed line in Fig. 2-31 (i.e., those in the catalytic cycle) were not observable; they were "invisible" during the catalysis because their concentrations were too low to be measured. The cycle was pieced together from results of separate experiments determining kinetics and equilibria under conditions chosen so that the catalytic cycle proceeded only very slowly or not at all.

The Wilkinson hydrogenation cycle illustrates the occurrence of the familiar oxidative addition, reductive elimination, and insertion reactions. Keys to the success of the rhodium phosphine complexes in olefin hydrogenation catalysis are the following:

1. Rh exists in two oxidation states separated by two units, allowing the oxidative addition and reductive elimination to occur readily.

2. There are no intermediates that are so stable as to form bottlenecks in the cycle. The intermediates are in delicate balance. They are all present in low concentrations, and they react predominantly within the cycle rather than to give dead-end complexes. When the phosphine concentration is too high or the hydrogen concentration too low, most of the rhodium is present as $RhCl(PPh_3)_3$ and the dimeric complex

When olefins that are too tightly bound to the rhodium are used (e.g., ethylene or 1,3-butadiene), the catalysis is also slowed down. These compounds are competitive inhibitors. Competing ligands such as pyridine are bonded to the rhodium so tightly that they shut down the cycle; these ligands are called poisons.

The reactivities of the intermediate species are strongly influenced by those ligands bonded to the metal that are not reactants in the catalytic cycle, in this case the phosphines. The data of Table 2-4 illustrate this point, showing more than a fiftyfold variation in catalytic activity with changes in the substituent groups. The choice of these ligands provides a powerful lever in catalyst design. In the development of the industrial catalytic processes discussed below, much careful work has gone into the optimization of the ligands.

Table 2-4
RELATIVE ACTIVITIES OF RHODIUM COMPLEXES
AS CATALYST PRECURSORS IN THE WILKINSON
HYDROGENATION OF CYCLOHEXENE [45]

Rh Complex	Relative Activity
$RhCl[P(p\text{-}C_6H_4Cl)_3]_3$	1.7
$RhCl[P(p\text{-}C_6H_5)_3]_3$	41
$RhCl[P(p\text{-}C_6H_4\text{—}CH_3)_3]_3$	86
$RhCl[P(p\text{-}C_6H_4\text{—}OCH_3)_3]_3$	100

EXAMPLE 2-6 Kinetics of the Wilkinson Hydrogenation

Problem

Experimental kinetics data for the catalytic hydrogenation of cyclohexene with $RhCl(PPh_3)_3$ in benzene at 25°C are summarized in Table 2-5. Show that these data are consistent with the catalytic cycle of Fig. 2-31 and with Halpern's inference that the insertion step represented by k_6 is rate determining.

Solution

The data in the figure suggest that the insertion step characterized by k_6 can be regarded to a first approximation as rate determining, since the other steps are shown as equilibria or characterized by much larger rate constants than k_6. Consequently,

$$r = -\frac{dC_{\text{cyclohexene}}}{dt} = k_6 C_{RhCl(PPh_3)_2H_2\text{cyclohexene}} \tag{2-136}$$

$$K_5 = \frac{C_{PPh_3}C_{RhCl(PPh_3)_2H_2\text{cyclohexene}}}{C_{\text{cyclohexene}}C_{RhCl(PPh_3)_3H_2}} \tag{2-137}$$

Table 2-5
KINETICS OF CYCLOHEXENE HYDROGENATION IN THE PRESENCE
OF WILKINSON'S CATALYST $RhCl(PPh_3)_3$ IN BENZENE SOLUTION
AT 25°C [46][a]

$C_{\text{cyclohexene}}$, mol/L	P_{H_2}, atm	$10^6 \times$ Observed Reaction Rate, mol/(L·s)	$10^6 \times$ Calculated Reaction Rate, mol/(L·s)
0.37	0.79	2.2	2.1
0.62	0.79	3.3	3.2
0.83	0.79	4.1	4.1
1.24	0.79	5.6	5.4
0.62	0.81	3.3	3.2
0.62	0.54	2.9	3.2
0.62	0.36	2.5	3.2

[a] $(C_{Rh})_t = 7.6 \times 10^{-5}$ mol/L.

And, since the predominant forms of the rhodium are expected to be the stable coordinatively saturated (octahedral) complexes,

$$C_{RhCl(PPh_3)_3H_2} + C_{RhCl(PPh_3)_2H_2cyclohexene} = (C_{Rh})_t \qquad (2\text{-}138)$$

where $(C_{Rh})_t$ is the total concentration of rhodium complex, the conveniently measured catalyst concentration term introduced in Chapter 1.

Solving Eq. 2-137 for $C_{RhCl(PPh_3)_3H_2}$ and combining it with Eq. 2-138 gives

$$C_{RhCl(PPh_3)_2H_2cyclohexene} = \frac{K_5 C_{cyclohexene}(C_{Rh})_t}{K_5 C_{cyclohexene} + C_{PPh_3}} \qquad (2\text{-}139)$$

Combining this result with Eq. 2-137 gives

$$-\frac{dC_{cyclohexene}}{dt} = \frac{dC_{RhCl(PPh_3)_3H_2}}{dt}$$

$$= \frac{k_6 K_5 C_{cyclohexene}(C_{Rh})_t}{K_5 C_{cyclohexene} + C_{PPh_3}} \qquad (2\text{-}140)$$

The predictions of this equation are compared with the experimental rates in Table 2-5. The agreement is good.

The form of the equation is suggestive of those encountered in Chapter 1; the reaction is first order in catalyst concentration, and the dependence of rate on cyclohexene concentration indicates saturation kinetics (a dependence like that of Fig. 1-4). The rate is independent of hydrogen concentration, as the rate-determining step does not involve H_2 [and the data approximately confirm this result (Table 2-5)]. The result also shows that triphenylphosphine is a competitive inhibitor, because its concentration appears only in the denominator. This result means that triphenylphosphine drains away the complex involved in the rate-determining step, [RhCl(PPh$_3$)$_2$H$_2$cyclohexene]. But it is important to recognize that triphenylphosphine is more than just an inhibitor as these ligands are present in all the rhodium complexes in the cycle, and without them the cycle could not occur. Somewhere outside the observed range of concentrations, Eq. 2-140 must no longer be valid.   ∎

The results of this example shed more light on the previously mentioned ligand effects summarized in Table 2-4. The slow step is an insertion, and the rate of this step is increased from top to bottom in the list, that is, as the electron donor strength (basicity) of the ligand is increased. This pattern has been confirmed more generally in organometallic chemistry [37]. However, the cycle prevailing with one donor ligand may not be the same as that prevailing with another, and only for triphenylphosphine has a cycle been established.

The Halpern cycle for the Wilkinson hydrogenation is one of the best-understood examples of a catalytic reaction, and it has challenged theoretical chemists to examine it in detail. Molecular orbital calculations have been performed to provide evidence of the cycle, as discussed below. But even with modern computers, these calculations are slow, and calculations for the full Halpern cycle are still not feasible. They have instead been performed for a simplified cycle, with the triphenylphosphine ligand being replaced by the simpler phosphine (PH_3), although it does not give as active a catalyst, and the reactant olefin being taken to be ethylene, although this is not converted significantly with the Wilkinson catalyst (only higher olefins such as cyclohexene are converted significantly).

The key structures in the proposed cycle are shown in Fig. 2-32 [47]; transition states are included. The molecular orbital calculations yielded the potential energy profile shown in Fig. 2-33, where the structures in Fig. 2-32 are referred to by number and the elementary steps (oxidative addition, olefin association, etc.) are designated on the horizontal axis. The calculations are consistent with the inference that the olefin insertion step is rate determining (what is the activation energy for this step?). The oxidative addition of H_2 and the olefin coordination proceed virtually without activation energies. The reductive elimination step is characterized by a significant activation energy, but it is small enough that the olefin insertion is rate determining.

This is the first theoretical study of a full catalytic cycle, and it substantially confirms the experimental results for the similar Wilkinson cycle, verifying the mechanism and providing insights into the details of the mechanism. It is instructive to read the original paper [47]. This work is likely to be just part of the beginning of major contributions of theoretical chemistry to catalysis.

Chiral Catalysis: The Synthesis of L-Dopa

The role of the phosphine ligand in directing the catalysis of olefin hydrogenation inspired Knowles and colleagues [48] to design phosphine ligands to control selectivity in the hydrogenation of a prochiral olefin, that is, one that upon hydrogenation would give a chiral product. The goal was the chiral hydrogenation of acetamidocinnamic acid to give the precursor of L-dopa (3,4-dihydroxyphenylalanine), a drug used in the treatment of Parkinson's disease. Only the L isomer is biologically active. The goal was to prepare a chiral catalyst that would direct the stereochemical course of the reaction.

The strategy was to use a rhodium complex similar to those of the Wilkinson hydrogenation cycle, but with bulky bidentate phosphine ligands, in an attempt to direct the stereochemistry of the catalytic reaction to favor the desired L isomer of the product. The result was highly active and selective catalysts that have been used in commercial processes. (Today, however, there is a more economical route to L-dopa than through hydrogenation of the prochiral precursor.) A number of bidentate phosphine ligands, some chiral, give catalysts that are highly selective for the desired product.

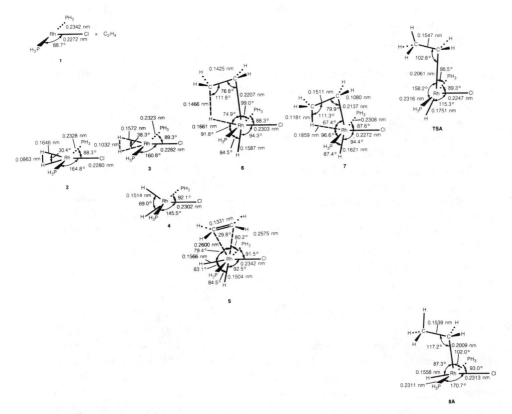

Figure 2-32
Intermediates in the olefin hydrogenation cycle characterized in Fig. 2-33. [47].

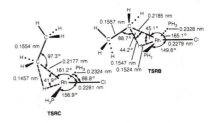

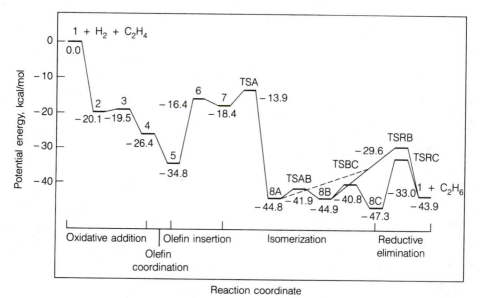

Figure 2-33
Potential energy diagram calculated for the ethylene hydrogenation cycle involving species shown in Fig. 2-32 [47].

The details of this chemistry have been investigated to provide a mechanistic interpretation of the selectivity [49, 50]. Detailed kinetics experiments comparable to those described above for the Wilkinson hydrogenation were carried out to determine the catalytic cycle for hydrogenation with the ligand represented as DIPHOS (Fig. 2-34), the structure of which is the following:

The results are quantitative, as shown in Fig. 2-34. Further experiments were done with a rhodium complex having a chiral diphosphine ligand, DIPAMP,

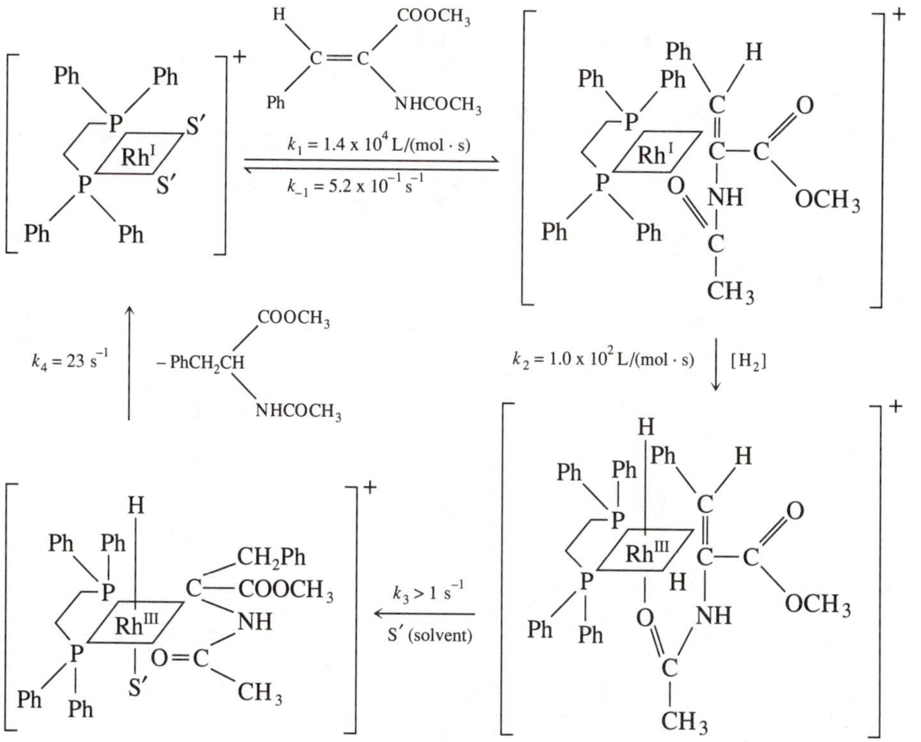

Figure 2-34
Catalytic cycle for hydrogenation of methyl-(Z)-α-acetamidocinnamate catalyzed in the presence of a rhodium complex with bidenate DIPHOS ligands. The rate constants were measured at 25°C [49, 50].

having the following structure:

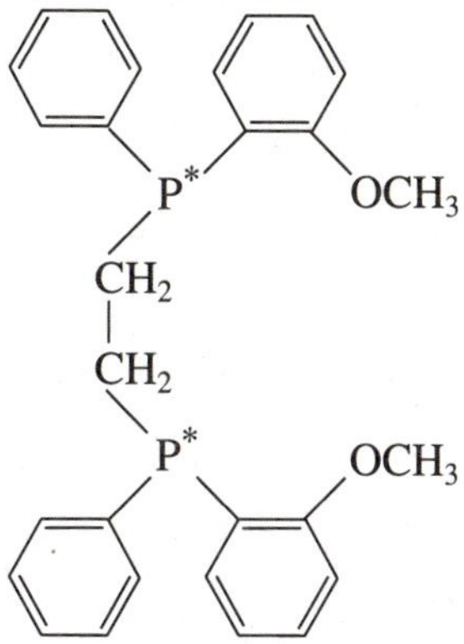

(The asterisks denote the chiral centers.) The catalytic cycle giving the chiral product was found to be closely similar to that of Fig. 2-34.

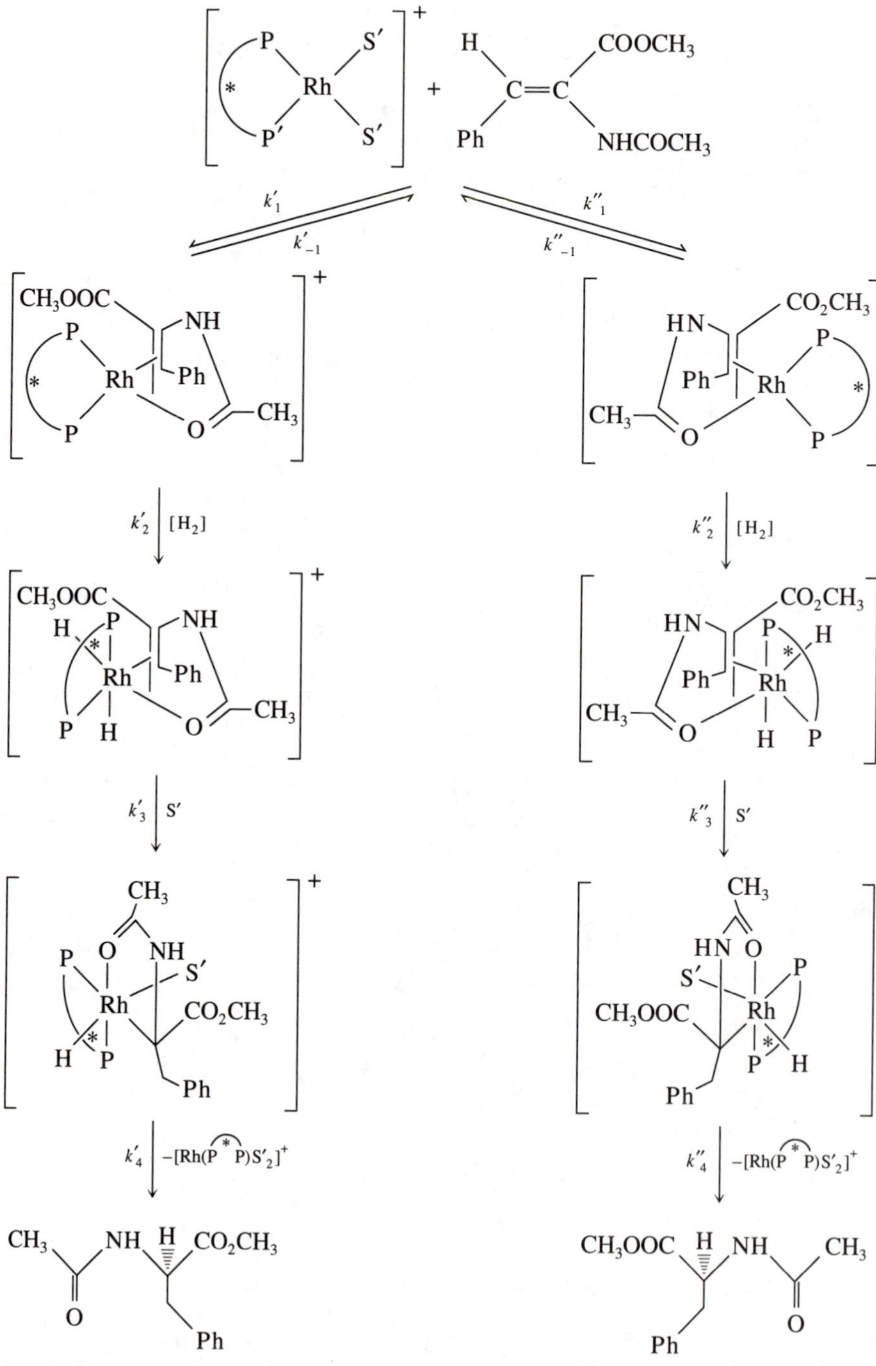

Figure 2-35
Pathways for the hydrogenation of a prochiral reactant, methyl-(Z)-α-ace-tamidocinnamate, catalyzed by a rhodium complex with a chiral diphosphine ligand (e.g., DIPAMP, symbolized as P*P). The solvent S is methanol [50].

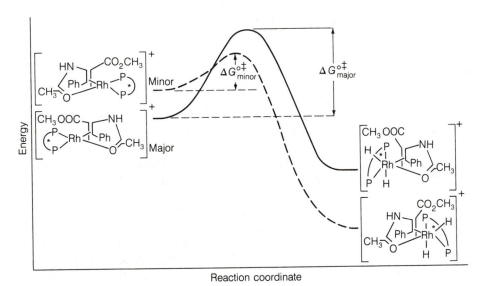

Figure 2-36
Schematic representation of energy profiles for the pathways of Fig. 2-35 for
the reactions of rhodium DIPAMP complexes with H_2 [50].

Two pathways were found for the chiral hydrogenation, as shown in Fig.
2-35. The pathway at the left of the illustration shows the preferred mode of
initial binding of the reactant to the catalyst. The pathway shown at the right
involves a minor isomer of the reactant–catalyst complex. These pathways
give products with different stereochemistries.

The results of the investigation showed that, contrary to the first expectation, the chirality of the product was not determined by the preferred mode
of initial binding of the reactant. Rather, the predominant product resulted from
the other pathway by virtue of a much higher reactivity of the reactant–catalyst
adduct with H_2, a result that had not been anticipated and could not easily
have been predicted.

The schematic energy profile for the reaction shown in Fig. 2-36 indicates
the essence of the catalytic chemistry. This example represents some of the
most subtle and efficient control of a catalytic reaction by the intervention of
a man-made catalyst. It is an indication of excellent opportunities for delicate
control of reactions with precisely structured catalysts. It gives a hint of the
subtlety of the chemistry of nature's catalysts, the enzymes, which are addressed in Chapter 3.

OLEFIN HYDROFORMYLATION

One of the most important industrial processes involving transition-metal complex catalysis is the hydroformylation of olefins (also known as the oxo process), exemplified below for propylene:

$$CH_3-CH{=}CH_2 + H_2 + CO \rightarrow CH_3-CH_2-CH_2-CHO$$

(2-141)

$$\searrow \quad CH_3-\overset{\displaystyle CHO}{\underset{\displaystyle |}{CH}}-CH_3$$

In the process for the manufacture of n-butyraldehyde, a rhodium phosphine catalyst is used under mild conditions, for example, 80°C and 15 atm. The reactants CO and H_2 (synthesis gas) can be produced from coal or natural gas and are expected to become increasingly important industrial feedstocks as petroleum is eventually replaced as the primary source of building blocks of organic chemicals.

The catalytic cycle for olefin hydroformylation has not been elucidated fully, but the available evidence is consistent with the suggestion of Fig. 2-37. This cycle is again characterized by the familiar steps: ligand association–dissociation (①, ②), oxidative addition (③, ④), reductive elimination (⑤), and the cis insertion (⑥, ⑦). (What are the steps not identified with numbers in the figure?)

Parallel paths in the cycle account for the formation of both the major products, n-butyraldehyde and isobutyraldehyde. Since the straight-chain aldehyde is the more valuable product, much effort has gone into developing catalysts with maximum selectivity for formation of n-butyraldehyde. To simplify Fig. 2-37, only the formation of this product is shown there.

The activity of the catalyst depends on the concentration of the ligands. The data of Fig. 2-38 show the dependence of the rate of hydroformylation of propylene on the ratio of the concentration of triphenylphosphine to the concentration of rhodium. For low values of this ratio, increasing the ratio increases the catalytic activity, since increasing amounts of the catalytically active rhodium phosphine complex are formed. But increasing the concentration ratio further leads to a decrease in the catalytic activity. At high concentrations, the phosphine is a reaction inhibitor; increasingly high concentrations of the phosphine remove an increasingly large fraction of the rhodium from the cycle, taking it into a "blind alley" rather than allowing it to pass along the cyclic route.

A general conclusion can be inferred from this observation: Any group that can bond to the catalyst in competition with the reactants can be a reaction inhibitor (but, as in the Wilkinson hydrogenation, the role of the phosphine is more complicated than that of a simple inhibitor). Many products of catalytic reactions are inhibitors, because they are often chemically similar to reactants and therefore are able to bond to the catalyst. Even reactants can be inhibitors, that is, they can be present in concentrations so high that they usurp the catalyst and hinder the bonding of coreactants. This point is illustrated by the hydroformylation cycle of Fig. 2-37: an excess of reactant CO leads the rhodium into another blind alley. In principle, the same kind of behavior is expected with any one reactant when there are more than one.

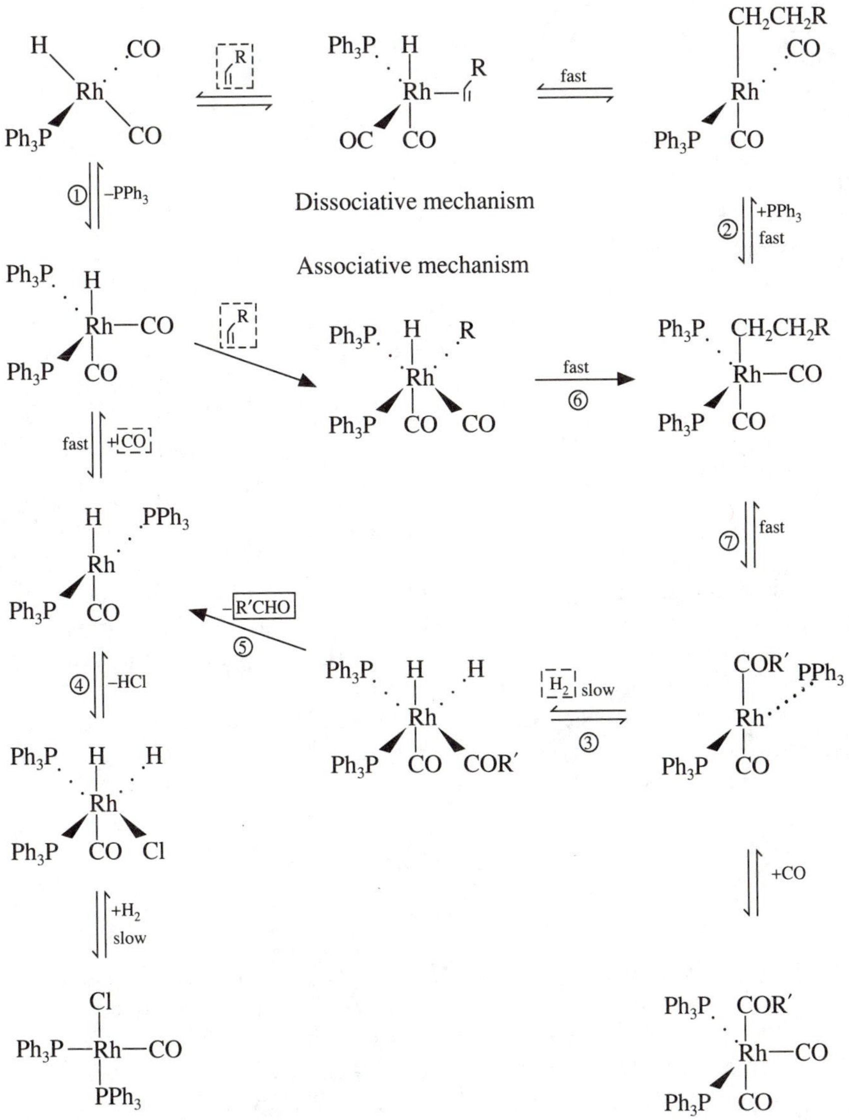

Figure 2-37
Catalytic cycles proposed for the propylene hydroformylation reaction.

Rhodium complexes are industrial catalysts for propylene hydroformylation [52–54]; but cobalt complexes, although they are much less active, have been applied for many more years and still predominate. The cobalt catalysts are even used for long-chain olefins, producing aldehydes that are reduced to give the so-called oxo alcohols; conversion of these into C_{12}–C_{15} sulfonates

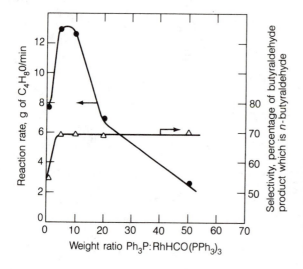

Figure 2-38

Dependence of rate and selectivity of hydroformylation of propylene on the phosphine:rhodium concentration ratio at 100°C and 35 atm. The reactant consisted of a $1:1:1$ molar ratio of $CO:H_2:C_3H_6$ [51].

produces detergents. The proposed elementary steps and the catalytic cycle are similar to those discussed for the rhodium catalyst, but the ligands surrounding the metal are now H, CO, and hydrocarbon; this cycle is not so well established as that involving the rhodium complexes.

A rate equation has been determined for the hydroformylation of diisobutylene in the presence of cobalt complexes at 150°C, $P_{CO} = 20$ to 150 atm, and $P_{H_2} = 50$ to 275 atm [55]:

$$r = \frac{kC_{catalyst}C_{olefin}P_{H_2}}{P_{CO} + KP_{H_2}} \tag{2-142}$$

This equation shows that CO and H_2 are inhibitors as well as reactants, since the partial pressures of CO and H_2 appear in the denominator of the equation. The special case for which $P_{CO} \ll KP_{H_2}$ shows saturation kinetics in H_2; the dependence of the rate on P_{H_2} is as shown in Fig. 1-4.

The role of CO in this example is intriguing. In the partial-pressure range for which the rate equation is valid, an increase in the CO partial pressure decreases the rate; but since CO is a reactant, it follows that at some lower partial pressures, increasing the CO partial pressure must increase the rate, but a rate equation is lacking for that range.

The complicated role of CO and the need for a high pressure (typically 200 atm) in the industrial process with cobalt catalysts is explained as follows: The CO stabilizes the catalyst because the Co—CO bonds form in competition not only with the other Co–reactant bonds but in competition with Co—Co bonds. A number of compounds with metal–metal bonds can form, including the following, in which both terminal and bridging CO ligands are depicted:

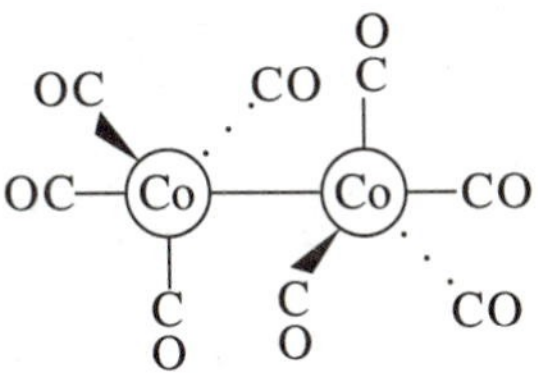

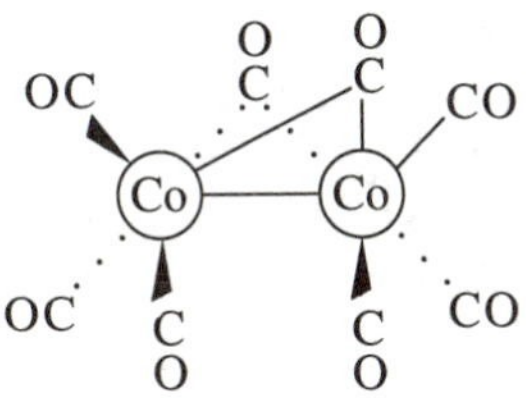

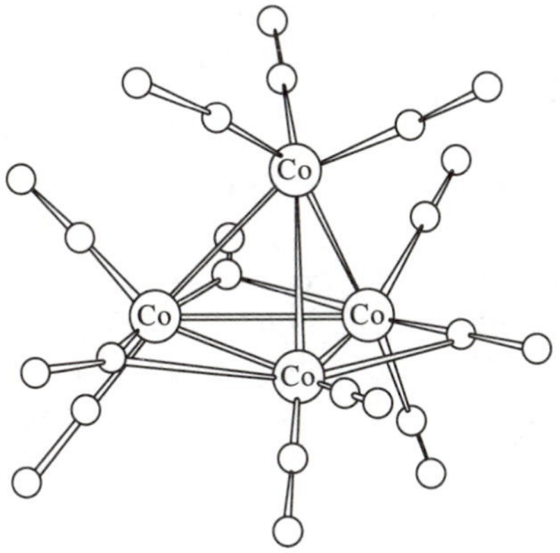

(Such compounds with metal–metal bonds are discussed subsequently in this chapter.) Ultimately, metallic Co can plate out on the reactor walls if the CO partial pressure becomes too low. In other words, the role of CO in maintaining the catalyst stability is to keep the metal dispersed in solution; the CO is simultaneously a reactant, an inhibitor, and a catalyst stabilizer. With the Rh catalyst, the role of stabilization is taken over by the phosphine. When the phosphine is absent, compounds such as $Rh_4(CO)_{12}$ and $Rh_6(CO)_{16}$ can form and complicate matters.

The hydroformylation process suggests some reactor design issues that are generally applicable to catalysis by transition metal complexes. The reactants are present in both gas and liquid phases, and the design must allow for efficient transfer of reactants from the gas phase into the liquid phase. The transfer rate can be maximized by (1) maximizing the interfacial area between phases and (2) increasing the rate of transport of the molecules across the interface. A high interfacial area can be obtained by sparging, that is, introducing a swarm of small bubbles of reactant into the liquid in a tank or by bringing the gas and liquid into contact in a column containing solid packing designed to cause mixing of the two phases. The rate of transport between phases can be increased by increasing convective motion, for example, by stirring the contents of the tank or by increasing the flow rate of the gas and/ or liquid in the packed column. Both stirred-tank and packed-column reactors are used in industrial hydroformylation processes.

Another important concern in the industrial processes is corrosion, which has already been mentioned in the discussion of acid catalysis. The problems can be countered by the use of corrosion-resistant materials, including stainless steels like Hastelloy C, but these are invariably expensive. Cobalt can plate out on vessel walls, and rhodium (and many other transition metals) can also react with metal surfaces. The consequences are not only fouling of metal surfaces but loss of the catalyst, and substantial losses cannot be tolerated when extremely rare and expensive metals like rhodium are used.

The industrial processes using soluble catalysts are all beset with the problems of separating the catalyst from the reaction products and returning (recycling) it to the reactor. As mentioned in the discussion of acid catalysis, this difficulty points to the application of solid catalysts, and in the following chapters transition metal complexes attached to solids are introduced.

EXAMPLE 2-7 A Two-Phase Hydroformylation Process
Problem

One of the complications of the processes used commercially for hydroformylation of propylene with rhodium complex catalysts is the difficulty of separation of the catalyst from the products. The separation is accomplished by distillation, but high-boiling products remain with the catalyst and lead to deactivation. Conceive an alternate process whereby the distillation is eliminated.

Solution

The problem can be overcome by design of a catalyst that is present in a separate phase. This phase has been chosen to be an aqueous phase, and a water-soluble catalyst has been prepared by sulfonating the phenyl groups in triphenyl phosphine:

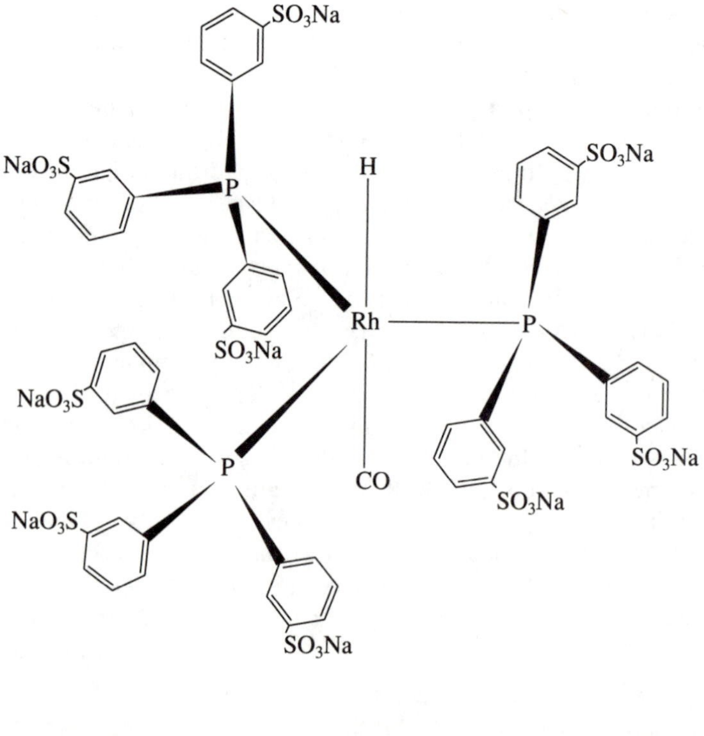

This catalyst can be used in a process with two liquid phases, one containing the organics, including reactants and products, and the other containing water and the catalyst [56]. The catalyst is likely present predominantly at the liquid–liquid interface. A process flow diagram is suggested in Fig. 2-39. The gas is introduced into a mixed reactor with two liquid phases, and these two phases are separated downstream in a settler (it works like a separatory funnel) and are recycled. Such processes have found industrial application and appear to be economically competitive with the single liquid-phase process. ∎

EXAMPLE 2-8 The Aldox Process

Problem

An industrially important alcohol used in making plasticizers (softeners for plastics) is 2-ethylhexanol:

$$CH_3-CH_2-CH_2-CH_2-CH-CH_2-OH$$
$$\overset{|}{CH_2}$$
$$\overset{|}{CH_3}$$

Suggest a process for manufacturing this from propylene, CO, and H_2.

Solution

Hydroformylation of propylene gives *n*-butylraldehyde, as described above. Soluble Rh or Co complexes are good catalysts.

Aldol condensation of the aldehyde gives a C_8 product with a carbon

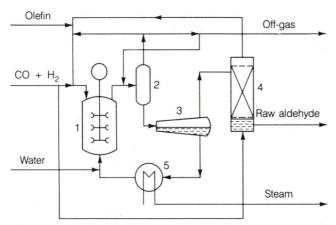

1 reactor; 2 separator; 3 phase separator; 4 stripping column; 5 heat exchanger

Figure 2-39
Schematic flow diagram for a hydroformylation process with a gas–liquid–liquid reactor [56].

skeleton like that of the desired product:

$$2CH_3-CH_2-CH_2-CHO \rightarrow CH_3-CH_2-CH_2-CH=\underset{\underset{\underset{CH_3}{|}}{\underset{CH_2}{|}}}{C}-CHO + H_2O$$

$$(2\text{-}143)$$

The reaction is catalyzed by acidic or basic solutions; a good choice is aqueous KOH.

The product, upon hydrogenation, gives 2-ethylhexanol:

$$CH_3-CH_2-CH_2-CH=\underset{\underset{\underset{CH_3}{|}}{\underset{CH_2}{|}}}{C}-CHO + 2H_2 \rightarrow$$

$$CH_3-CH_2-CH_2-CH_2-\underset{\underset{\underset{CH_3}{|}}{\underset{CH_2}{|}}}{CH}-CH_2-OH \qquad (2\text{-}144)$$

One might expect a catalyst like the Wilkinson catalyst to be active for the reaction. But nickel metal is also a good hydrogenation catalyst, and it is a preferred choice in industry.

The process could be carried out directly in three separate reactors, each containing the appropriate catalyst. The industrial process is similar to this conception [54]. ∎

METHANOL CARBONYLATION [57]

Another important industrial process involving CO as a reactant is the methanol carbonylation to give acetic acid:

$$CH_3OH + CO \rightarrow CH_3COOH \qquad (2\text{-}145)$$

Since the methanol is produced from synthesis gas (CO + H_2), acetic acid can be manufactured from coal and many other hydrocarbons which, upon gasification (reaction with steam at high temperatures), give synthesis gas. Today, the principal source of synthesis gas is natural gas, which is mostly methane.

As the stoichiometry of methanol carbonylation suggests an insertion of CO between CH_3 and OH, an insertion step might be expected in the catalytic cycle. The cycle is well established (Fig. 2-40); spectroscopic evidence of the intermediates has been obtained and independent experiments have been done with a key intermediate, $[Rh(CO)_2I_2]^-$. There is an insertion step, involving, as expected, the metal and a reactant ligand (CH_3). Attachment of a CH_3 ligand

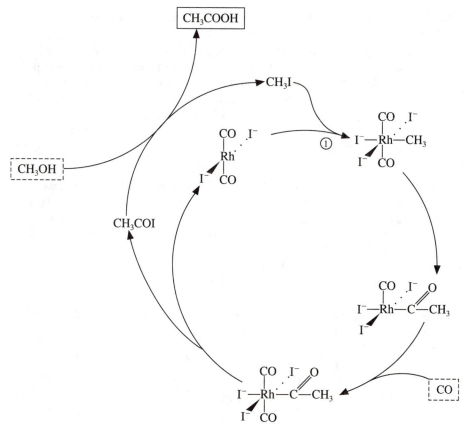

Figure 2-40
Catalytic cycle for methanol carbonylation in the presence of a rhodium complex catalyst and methyl iodide cocatalyst [57].

to the Rh requires a compound incorporating CH_3 that readily undergoes an oxidative addition reaction with the metal. Such a compound is methyl iodide; it is called a cocatalyst (or promoter) in this cycle. It is not consumed in the process, and the cycle cannot be closed without it.

The step marked ① in Fig. 2-40 is rate determining, and therefore the overall kinetics is [58]

$$r = kC_{CH3I}C_{Rh\ complex} \qquad (2\text{-}146)$$

where $k = (3.5 \times 10^6)\, e^{-14.7/RT}$ L/(mol·s), and the activation energy is in kcal/ mol. The rate is independent of the reactant concentrations over the range of applicability of the equation, which extends to partial pressures of CO as low as about 1 atm. In industrial practice, the reaction takes place at about 180°C and 15 atm.

EXAMPLE 2-9 A Noniodide Cocatalyst for Methanol Carbonylation

Problem

The disadvantages of the methyl iodide cocatalyst in the methanol carbonylation process are its high volatility, which makes it difficult to separate from the reaction products, and its corrosiveness. A substitute has been suggested, pentachlorothiophenol, which is much less volatile and corrosive [59]. But the reaction rate with this substitute is only about 3% of that observed for CH_3I, and it is not applied. Suggest an explanation for the role of the new cocatalyst.

Solution

The mechanism of reaction with pentachlorothiophenol has not been determined, but one might assume that it is the precursor of a cocatalyst that functions like CH_3I, since it is a "pseudohalide" with a reactivity similar to that of CH_3I. An equilibrium is set up in the industrial process involving CH_3I and HI:

$$CH_3OH + HI \rightleftharpoons CH_3I + H_2O \tag{2-147}$$

It has been suggested [59] that the role of pentachlorothiophenol is analogous to that of HI:

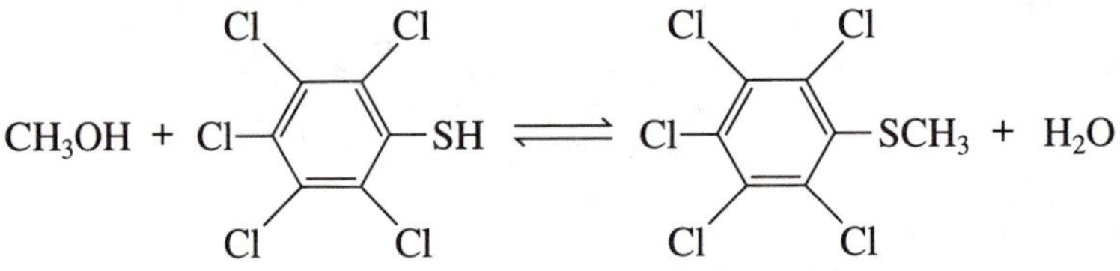

$$\tag{2-148}$$

One could envision an oxidative addition like the rate-determining step and other steps like those in Fig. 2-40, with the I^- ligands replaced by the pseudohalide anion

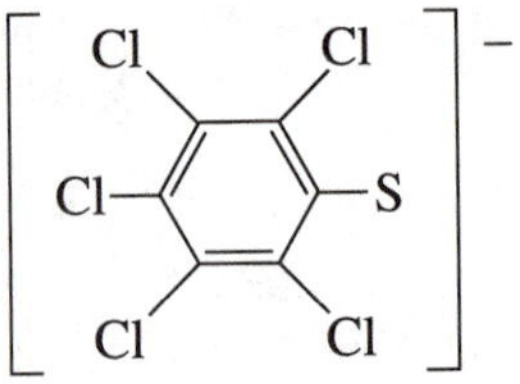

■

Methanol carbonylation is also catalyzed industrially by cobalt complexes in the presence of an iodide cocatalyst. But the selectivity is less than that of the newer processes using rhodium, and the pressure is much higher (hundreds of atmospheres). The requirement of high pressure, as in the hydroformylation process, is a reflection of the need for CO ligands to minimize Co—Co bond formation and to keep the cobalt in solution.

EXAMPLE 2-10 Process Conception for Methanol Carbonylation

Problem

Select a preliminary set of processing conditions and identify the process engineering problems for methanol carbonylation.

Solution

Consider the components present in the reactor and product stream. These include the reactants CO and CH_3OH, the product CH_3COOH, and the catalyst (Rh complex) and cocatalyst (CH_3I). Under typical reaction conditions cited in the literature [57, 58] (180°C and 15 atm), several side reactions are expected:

$$CH_3OH + HI \rightleftharpoons CH_3I + H_2O \qquad (2\text{-}149)$$

$$CH_3COOH + CH_3OH \rightleftharpoons CH_3COOCH_3 + H_2O \qquad (2\text{-}150)$$

$$2CH_3OH \rightleftharpoons CH_3\text{—}O\text{—}CH_3 + H_2O \qquad (2\text{-}151)$$

The first of these is mentioned in Example 2-9; the others are the familiar acid-catalyzed esterification and alcohol dehydration reactions (why is the solution acidic?). These reactions can, as a first approximation, be regarded as fast and in virtual equilibrium. Since small yields of CO_2 and H_2 have been observed, one might also infer the occurrence of the water–gas shift reaction [57b]:

$$CO + H_2O \rightarrow CO_2 + H_2 \qquad (2\text{-}152)$$

The following additional components of the product stream are now evident: H_2, CO_2, $CH_3\text{—}O\text{—}CH_3$, H_2O, HI, and CH_3COOCH_3.

Dominant issues in the process design therefore include the following:

1. Corrosion: Since the product is an acidic aqueous solution containing iodide, expensive materials such as stainless steels will be required in all the equipment coming in contact with this solution.

2. Product purification: The product mixture is complex, and the rhodium in solution must be recovered with an extremely high efficiency and returned to the reactor, because rhodium is so expensive. Distillation is the suggested method of product purification, since it is likely the most economical. As there is a wide range of volatilities of the components, and the product is near the middle, a complex distillation scheme, probably with multiple columns, would be required.

3. Calculations beyond the scope of this textbook suggest that at the temperature and pressure indicated by the literature, the rate of reaction in a well-mixed stirred-tank reactor may be near the edge of gas–liquid mass transfer influence; this issue (discussed later) may be important in the choice of temperature and pressure. How does the rate under these conditions compare with the criterion of Weisz stated in Chapter 1?

OLEFIN POLYMERIZATION

The polymerization of olefins, already described as a reaction proceeding through carbenium ion intermediates, is also catalyzed by complexes of many transition metals. The essential steps in the catalytic cycles are (1) insertion of an olefin between the metal and an alkyl ligand,

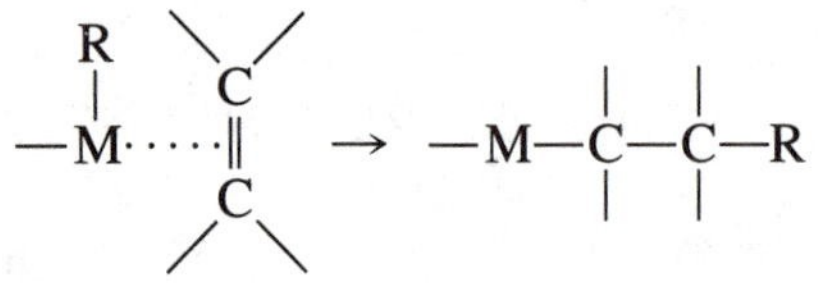

$$
\begin{array}{c} R \\ | \\ -M\cdots\| \\ \ \ \ \ C \end{array} \quad \rightarrow \quad -M-\overset{|}{C}-\overset{|}{C}-R \tag{2-153}
$$

and (2) a β-hydrogen transfer, which is the reverse of an insertion reaction that involves a metal hydride as the starting complex,

$$
-\overset{|}{M}-CH_2CH_2R \quad \rightarrow \quad -\overset{|}{M}H + CH_2=CHR \tag{2-154}
$$

A simple example is the dimerization of ethylene in the presence of π-allylnickel halide complexes:

$$
2CH_2=CH_2 \rightarrow CH_2=CHCH_2CH_3 \tag{2-155}
$$

An active form of the catalyst, a nickel hydride, is believed to form by addition of an olefin and displacement of the π-allyl group [60]:

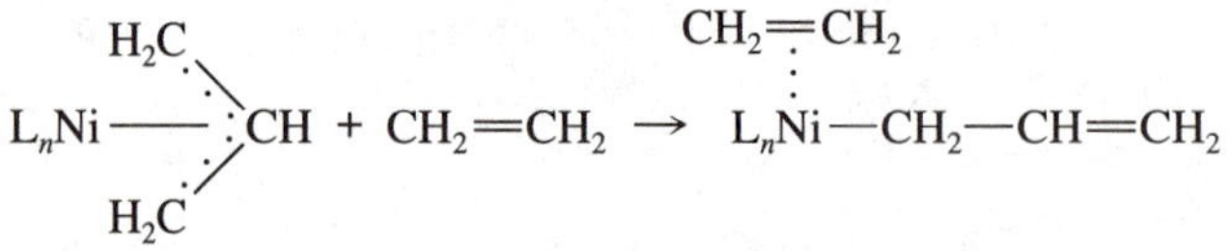

$$
\underset{\substack{H_2C}}{\overset{\substack{H_2C}}{L_nNi\text{------}CH}} + CH_2=CH_2 \quad \rightarrow \quad L_nNi-CH_2-CH=CH_2 \tag{2-156}
$$

This reaction involves a conversion of the fluxional π-bonded allyl (with three carbons bonded to the metal) into a σ-bonded species (with only one carbon bonded to the metal). The following steps are the insertion and the β-hydrogen transfer [60]:

$$CH_2{=}CH_2$$
$$\vdots$$
$$L_nNi{-}CH_2{-}CH{=}CH_2 \;\rightarrow\; L_nNi{-}CH_2CH_2CH_2CH{=}CH_2 \qquad (2\text{-}157)$$

$$L_nNi{-}CH_2CH_2CH_2CH{=}CH_2 \;\rightarrow\; L_nNiH + CH_2{=}CHCH_2CH{=}CH_2 \qquad (2\text{-}158)$$

The nickel hydride now enters the catalytic cycle (Fig. 2-41); it is consumed in step ① and regenerated in step ③.

An important complication is anticipated in the cycle of Fig. 2-41—not only ethylene but the product butene can coordinate to the metal in step ①. Such a step would lead to the formation of trimer; routes for the formation of higher oligomers and polymers are obvious.

The π-allylnickel chlorides exist as dimers with bridging chloride ligands:

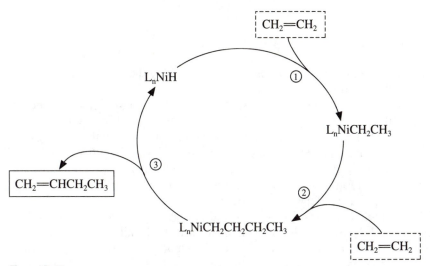

Here is an opportunity for self-inhibition; the nickel catalysts alone are not highly active. But when Lewis acid promoters such as $AlCl_3$ are added, the nickel complexes become some of the most active known catalysts for dimerization of ethylene. The role of the promoter is believed to be the opening of the bridging bonds, with one electron pair of the chloride being shared with

Figure 2-41
Catalytic cycle for the dimerization of ethylene in the presence of π-allylnickel halide [60].

the aluminum:

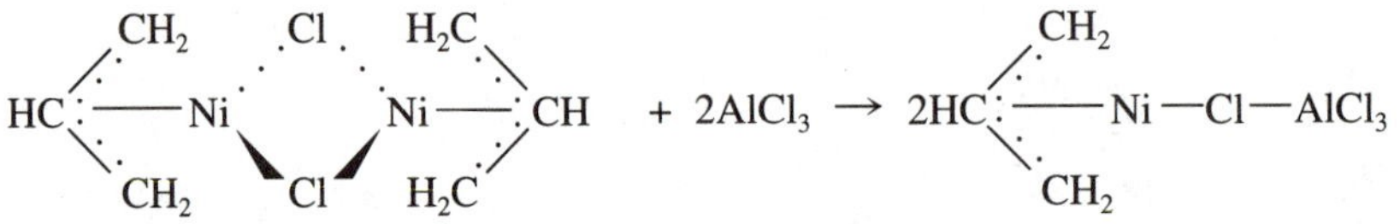

$$\text{(2-159)}$$

This reaction creates open coordination sites on the nickel and allows it to enter into a catalytic cycle like that of Fig. 2-41.

Nickel complexes are applied as catalysts for a variety of reactions of olefins [61].

The best catalysts for oligomerization and polymerization of olefins are derived from Group 4–6 transition metals, the most common being Ti, V, Cr, and Zr. Often they are used in combination with organometallic promoters, for example, $TiCl_4$ + $(C_2H_5)_3Al$. (Some catalysts work without the promoters.) These combinations are referred to as Ziegler catalysts. Similar solid polymerization catalysts have found wide commercial application, as described in Chapter 6.

The Ziegler catalysts form active species with structures that may resemble the following:

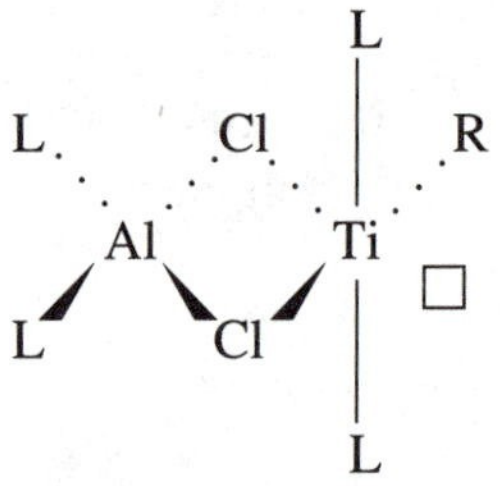

The square □ denotes a vacant coordination site where the olefin can be bonded cis to the alkyl group R.

Some polymerizations of ethylene catalyzed by complexes of Ti and Zr proceed with high rates and selectivities to give linear high-molecular-weight polyethylene. Polymerization of propylene also occurs, and with some ligands chosen to give a chiral catalyst, the polymerization proceeds with a high degree of stereospecificity [62–64]. Solid catalysts are also active and selective for these polymerizations, and they are the ones used industrially to make polyethylene and polypropylene, as described in Chapter 6.

The soluble catalysts are simpler and better understood than the solids, but details of the polymerization mechanisms are still lacking. Numerous complexes of Ti and Zr catalyze ethylene and propylene polymerization, and some have high activities even at temperatures of 0°C and less. The most remarkable of these catalysts are the ones that give high-molecular-weight polymers and stereoregular polypropylene. Most of the known catalysts give polypropylene that lacks stereoregularity, called *atactic* polypropylene. When some chiral catalysts are used, however, high yields of *isotactic* polypropylene are formed, and these are solids with much more desirable physical properties than the gooey atactic polymer. Isotactic polypropylene has all the methyl groups oriented in the same fashion:

$$CH_3 \quad CH_3 \quad CH_3$$

isotactic polypropylene

This depiction represents a stretched form of the polymer chain. In the typical solid, the structure is more complicated; for example, coils might be formed. Another stereoregular polypropylene has also been formed, although it is not commercially important; the methyl groups in this *syndiotactic* polypropylene alternate in the in and out positions:

$$CH_3 \qquad CH_3\,CH_3$$

syndiotactic polypropylene

The atactic polymer has a random distribution of methyl groups.

The soluble catalysts that are selective for the stereospecific polymerization to give isotactic polypropylene are chiral and made, for example, from ethylenebis(tetrahydro-1-idenyl)zirconocene and aluminooxane. Ion pairs of the type $[Cp_2M{-}R]^+[X(Al(CH_3){-}O)_n]^-$ are implicated [64], but the structures of the intermediates are not well understood.

A simplified stereochemical model for the intermediate undergoing the olefin insertion step in the polymerization is suggested in Fig. 2-42 [65]. The

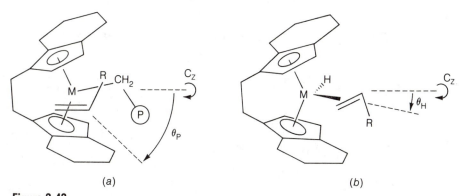

(a)   (b)

Figure 2-42
Simplified stereochemical models for the olefin insertion step in stereospecific polymerization (a) and hydrogenation (b) [65]. The angle θ_P is greater than θ_H.

alkyl ligand in the complex shown as (*a*), which is the growing polymer chain —CH$_2$—Ⓟ, is inferred to affect the stereochemistry of bonding of propylene to the metal M. Evidently the olefin always bonds in the same way and is inserted in a way that consistently gives the isotactic product.

The catalyst is also active for hydrogenation of olefins, and the stereochemistry is different from that of the polymerization; structure (*b*) of Fig. 2-42 has been suggested [64]. The hydride ligand in this structure is so small that the olefin can approach the metal center without the constraints encountered in structure (*a*).

Kinetics of styrene polymerization has been investigated with Zr(benzyl)$_4$, which is not a stereospecific catalyst. The formation of catalytically active species appears to involve the following steps [66]:

$$(C_6H_5CH_2)_3ZrCH_2C_6H_5 \; + \; CH_2{=}CHC_6H_5 \; \rightleftharpoons \; \begin{matrix} CH_2{=}CHC_6H_5 \\ | \\ (C_6H_5CH_2)_3ZrCH_2C_6H_5 \end{matrix}$$

$$\text{C}' \text{ (catalyst precursor)} \qquad \text{M (monomer)} \qquad \qquad \text{C}'\text{I}_0 \text{ (intermediate)}$$

$$(2\text{-}160)$$

$$\begin{matrix} CH_2{=}CHC_6H_5 \\ | \\ (C_6H_5CH_2)_3ZrCH_2C_6H_5 \end{matrix} \; \xrightarrow{\text{slow}} \; (C_6H_5CH_2)_3Zr(\underset{\underset{C_6H_5}{|}}{CHCH_2})CH_2C_6H_5$$

$$\text{C}'\text{I}_0 \qquad\qquad\qquad\qquad\qquad\qquad \text{C}'\text{P}$$

$$(2\text{-}161)$$

The suggested propagation (polymer-chain-growth) steps are as follows [65]:

$$(C_6H_5CH_2)_3Zr(\underset{\underset{C_6H_5}{|}}{CH{-}CH_2})CH_2C_6H_5 \; + \; CH_2{=}CHC_6H_5$$

$$\text{C}'\text{P}$$

$$\downarrow$$

$$\begin{matrix} CH_2{=}CHC_6H_5 \\ | \\ (C_6H_5CH_2)_3Zr(\underset{\underset{C_6H_5}{|}}{CHCH_2})CH_2C_6H_5 \end{matrix}$$

$$\text{C}'\text{I}$$

$$\downarrow$$

$$(2\text{-}162)$$

$$(C_6H_5CH_2)Zr(\overset{|}{\underset{C_6H_5}{CHCH_2}})_2CH_2C_6H_5$$

$$C'P_2$$

$$\Big\downarrow CH_2{=}CHC_6H_5$$

$$\vdots$$

$$\Big\downarrow CH_2{=}CHC_6H_5$$

$$(C_6H_5CH_2)_3Zr(\overset{|}{\underset{C_6H_5}{CHCH_2}})_nCH_2C_6H_5$$

$$C'P_n$$

β-Hydrogen transfer terminates the growth of the polymer chains and disengages them from the metal centers.

The kinetics and equilibrium information can be summarized in the following compact form:

$$C' + M \overset{K_{163}}{\rightleftharpoons} C'I_0 \tag{2-163}$$

$$C'I_0 \overset{k_{164}}{\longrightarrow} C'P \tag{2-164}$$

$$C'P_n + M \overset{K_{165}}{\rightleftharpoons} C'I_n \tag{2-165}$$

$$C'I_n \overset{k_{166}}{\longrightarrow} C'P_{n+1} \tag{2-166}$$

$$C'P_n \overset{k_{167}}{\longrightarrow} P_n + C'' \tag{2-167}$$

$$C'P_n + M \overset{k_{168}}{\longrightarrow} Z_n \tag{2-168}$$

The final reaction indicates the formation of polymers attached to inactive metal atoms (Z_n); it is a blind alley in the catalytic cycle.

The kinetics can be worked into a form convenient for comparison with the experimental results, as follows: From a mass balance on the Zr,

$$(C_C)_t = C_{C'} + \sum_{n=1}^{\infty} C_{C'P_n} + \sum_{n=1}^{\infty} Z_n + C_{C''} + \sum_{n=1}^{\infty} C_{C'I_n} \tag{2-169}$$

Furthermore, it has been shown experimentally [66] that $C_{C'} \approx (C_C)_t$ at low conversions, that is, that the catalyst is predominantly present as $(C_6H_5CH_2)_3ZrCH_2C_6H_5$ and that $C_{C'I_0} \ll (C_C)_t$, which suggests the plausibility of the assumption that $\sum_{n=1}^{\infty} C_{C'I_n} \ll \sum_{n=1}^{\infty} C_{C'P_n}$. Therefore, the expression for equilibrium in steps 2-163 and 2-165 is written as follows:

$$C_{C'I_0} = K_{163}C_{C'}C_M \cong K_{163}(C_C)_0(C_M)_0 \tag{2-170}$$

and

$$\sum_{n=1}^{\infty} C_{C'I_n} = K_{165} \sum_{n=1}^{\infty} C_{C'P_{n+1}} C_M \tag{2-171}$$

Now, expressions are written for the following:

1. The rate of initiation of the polymerization, that is, the rate of formation of C'P,

$$r_i = k_{164} C_{C'I_0} = k_{164} K_{163} (C_C)_0 (C_M)_0 \tag{2-172}$$

2. The rate of propagation, that is, the rate of growth of the polymer chains,

$$r_p = k_{166} \sum_{n=1}^{\infty} C_{C'I_n} = k_{166} K_{165} \sum_{n=1}^{\infty} C_{C'P_n} (C_M)_0 \tag{2-173}$$

and

3. The rate of termination of polymerization,

$$r_t = [k_{167} + k_{168}(C_M)_0] \sum_{n=1}^{\infty} C_{C'P_n} \tag{2-174}$$

The rate of polymerization (i.e., the rate of monomer consumption) is almost identical to the rate of propagation r_p, since the amount of monomer consumed in the initiation and termination reactions is negligible in comparison with that consumed in the polymerization.

The further analysis of the kinetics is simplified by invoking the steady-state approximation illustrated in Chapter 1. Assume that after a short induction period, the term $\sum_{n=1}^{\infty} C_{C'P_n}$ achieves a nearly constant (time-invariant) value, since these species are highly reactive and the sum of their concentrations is small. Therefore, $r_i = r_t$, and the term $\sum_{n=1}^{\infty} C_{C'P_n}$ can be solved for and combined with Eq. 2-173 to give an equation for the polymerization rate in terms of conveniently measurable concentrations:

$$r_p = \frac{k_{164} k_{166} K_{163} K_{165} (C_C)_t (C_M)_0^2}{k_{167} + k_{168}(C_M)_0} \tag{2-175}$$

This equation predicts a linear dependence of rate on total catalyst concentration; the prediction is consistent with the experimental results of Fig. 2-43.

Equation 2-175 can be rearranged to the form

$$\frac{1}{(C_M)_0} + A = \frac{B(C_C)_t (C_M)_0}{r_p} \tag{2-176}$$

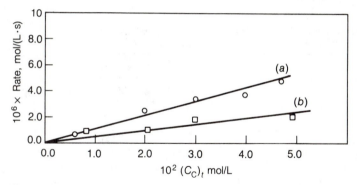

Figure 2-43
Linear dependence on catalyst concentration of initial rate of polymerization of styrene initiated by Zr(benzyl)$_4$ in toluene solvent at 30°C. Curve (a), $(C_M)_0$ = 5.0 M; curve (b) $(C_M)_0$ = 3.0 M [66b].

where $A = k_{168}/k_{167}$ and $B = k_{164}k_{166}K_{163}K_{165}/k_{167}$. The data of Fig. 2-44 are consistent with the equation—plots of $(C_M)_0/r_p$ vs. $1/(C_M)_0$ are good straight lines at given values of temperature and $(C_C)_t$. The experimental kinetics therefore confirms the suggested sequence of steps. The rate equation parameters are summarized in Table 2-6.

OLEFIN METATHESIS

The olefin metathesis reaction (sometimes referred to as a dismutation, or a disproportionation reaction), recently discovered by H. S. Eleuterio [67], is an industrially applied route for breaking of two carbon–carbon double bonds with formation of two new carbon–carbon double bonds:

$$
\begin{array}{ccc}
RCH{=}CHR' & & RCH \quad HCR' \\
+ & \rightarrow & \| \quad + \quad \| \\
RCH{=}CHR' & & RCH \quad HCR'
\end{array}
\qquad (2\text{-}177)
$$

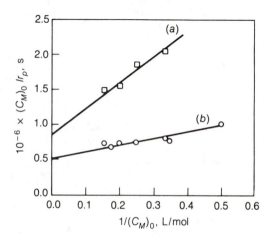

Figure 2-44
Dependence of initial rate of polymerization of styrene on the styrene concentration. Polymerization initiated by Zr(benzyl)$_4$ in toluene solution at an initial Zr(benzyl) concentration of 0.03 M. Curve (a), 30°C; curve (b), 40°C [66b].

Table 2-6

KINETICS OF POLYMERIZATION OF STYRENE INITIATED BY Zr(benzyl)$_4$ IN TOLUENE SOLUTION: THE PARAMETER VALUES FOR EQ. 2-175 [66]

	Parameter Value	
	At 30°C	**At 40°C**
$10^8 \times k_{164}K_{163}$, L/(mol·s)	3	6.7
$k_{166}K_{165}$, L/(mol·s)	0.07	0.2
$10^4 \times k_{167}$, s^{-1}	1.7	4.0
$10^4 \times k_{168}$, L/(mol·s)	0.44	2.0
$10^5 \times \sum_{i=1}^{\infty} C'C_{P_n}$, mol/L	0.7	0.6
$[(C_M)_0 = 3.00$ mol/L, $(C_C)_t = 0.03$ mol/L]		

The first known catalysts were solids (mentioned in Chapter 6); catalysts discovered more recently include soluble Ziegler combinations consisting of a transition metal halide with an alkylaluminum halide [68, 69]. Most of the catalysts include compounds of tungsten, molybdenum, or rhenium in high oxidation states, for example, Re^{7+}. One highly active catalyst–cocatalyst combination is W(OCH$_3$)$_6$ plus C$_2$H$_5$AlCl$_2$; other combinations are PhWCl$_3$ plus AlCl$_3$ and (Ph$_3$P)$_2$MoCl$_2$(NO)$_2$ plus CH$_3$Al$_2$Cl$_3$. Some highly active catalysts do not require a cocatalyst, examples being W(CH—t-Bu)(OR)$_2$X$_2$ (where X is halide) [70] and W(CH—t-Bu)(NAr)(OR)$_2$ (where Ar is aromatic) [71].

The reactants that undergo metathesis include most simple olefins, substituted olefins, cyclopropanes, and acetylenes. Cyclic olefins can react to give linear high-molecular-weight polymers; this result and the nature of the catalyst–cocatalyst combinations point to a similarity between the metathesis reaction and the olefin polymerizations discussed in the preceding section.

A basic question is which bonds are broken and which are formed in the metathesis reaction? There are two plausible possibilities: (1) cleavage and transfer of alkyl groups, and (2) cleavage of the C—C double bonds and transfer of the alkylidene groups. These possibilities can be distinguished with reactants containing deuterium labels. Consider the reaction of but-2-ene with perdeutero-but-2-ene:

$$H_3CCH{=}CHCH_3$$
$$+$$
$$D_3CCD{=}CDCD_3$$

The metathesis reaction of these gives the following product [72]:

$$D_3CCD{=}CHCH_3$$

without any of the following:

$$H_3CCD{=}CDCD_3$$

$$H_3CCH\!=\!CHCD_3$$

$$H_3CCD\!=\!CDCH_3$$

$$D_3CCH\!=\!CHCD_3$$

$$D_3CCH\!=\!CDCD_3$$

It follows that metathesis involves double-bond cleavage and not alkyl group transfer [71]. A mechanism to explain metathesis must therefore account for breaking of the carbon–carbon double bond and for the randomization of the alkylidene units.

A proposed mechanism involves a carbene complex, which is known to be capable of reacting with an olefin and forming a new olefin and another carbene complex:

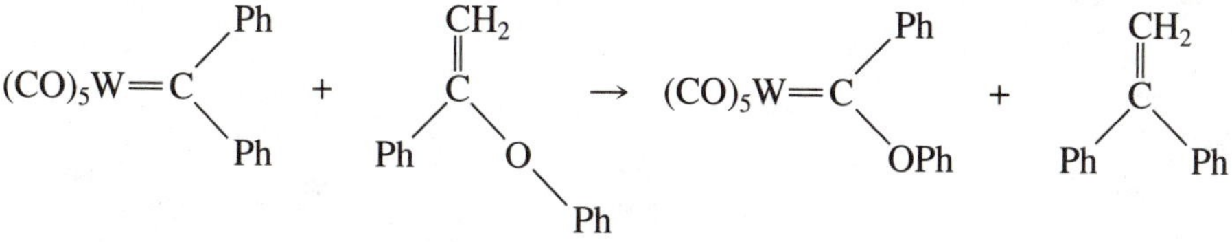

$$(2\text{-}178)$$

This reaction provides part of the basis for the following suggested steps in metathesis:

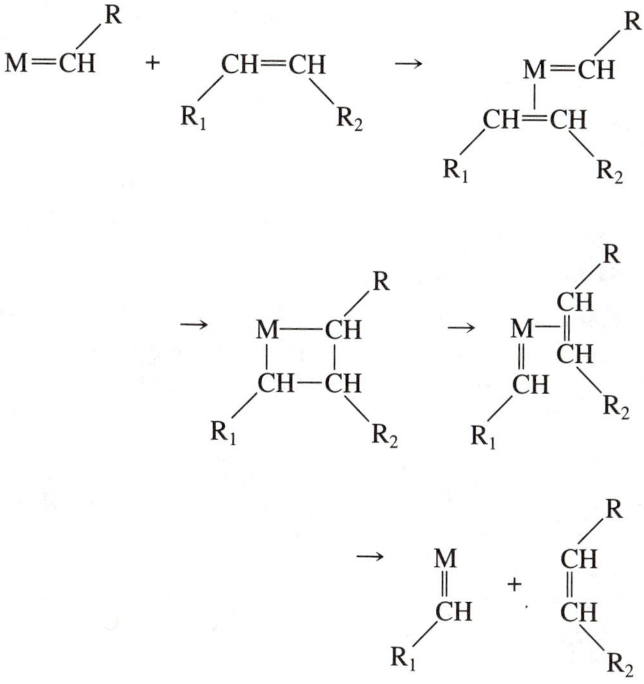

$$(2\text{-}179)$$

Here, the other ligands on the metal M have been omitted. The cyclic intermediate shown above is a metallacycle.

This sequence of steps meets the criteria mentioned above. Each of the
two fragments (alkylidene units) of each of the two reactant olefins can in this
way become bonded to any of the other fragments. A representation of the
possibilities is shown in the schematic cycle of Fig. 2-45; only some of the
imaginable combinations are included.

Researchers investigating the reactivities and structures of likely precursors of reactive intermediates in metathesis cycles [71, 73] have postulated
details of the mechanisms. A crystal structure of $W[\overline{C(CH_2)_3CH_2}](OCH_2\text{-}t\text{-Bu})_2\text{-}$
$Br_2 \cdot GaBr_3$ [73] led to the suggestion of structure (a) below, in which an olefin
is coordinated in a position cis to the carbene ligand. This could be converted
into a metallacyclobutane, structure (b), postulated to be an intermediate in
the metathesis cycle:

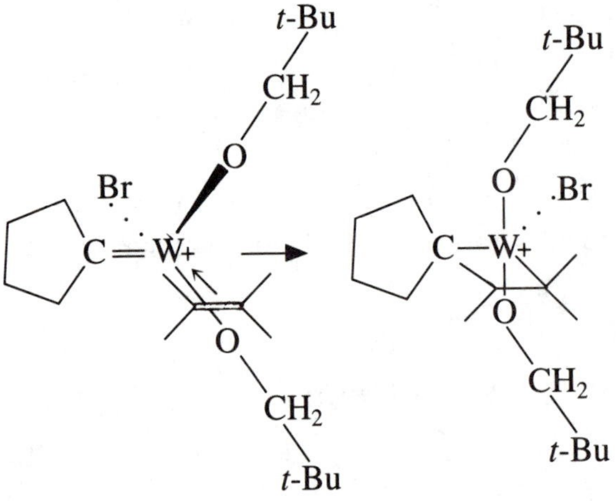

(a)   (b)

PARTIAL OXIDATION

Many useful conversions of organic compounds involve reactions with oxygen
to give products other than CO and CO_2. The partial oxidation of hydrocarbons,
which is especially important in technology, incorporates functional groups
such as —CHO, —COOH, and $-\overset{\displaystyle O}{\overset{\displaystyle \diagup\!\diagdown}{C-C}}-$ into the products. Most partial oxidations, both catalytic and noncatalytic, proceed via free-radical intermedi-
ates, as described in Section 2.3.2. But there are simpler oxidations involving
the now familiar reactions taking place in the coordination spheres of transition
metal ions. The metals may bond to the organic reactant and/or the oxygen
itself.

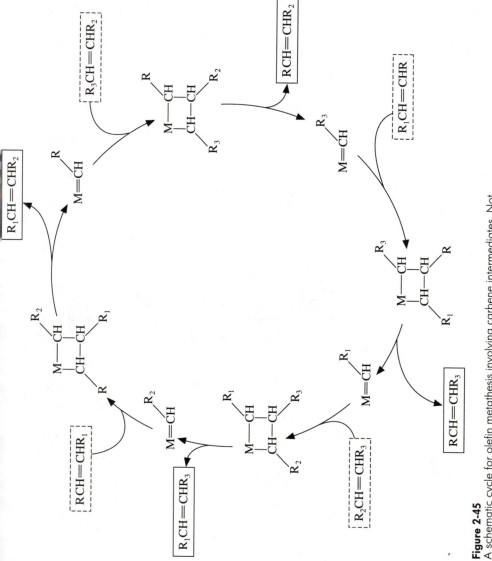

Figure 2-45
A schematic cycle for olefin metathesis involving carbene intermediates. Not all ligands around the metal M are shown.

Triphenylphosphine Oxidation

The oxidation of triphenylphosphine to triphenylphosphine oxide is catalyzed by complexes of several transition metals, including Ni, Rh, Ir, Ru, Pd, and Pt. The reaction involving Pt has been investigated most thoroughly [74]; the spectroscopic methods of determining intermediates in the catalytic cycles and the kinetics provide another good example of the detailed workings of a catalyst. Halpern et al. [74] observed spectra of a benzene solution to which $Pt(PPh_3)_4$ had been added, concluding that the predominant metal complex was $Pt(PPh_3)_3$, formed by a ligand dissociation:

$$Pt(PPh_3)_4 \rightarrow Pt(PPh_3)_3 + PPh_3 \tag{2-180}$$

The reaction of $Pt(PPh_3)_3$ with O_2 in solution was followed spectrophotometrically by monitoring an absorption peak indicative of $Pt(PPh_3)_3$; the following reaction took place in the near absence of side reactions when C_{O_2} was much less than $(C_{Pt\ complex})_t$:

$$Pt(PPh_3)_3 + O_2 \xrightarrow{k_{181}} Pt(PPh_3)_2O_2 + PPh_3 \tag{2-181}$$

and the kinetics was found to be

$$r = k_{182}C_{Pt(PPh_3)_3}C_{O_2} \tag{2-182}$$

where $k_{182} = 2.6$ L/(mol·s) at 25°C.

When a large excess of PPh_3 was present in the solution with $Pt(PPh_3)_2O_2$, then the next two reactions in the catalytic cycle, taking place almost in the absence of any side reactions, were characterized by spectrophotometric observations of $Pt(PPh_3)_3$:

$$Pt(PPh_3)_2O_2 + PPh_3 \xrightarrow{k_{183}} Pt(PPh_3)_3O_2 \tag{2-183}$$

$$2PPh_3 + Pt(PPh_3)_3O_2 \xrightarrow{fast} Pt(PPh_3)_3 + 2\ Ph_3PO \tag{2-184}$$

The second of these reactions was fast, and the rate constant of the former (k_{183}) was determined to be 0.15 L/(mol·s) at 25°C. The catalytic cycle is summarized in Fig. 2-46.

When all reactions in the cycle take place simultaneously, the rate of conversion of O_2 is expressed by Eq. 2-182; the prevailing concentration of $Pt(PPh_3)_3$ can be established if the steady-state approximation is made for this intermediate:

$$0 = -\frac{dC_{Pt(PPh_3)_3}}{dt} = k_{181}C_{Pt(PPh_3)_3}C_{O_2} - k_{183}C_{Pt(PPh_3)_2O_2}C_{PPh_3} \tag{2-185}$$

Rearrangement and use of the result that the platinum is present almost ex-

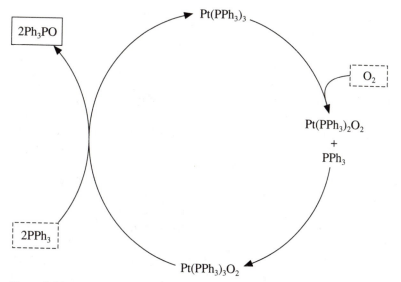

Figure 2-46
Catalytic cycle for the oxidation of triphenyl phosphine [74].

clusively as $Pt(PPh_3)_3$ and $Pt(PPh_3)_2O_2$ gives

$$r = -\frac{dC_{O_2}}{dt} = \frac{k_{181}k_{183}(C_{Pt\ complex})_t C_{PPh_3} C_{O_2}}{k_{181}C_{O_2} + k_{183}C_{PPh_3}} \qquad (2\text{-}186)$$

The appropriateness of this equation for describing the rate of the overall catalytic reaction has been confirmed in kinetics experiments independent of those mentioned in the foregoing discussion. The detailed mechanism of the reaction involving oxygen in the coordination sphere of Pt is still not known.

Oxidation of Ethylene to Acetaldehyde (the Wacker Reaction)

One of the oldest processes for partial oxidation of a hydrocarbon is the Wacker process for oxidation of ethylene to acetaldehyde. The stoichiometric oxidation of ethylene in the presence of $PdCl_2$ in aqueous solution has been known for many years:

$$CH_2{=}CH_2 + H_2O + PdCl_2 \rightarrow CH_3CHO + Pd + 2HCl \qquad (2\text{-}187)$$

Smidt and co-workers [75] recognized that combining this reaction with a rapid reoxidation could give an economical cycle:

$$Pd + 2CuCl_2 \rightarrow PdCl_2 + 2CuCl \qquad (2\text{-}188)$$

$$2CuCl + \tfrac{1}{2}O_2 + 2HCl \rightarrow 2CuCl_2 + H_2O \qquad (2\text{-}189)$$

These reactions close the cycle. Summing Eqs. 2-187–2-189 gives the simple stoichiometry of the ethylene oxidation:

$$CH_2{=}CH_2 + \tfrac{1}{2}O_2 \rightarrow CH_3CHO \tag{2-190}$$

The redox reactions involving Pd and Cu provide the opportunity for catalysis. The strategy illustrated by the work of Smidt et al. [75] is potentially valuable in other industrial situations; if a reaction is recognized that converts a reactant into a desired product, then it may form the basis for a catalytic cycle. The challenge here is to find the chemistry that will regenerate the expensive reactants (Pd in this case) rapidly, with the expenditure of only relatively cheap reactants (O_2 or air in this case). The combination of Pd and Cu has been found to be active for a number of partial oxidation reactions.

The industrial processes for ethylene oxidation fall into two categories: (1) those with one reactor, in which all three reactions [2-187 to 2-189] take place; and (2) those with two reactors, with the ethylene oxidation, reaction 2-187, and the Pd reoxidation, reaction 2-188, taking place in one reactor and the reoxidation of Cu, reaction 2-189, taking place in the other. The processes are shown schematically in Fig. 2-47. These depictions are called process flowsheets; they define the principal pieces of equipment, such as reactors and distillation columns, without details such as valves, pumps, and control devices; and they indicate the patterns of flow of the reactants, products, solvents, and so forth. The creation of new processes involves an integration of the chemistry with principles of process engineering; alternative flowsheets are

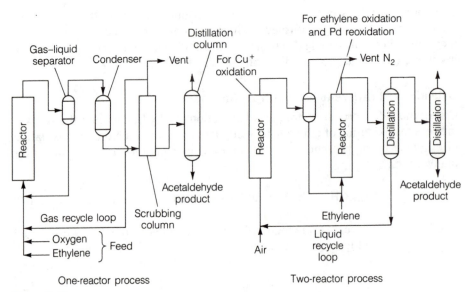

Figure 2-47
Simplified flowsheets of the one-reactor and two-reactor Wacker processes for ethylene oxidation to give acetaldehyde.

conceived at an early stage of the development. Selection among them is based largely on cost estimates.

The comparison between the two Wacker processes illustrates some of the important engineering concerns in catalysis by soluble transition metal complexes. One is the materials of construction, since the chlorides of Pd and Cu in an acidic solution are highly corrosive. Expensive corrosion-resistant materials such as titanium must be used. An advantage of the one-reactor process is that the corrosive liquid is confined to the reactor; in the other process, two reactors, pipes, and a pump must be constructed of the expensive material. But in the two-reactor process, air (fed to the reactor where Cu^+ is reoxidized) can be used instead of the expensive oxygen, since the inert components can flow once through, exiting to a vent. In contrast, the one-reactor process uses a gas recycle loop, and therefore nearly pure oxygen is needed in the feed instead of air. If air were used, the nitrogen and other inert components would accumulate and would dominate the recycle stream. Even with a feed of nearly pure oxygen, a bleed stream is drawn off the gas recycle loop to prevent buildup of the inerts introduced as impurities. Another concern is the possibility of forming explosive mixtures of oxygen and ethylene; the need to operate with safe mixtures constrains the feed composition and the conversion in the ethylene oxidation reactor. The reactor design also requires efficient gas–liquid contacting, so that the gas phase reactants are efficiently transferred to the liquid phase, where the reaction occurs. The issues are those mentioned in the discussion of hydroformylation of olefins (p. 95); they pertain to a majority of the industrial processes involving catalysis by transition metal complexes in solution.

The chemistry of the ethylene oxidation, reaction 2-187, is similar to that of the previously discussed reactions of transition metal complexes. Ethylene is π-bonded to the metal, where it is activated, but the details of the cycle are not well understood. In contrast, the chemistry of the oxidation of CuCl, reaction 2-189, involves free radical intermediates.

An important distinction between the ethylene oxidation reaction and the previously discussed triphenylphosphine oxidation reaction is the nature of the species donating oxygen to the organic ligand; the oxygen donor is believed to be water in the Wacker reaction and coordinated dioxygen in the triphenylphosphine oxidation. In the Wacker cycle, water is an intermediate oxidation product formed from O_2 in a separate reaction. In catalytic oxidation the common pattern is that a species other than the O_2 itself donates the oxygen to the organic reactant; this pattern is encountered again in Chapter 6.

C—H BOND ACTIVATION

The industrially important examples of organometallic catalysis described here involve reactants such as olefins, carbon monoxide, oxygen, and other compounds having nonbonding electrons, which provide the reactivity for interaction with the metal. An important class of compounds is missing from this list—saturated hydrocarbons, paraffins—which are major constituents of petroleum and natural gas and therefore of great importance as raw materials in the organic chemicals industry. Some reactions of these compounds having nonactivated C—H bonds were encountered, for example, in strong-acid-catalyzed paraffin isomerization (with the highly reactive R^+ ab-

stracting a hydride ion from $-\overset{|}{\underset{|}{C}}-H$ and generating another carbenium ion) and

in autoxidation (with RO_2^- abstracting a hydrogen from $-\overset{|}{\underset{|}{C}}-H$ and generating

another free radical). But reactions in which a simple C—H bond is activated by a transition metal complex are rare, and there are no industrially important examples.

In recent years much effort has gone into research with the goal of activating paraffinic C—H bonds with complexes of rhenium, iridium, and others. Thus far the results are only some slow catalytic reactions, with most characterized by few turnovers (reaction events per metal atom). Much remains to be done, and for now the only applications of catalytic C—H bond activation involve surfaces. The subject reappears in Chapter 6 in the context of important paraffin conversion processes catalyzed by surfaces of transition metals.

2.4.3 Catalysis by Metal Clusters

The metal complex catalysts considered thus far have been *mononuclear*, containing a single metal atom, and complexes containing two or more metal atoms held together by bridging ligands. There is a class of metal complexes that contain more than one metal atom and are held together by metal–metal bonds, exemplified by $Co_4(CO)_{12}$ (p. 95). These have frameworks of three or more metal atoms or ions; they are called *metal clusters*. Hundreds of these compounds are known, many having well-defined structures that have been determined by X-ray crystallography [76]. The stable compounds incorporate familiar ligands, commonly carbonyls, hydrides, phosphines, chlorides, and some hydrocarbons.

One of the simplest and most thoroughly investigated metal clusters is $Os_3(CO)_{12}$, which is quite stable and has the following structure:

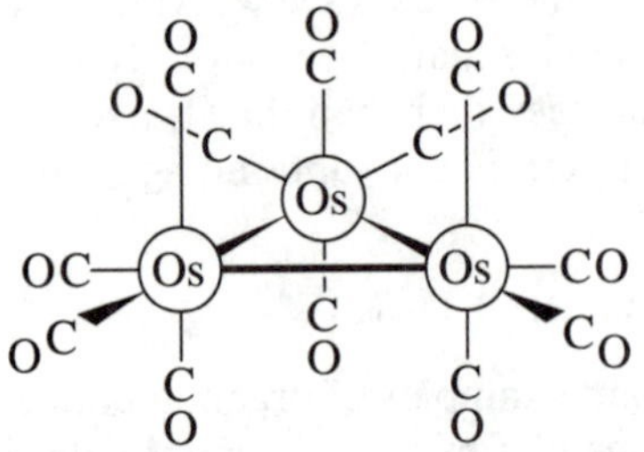

The cluster consists of a triangle of metal atoms, each bearing four terminally bonded CO ligands. $Ru_3(CO)_{12}$ has the same structure.

There are also many tetranuclear structures, for example, $Ir_4(CO)_{12}$, shown below; the CO ligands here are represented by connected full circles and the Ir atoms are represented by open circles:

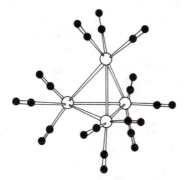

The metal atoms define a tetrahedron, and again the CO ligands are terminally bonded. The structure of $Co_4(CO)_{12}$ is slightly different; the metal atoms again define a tetrahedron, but there are bridging as well as terminal CO ligands.

Many cluster compounds are of metals important in catalysis, including Fe, Ru, Rh, Re, Os, Ir, and Pt. There is a family of platinum cluster anions that are oligomers of triplatinum units (Fig. 2-48) and a rhodium cluster (Fig.

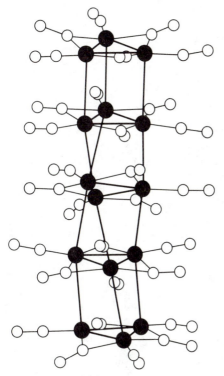

Figure 2-48
Structure of $[Pt_3(CO)_6]_5^{2-}$ anion [77]. Full circles: platinum atoms; open circles: C and O atoms.

2-49) in which the metal atoms have a hexagonal close packing typical of many crystalline solids. There is even a platinum carbonyl incorporating 19 metal atoms (Fig. 2-50), and still larger clusters are known. The clusters provide a transition between mononuclear metal compounds and particles of solid metals, important catalysts which are discussed in Chapter 6. One might anticipate that the clusters will also have catalytic properties intermediate between those of mononuclear metal complexes and surfaces of metal particles; but it is important to remember that the ligands on the clusters help hold the metal frameworks together. These ligands typically act as catalyst poisons on surfaces of metal particles.

Catalysis by metal clusters is difficult to characterize because the clusters often fragment in solution to give mononuclear metal complexes, which may be responsible for the catalysis, and they also often aggregate to give colloidal metal particles or crystallites, which may also be responsible for the catalysis. There are, therefore, only a few well documented examples of catalysis by intact metal clusters [80]. One example involves a triosmium cluster catalyst, the reaction being olefin isomerization. The proposed cycle is shown in Fig. 2-51. The CO ligands on the cluster are omitted to simplify the depiction and

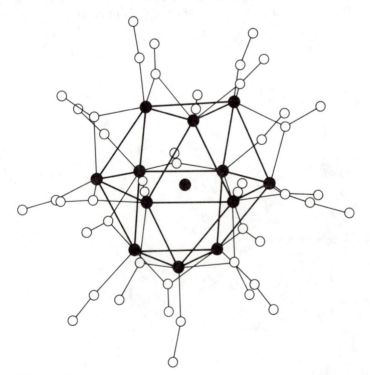

Figure 2-49
Structure of $[Rh_{13}(CO)_{25}H_{5-n}]^{n-}$ anions, where $n = 2$ or 3 [78]. Full circles: Rh atoms; open circles: CO ligands.

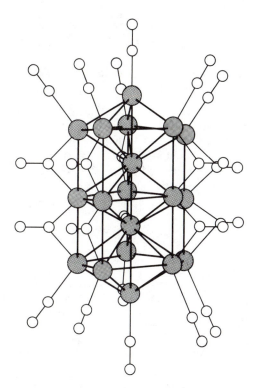

Figure 2-50
Structure of $[Pt_{19}(CO)_{22}]^{4-}$ [79]. Full circles: platinum atoms; open circles: C and O atoms.

to emphasize the interactions of the hydride and hydrocarbon ligands. The triosmium cluster is initially coordinatively unsaturated; the olefin is bonded to Os to begin the catalytic cycle.

Metal clusters present unique opportunities in catalysis [83, 84]:

1. As illustrated by the triosmium example, the clusters allow multiple bonding of a single reactant molecule, offering opportunities for conversions not possible with mononuclear species. Some of the opportunities for bonding are illustrated in Fig. 2-52.

2. Having neighboring metal centers, the clusters allow bonding of different reactants in neighboring positions. Since both CO and H are common ligands in clusters, there are prospects for development of cluster catalysts for synthesis gas (CO + H_2) conversion, the importance of which was mentioned in connection with the olefin hydroformylation process. There is evidence that rhodium carbonyl clusters in basic solutions are catalytically active for CO hydrogenation to give ethylene glycol, but pressures of hundreds of atmospheres are required, possibly to stabilize the clusters [80]. An opportunity exists for preparation of bimetallic clusters that have one metal specifically chosen to bond to one reactant and the other metal chosen to bond to the other reactant.

3. As effects of ligands on reactivity can be transmitted through the metal–metal bonds, the opportunities for designing metal cluster catalysts with appropriate activity are more varied than with mononuclear metal catalysts. One could adjust the ligands on the metal centers not holding reactants, which would leave open more bonding positions for reactants on the reactive center.

4. Ligands bonded to metal clusters can migrate rapidly from one metal center to another, as has been demonstrated for CO and hydride ligands by NMR spectroscopy. A suggested mechanism for migration of terminal CO ligands involves formation of intermediates with bridging CO ligands. In catalysis, therefore, reactants may have the opportunity to be bonded at one metal center and then to migrate to a neighboring center where they react.

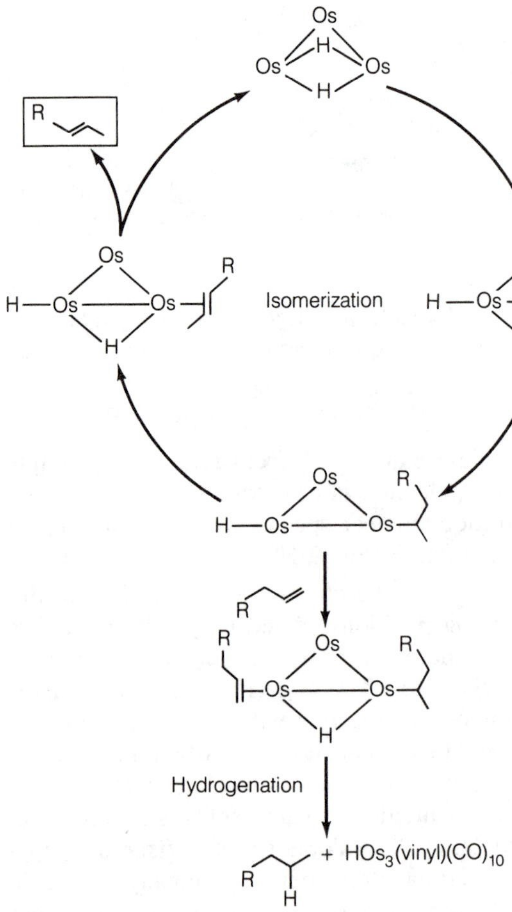

Figure 2-51
Catalytic cycle for olefin hydrogenation catalyzed by triosmium clusters [81, 82]. The carbonyl ligands on Os are omitted for simplicity.

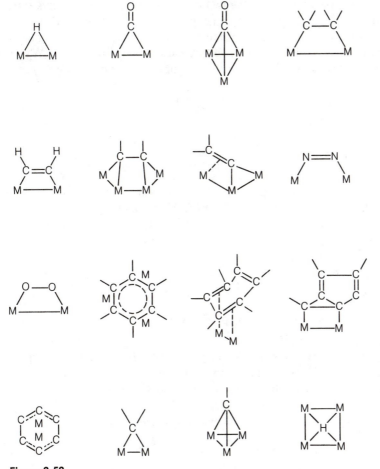

Figure 2-52
Some metal–ligand combinations in metal cluster compounds [83].

Nonetheless, the fragility of the clusters is a severe limitation to their application as catalysts. They are most important for the conceptual link that they offer to solid metal catalysts; consequently, they are mentioned again in Chapters 5 and 6.

2.5
CATALYSIS BY MACROMOLECULES

Some macromolecules may have molecular weights of hundreds of thousands or more and may still be soluble, and it is of interest to ask how the catalytic nature of a macromolecule incorporating a catalytic group compares with that of a small molecule incorporating the same group. The interest in macromolecular catalysts stems chiefly from the recognition that biological catalysts are macromolecules, and they function with amazing efficiency (Chapter 3).

It is not difficult to prepare long polymer chains incorporating catalytic groups like acids, bases, redox groups, and transition metal complexes. A simple polymer backbone that lends itself easily to functionalization is polystyrene:

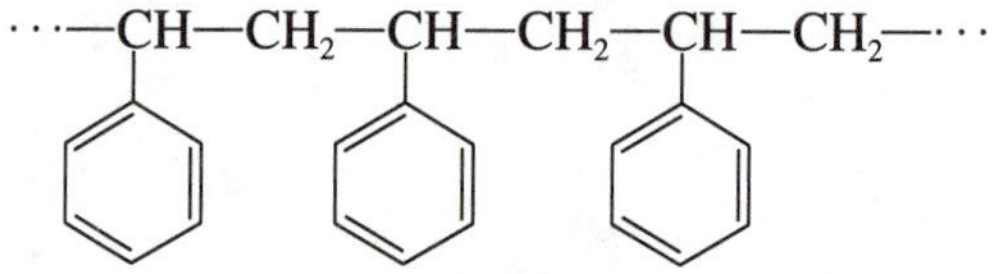

The benzene rings can be sulfonated, giving the water-soluble poly(styrenesulfonic acid), with most of the SO_3H groups in the para position:

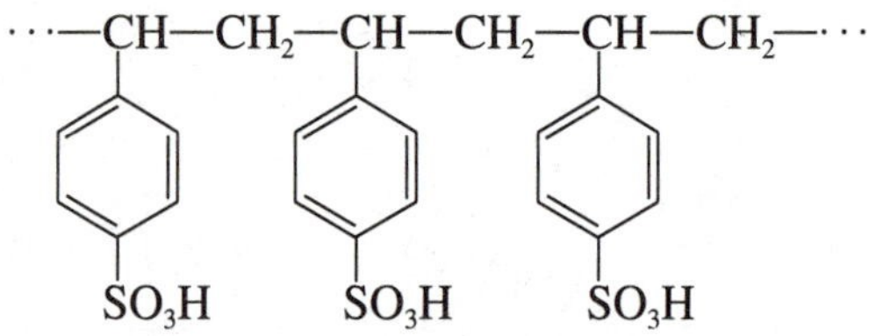

Catalytically, this polymer acts much like p-toluenesulfonic acid. Among the other polymers in this class is the following rhodium complex, which is, as expected, an olefin hydrogenation and hydroformylation catalyst [85]:

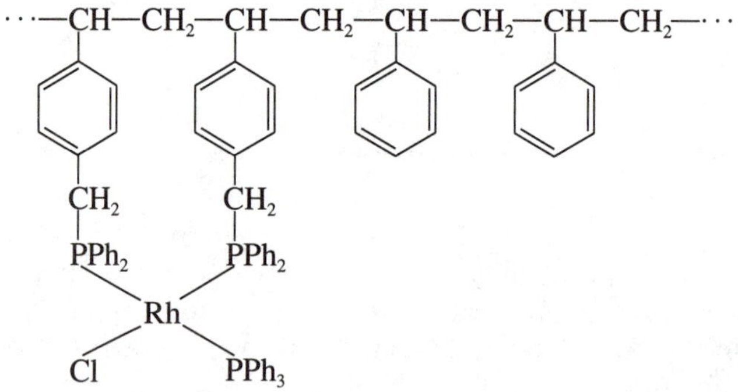

Poly(styrenesulfonic acid) in aqueous solution has been used for ester hydrolysis and has been compared with solutions of HCl of the same normality. When the ester was butyl acetate, the reaction was about an order of magnitude faster in the polymer solution than in the HCl solution; the difference between the activities of the two catalysts was negligible when the reactant was ethyl acetate [86]. An explanation of the influence of the hydrocarbon chain holding the catalytic groups rests on the notion of *hydrophobic* (or *apolar*) *bonding*:

like attracts like, and the butyl groups of the reactant interact with the hydro-carbon chain of the polymer, giving a high local concentration of reactant in the immediate vicinity of the catalytic groups. With the smaller ethyl groups, the effectiveness of the hydrophobic bonding is much less.

The important idea here concerns segregation of the ester reactant; it is concentrated near where it needs to be for catalysis to occur, that is, near the $SO_3^- H^+$ groups, because of its affinity for the hydrocarbon polymer chain. The hydrocarbon segments are a second kind of functional group. They tend to anchor the reactant ester molecule (although weakly) and to increase the prob-ability of its interaction (through the polar carbonyl group) with the $SO_3^- H^+$ group. A small extension of this idea leads to an understanding of the impor-tance of the distribution of reactants and catalyst between two liquid phases. With a proper choice of solvents in a two-phase mixture, it is possible to max-imize the concentration of a reactant in the phase where the catalyst is con-centrated. Alternatively (as in the example of paraffin alkylation, p. 60), surf-actants can be used to maximize the concentrations of reactants and catalyst at an interface, which can be considered a third phase.

The idea of anchoring a reactant molecule at one end to facilitate a cat-alytic reaction occurring at the other end has been demonstrated with tailor-made, multifunctional, donut-shaped molecules in solution (Fig. 2-53). The catalyst molecule, a cycloamylose (cyclodextrin), has a hydrophobic center (the donut hole) and polar catalytic groups at the ends. The reactant molecule becomes inserted in the cycloamylose so that the reactive ends are held in proximity to the functional groups at the ends of the cyclodextrin (e.g., acidic, basic, or metal complex groups). Consequently, the catalytic reaction is facil-itated. The multifunctional bonding is illustrated in Fig. 2-53.

It is not only apolar regions of macromolecules that can serve as a second catalyst function; many other kinds of functional groups besides the catalytic groups can be incorporated. If two monomers with different functional groups form an alternating copolymer, then the two functions will be positioned next to each other on a chain. The concept is illustrated by results obtained with a copolymer of 4(5)-vinylimidazole and *p*-vinylphenol:

This polymer in aqueous solution catalyzes the hydrolysis of 3-nitro-4-acetox-ylbenzoic acid [88]. The copolymer is about 60 times as active as imidazole for the hydrolysis reaction at pH 9. Phenol and poly(vinylphenol) have no effect on the solvolytic rate under these conditions. These results demonstrate bi-functional catalysis by the polymer, which possibly involves the imidazole group as a nucleophile attacking the reactant, and the phenolate ion acting as a base on the resulting tetrahedral intermediate [88]:

$$\cdots\!-\!CH\!\!-\!\!-\!CH_2\!\!-\!\!-\!CH\!\!-\!\!-\!CH_2\!\!-\!\cdots$$

The spacing of the functional groups is important for the efficient operation of such a (possibly concerted) mechanism. Because the macromolecule is free to bend, twist, and coil in solution, it may have a tendency to conform to the reactant, and there is no need for the pair of participating functional groups to be bonded to neighboring monomer units in the chain.

There are many related examples of macromolecular catalysis in solution [85, 89] that reinforce the ideas presented above. Another prevalent pattern in macromolecular catalysis is the competitive inhibition by compounds that are similar to the reactants in their tendencies to bind to the catalytic groups or to those groups that assist in the catalysis by bonding to a reactant and by holding it in the proximity of a catalytic group. Since either of these types of groups can become saturated with reactants, saturation kinetics, exemplified in a simple form by Eq. 1-24, is often observed.

Intriguing complications arise when the macromolecules assume secondary structures in solution. Instead of being randomly arranged, the long chains may arrange into fairly well-defined structures such as coils; polar functional groups on the inside of the coil may interact with each other to stabilize the coil. This kind of arrangement of the groups creates a distinct microenvironment within a coil; it can act like a separate phase, concentrating reactants for which it has a strong affinity, and consequently having a higher catalytic activity than a more rigid polymer that incorporates the same functional groups [90]. The microenvironment in a coil is comparable to that inside the "donut hole" of the cyclodextrin molecule of Fig. 2-53.

Hydrophobic pocket   Polar functional group

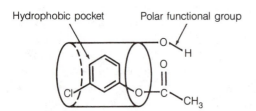

Figure 2-53
Schematic representation of a cyclodextrin inclusion complex [87].

2.6
PHASE TRANSFER CATALYSIS [91, 92]

The subjects of catalysis in two-phase liquid media and the influence of inter-facial phenomena have arisen in the discussion of alkylation and, again, in the preceding section on macromolecular catalysis. A unique class of catalyst has been designed to facilitate catalysis when two liquid phases are present by transporting reactants across the liquid–liquid interface; these *phase transfer catalysts* are used when the desired reaction involves a pair of reactants with far different solubility characteristics which, therefore, cannot both be present in a high concentration in any single fluid phase. For example, consider the reaction of an organic compound, such as an alkyl halide, RX (highly soluble in an organic phase but not in an aqueous phase), with an inorganic ion, such as OH^- or X^- (easily formed by dissolution of NaOH or NaX in water, but not formed in any appreciable concentration in an organic phase). The reaction can be facilitated near the interface between an aqueous and an organic phase, and a phase transfer catalyst can greatly accelerate the reaction by bringing the reaction partners into efficient contact.

The mechanism of action of a phase transfer catalyst is shown schemat-ically in Fig. 2-54. Consider the reaction

$$RX + Y^- \rightarrow RY + X^- \tag{2-191}$$

in the presence of QX. The phase transfer agent is Q^+, which is a highly lipophilic cation (one having a strong affinity for an organic solvent), such as a tetraalkyl phosphonium ion [e.g., $C_{16}H_{33}P^+(C_4H_9)_3$], a tetraalkyl ammonium ion, or a complexing agent like a crown ether. The following crown ethers are capable of solubilizing organic and inorganic alkali metal salts even in nonpolar organic solvents; they function by providing the cation with an "organic mask" by complex formation:

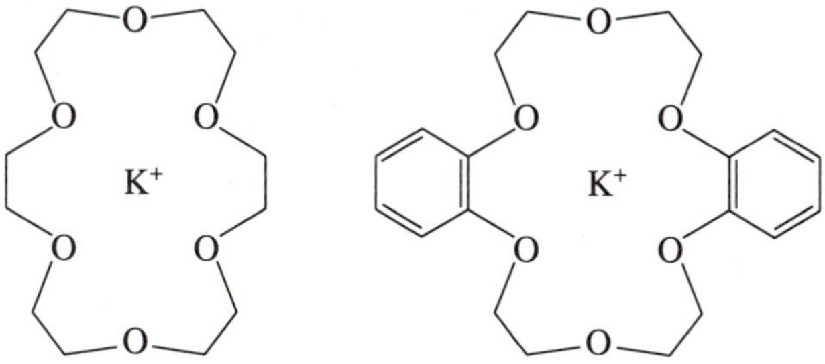

In the scheme of Fig. 2-54, Y^- is taken to be more lipophilic than X^-. The cation Q^+ migrates with the anion Y^- from the aqueous phase to the phase boundary (the liquid–liquid interface) and into the organic phase, where it comes in contact with RX. In the organic phase the ion pairs Q^+Y^- are subject

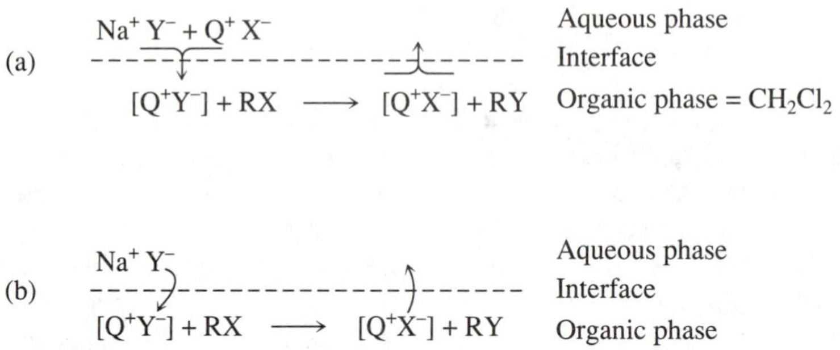

(a)

$Na^+ Y^- + Q^+ X^-$ Aqueous phase

-- Interface

$[Q^+Y^-] + RX \longrightarrow [Q^+X^-] + RY$ Organic phase = CH_2Cl_2

(b)

$Na^+ Y^-$ Aqueous phase

-- Interface

$[Q^+Y^-] + RX \longrightarrow [Q^+X^-] + RY$ Organic phase

Figure 2-54
Schematic representation of phase transfer catalysis. The reaction is RX + Y⁻ → RY + X⁻, and the phase transfer catalyst is Q⁺ [92].

to only slight solvation or aggregation, and the Y^- ions are therefore exposed and highly reactive with RX. After the rapid reaction, the catalytic cation Q^+ forms an ion pair with X^- (the reaction product) and migrates across the interface back into the polar phase. A properly chosen Q^+ has such a strong affinity for the organic phase that it does not exist in the aqueous phase in any appreciable concentration. Therefore, the simplified representation of the process, as shown in Fig. 2-54(b), is a good approximation of the role of Q^+. This ion associates with Y^- ions at the interface and transfers them in ion pairs to the organic phase, where they react with RX and generate X^-. Q^+ then associates with X^- and transfers it to the interface, where it exchanges X^- for another Y^-. Concentration gradients provide the driving force for the transfer processes and, therefore, for the catalysis. The key to the high reactivity of the ion pairs in weakly polar organic solvents is their weak solvation.

Industrial examples of phase transfer catalysis include polymerizations and substitution, condensation, and oxidation reactions, among others [91, 92], and also organometallic reactions [93]. A commonly used phase transfer catalyst is tetrabutylammonium bromide. Processing advantages include mild reaction conditions, relatively simple process flow diagrams, flexibility in the choice of solvents, and the opportunity to use NaOH as a base.

Some specific examples of phase transfer catalysis are the following:

1.

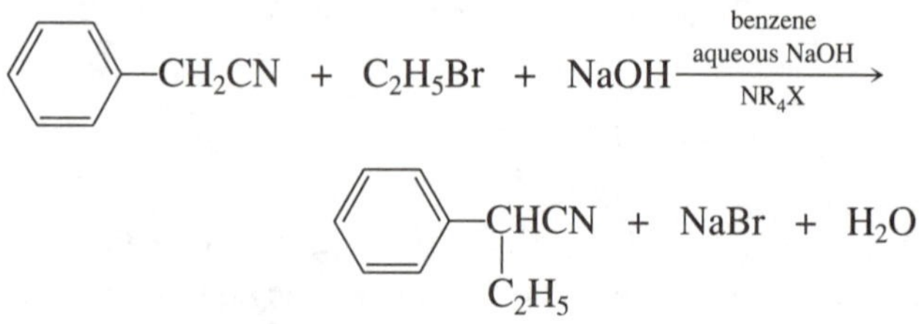

$$\text{C}_6\text{H}_5\text{—CH}_2\text{CN} + \text{C}_2\text{H}_5\text{Br} + \text{NaOH} \xrightarrow[\text{NR}_4\text{X}]{\overset{\text{benzene}}{\text{aqueous NaOH}}}$$

(2-192)

$$\text{C}_6\text{H}_5\text{—CHCN} + \text{NaBr} + \text{H}_2\text{O}$$
$$\qquad\qquad |$$
$$\qquad\quad \text{C}_2\text{H}_5$$

When this reaction is carried out in a single liquid phase, it requires anhydrous conditions and a base like sodium hydride or sodium amide.

2.

$$C_8H_{17}Cl + NaCN \xrightarrow[\text{NR}_4\text{X or PR}_4\text{X}]{\text{aqueous NaCN}} C_8H_{17}CN + NaCl \qquad (2\text{-}193)$$

This reaction gives a 99% conversion within hours under the influence of phase transfer catalysis, whereas without a catalyst, no product is formed even after weeks of contacting.

3.

$$\text{⟨Ph⟩}-CH_2Cl + CO + 2NaOH \xrightarrow[\text{PdCl}_2(\text{PPh}_3)_3,\ \text{PhCH}_2\text{N}(\text{C}_2\text{H}_5)_3\text{Cl, NR}_4\text{X}]{p\text{-xylene, aqueous NaOH}}$$

$$\text{⟨Ph⟩}-CH_2COONa + NaCl$$

$$(2\text{-}194)$$

2.7
CATALYSIS BY MICELLES

Depending on the composition and solubilities, a liquid mixture containing more than one component may consist of a single phase or more than one. The existence of a second liquid phase introduces the complications and opportunities described above, and there is an intriging variation on this theme that arises in liquids containing surfactant molecules such as soaps. The surfactant may be soluble in the liquid (e.g., water) up to a critical concentration, whereas at higher concentrations new microphases form; some of these are micelles, and they have been observed by electron microscopy [94].

A schematic representation of a micelle is shown in Fig. 2-55. The hydrophobic tails of the soap molecules form the center of the micelle, with the polar heads constituting an interface to the continuous aqueous phase. The interior region of the micelle is hydrophobic, being a much different chemical environment from that of the surrounding liquid. Reactant molecules with nonpolar groups may have a much greater affinity for the interior of the micelle than for the polar aqueous phase or the interfacial regions. Consequently, some reactants are concentrated in the micelles, and the high local concentrations may favor reactions between them. The rates of some reactions, therefore, increase markedly (perhaps one hundredfold) as the concentration of surfactant in the solution is increased beyond the critical value, and the micelles play a catalytic role [95].

The micelles may also play the role of concentrating an added catalyst along with reactants in the hydrophobic interior. They can also stabilize free-radical initiators by isolating single radicals and preventing the chain termination that results from the reaction of two radicals, at least, until a second

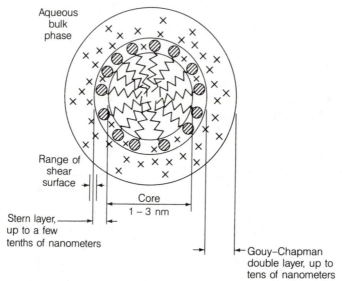

Figure 2-55
Schematic representation of the regions of a spherical ionic micelle. The counterions ($\times$), the polar head groups (⊘), and the hydrocarbon chains (〰) are indicated schematically to denote their location, but not their numbers, distributions, or configurations [95].

radical enters the micelle or is formed in it. This stabilization of an initiator is the essence of emulsion polymerization; high molecular weights of products [e.g., poly(vinyl acetate)] are formed in the presence of the micelle-isolated free radicals, whereas in solution the termination reactions are faster and the average molecular weight much less.

The micelle depicted in Fig. 2-55 has a hydrophobic interior and a polar surface; the reverse situation can also occur when the continuous liquid phase is nonpolar.

2.8
THE INFLUENCE OF DIFFUSION

In the discussion of kinetics up to this point, the rates of chemical reactions have been considered as though physical processes, such as diffusion, were of no importance. But a chemical reaction can proceed at a rate determined by intrinsic chemical reactivity only if the reactants are brought in contact with each other at this rate. At some stage of a catalytic cycle, the reactant molecules may be transported to the catalyst more slowly than they otherwise would react; in this situation, concentration gradients are set up, and the rate of the reaction may be influenced by both the intrinsic chemical reactivity (i.e., the catalytic activity) and the physical process of transport of the reactant to the catalyst (e.g., by diffusion, the rate of which is proportional to the concentration gradient of the diffusing species according to Fick's law).

When reactions take place in a homogeneous fluid phase, the influence of the physical processes is often negligible. But there are familiar reactions that are intrinsically so fast that the transport processes are important, for example, the burning of hydrocarbon vapors in a flame. In aqueous solutions, fast reactions involving proton transfer may be influenced by the physical transport processes.

A simple development helps provide insight into the interplay of the chemical and physical processes [96]. Consider a bimolecular elementary reaction involving molecules A and B, one of which might be a catalyst. In solution, each A molecule is surrounded by other molecules, including B, which diffuses toward it. At a great distance from A, the concentration of B will be the average concentration of B in the solution, C_B. If A and B react rapidly, then a concentration gradient will be set up, so that near A the concentration of B is less than C_B. The diffusion flux replaces the B molecules converted in the reaction with A.

If there were no resistance to diffusion, the rate would be written as proportional to the average concentrations C_A and C_B, just as has been done throughout this chapter. The rate per A molecule is then

$$\text{rate} = kC_B \tag{2-195}$$

In the more general case of a significant diffusion resistance, the rate of conversion of B—at steady state—is equal to the rate of transport of B from the surrounding fluid to A (this is a statement of conservation of mass). From Fick's law, this diffusion flux is written as

$$\text{rate} = -D(4\pi r^2)\frac{dC_B}{dr} \tag{2-196}$$

where D is the diffusion coefficient and r (not to be confused with rate) is the distance from A. Integrating this equation between $r = \infty$ and $r = (r_A + r_B)/2$ (the distance between A and B when they collide) gives

$$(C_B)_\infty - (C_B)_{(r_A + r_B)/2} = \frac{\text{rate}}{4\pi D\dfrac{(r_A + r_B)}{2}} \tag{2-197}$$

The equation for the intrinsic chemical kinetics is written in terms of the actual concentration of B in the immediate vicinity of A:

$$\text{rate} = k(C_B)_{(r_A + r_B)/2} \tag{2-198}$$

If there were no influence of diffusion, the rate would be the maximum attainable rate, defined as $rate_{max}$:

$$rate_{max} = k(C_B)_\infty \tag{2-199}$$

The dimensionless ratio of the actual to the maximum rate is defined as an *effectiveness factor* η, which is a measure of the diffusion influence:

$$\eta = \frac{rate}{rate_{max}} = \frac{(C_B)_{(r_A + r_B)/2}}{(C_B)_\infty} \tag{2-200}$$

Using these equations to eliminate the concentrations from Eq. 2-200 gives

$$\eta = \frac{1}{1 + \dfrac{k}{4\pi\left(\dfrac{r_A + r_B}{2}\right)D}} \tag{2-201}$$

The actual rate of reaction in the general case is

$$rate = \eta k(C_B)_\infty \tag{2-202}$$

The maximum rate, that is, the rate in the absence of a significant diffusion influence, is observed when η approaches its maximum value, unity. This limit is approached when

$$k \ll 4\pi\left(\frac{r_A + r_B}{2}\right)D \tag{2-203}$$

On the other hand, when this inequality is reversed the rate is dependent on the diffusion process:

$$rate = \eta k(C_B)_\infty = \frac{4\pi\left(\dfrac{r_A + r_B}{2}\right)D}{k} k(C_B)_\infty = 4\pi\left(\frac{r_A + r_B}{2}\right)D(C_B)_\infty \tag{2-204}$$

In this limiting case, reaction is said to be *diffusion controlled*; it depends on D but not on k.

There are transport processes besides diffusion, which is the result of random molecular motion; motion of whole fluid elements—*convection*—can be induced by stirring the contents of a reactor; natural convection can also occur. The typical liquid phase reaction is carried out in a mechanically stirred tank.

The more typical situation encountered in organometallic catalysis involves gas–liquid reactors, whereby the catalyst is present in a liquid, and one or more of the reactants is introduced into the reactor as a gas. The influence of transport processes may be much greater in these reactors, since the reactants must be transferred from the gas (usually a swarm of bubbles introduced through a sparger) through the gas–liquid interface and into the liquid. The resistances involving the gas phase are large in comparison with those in the liquid. Modeling and design of these reactors are complex and beyond the scope of this treatment.

2.9
PROCESS ENGINEERING

Some issues of the engineering of processes involving solution catalysis have arisen in this chapter. It is helpful to summarize them.

2.9.1 Separation

The products, catalyst, and unconverted reactants flowing from the reactor present problems in purification. Distillation is the most common separation process, but there are others used on a large scale, including extraction. If distillation is to be used, the catalyst must be stable enough to withstand the operation; one of the strong advantages of rhodium complexes over cobalt complexes in hydroformylation is that they have such stability. When the catalysts are expensive (e.g., rhodium), the efficiency of the separation must be exceedingly high, so that virtually all of the catalyst can be recycled to the reactor and reused. Often the separation devices dominate the flowsheet of a catalytic process that involves a soluble catalyst.

2.9.2 Corrosion

Almost all the solutions used in homogeneous catalysis are corrosive, and the need for expensive corrosion-resistant materials of construction extends beyond the reactor to the separation devices and recycle pumps and pipes. The importance of the issues of separation and corrosion often motivates the search for alternative processes using solid rather than soluble catalysts, since reaction products are easily separated from solids and corrosion is minimized. Solid catalysts are described in the chapters that follow.

2.9.3 Heat Transfer

The reactors used with soluble catalysts are typically designed for introduction of gaseous reactants by sparging and for removal of heat, because many of the reactions are exothermic. Heat may be removed through the reactor walls to cooling fluid in a surrounding jacket. Alternatively, cooling coils may be used with the flowing coolant, or a product may be vaporized, with the heat of vaporization absorbing the heat load of the reactor. The analysis and design of reactors, which rely on modeling of the chemical reaction processes and the physical processes of transport of mass and energy (heat), are described in chemical engineering textbooks [97, 98]; the subject is called *chemical reaction engineering*.

REFERENCES

1. Maccoll, A., and Stimson, V. R., *J. Chem. Soc.,* **1960,** 2836.
2. Wilhelmy, L., *Ann. Physik. Chem.* (Poggendorf), **81,** 413, 419 (1850).
3. Long, F. A., and Paul, M. A., *Chem. Rev.,* **57,** 935 (1957).
4. Wenthe, A. M., and Cordes, E. H., *J. Am. Chem. Soc.,* **87,** 3173 (1965).
5. Edward, J. T., and Meacock, S. C. R., *J. Chem. Soc.,* **1957,** 2000.
6. Ghosh, P. K., Guha, T., and Saha, A. N., *J. Appl. Chem.,* **17,** 239 (1967).
7. Tokio Kato, *Nippon Kagaku Zasshi,* **84,** 458 (1963).
8. de Jong, J. I., and Dethmers, F. H. D., *Rec. Trav. Chim.,* **84,** 460 (1965).
9. Williams, A., and Bender, M. L., *J. Am. Chem. Soc.,* **88,** 2502 (1966).
10. Bell, R. P., and Higginson, W. C. E., *Proc. R. Soc.,* **A197,** 141 (1949).
11. Hammett, L. P., and Deyrup, A. J., *J. Am. Chem. Soc.,* **54,** 2721 (1932).
12. Olah, G. A., *Friedel–Crafts Chemistry,* Wiley Interscience, New York, 1973, p. 368.
13. Hammett, L. P., *Physical Organic Chemistry,* Chap. 9, McGraw-Hill, New York, 1970.
14. Bender, M. L., *Mechanisms of Homogeneous Catalysis from Protons to Proteins,* Wiley Interscience, New York, 1971.
15. Zucker, L., and Hammett, L. P., *J. Am. Chem. Soc.,* **61,** 2785 (1939).
16. Reference 12, pp. 117–118.
17. Swain, C. G., Stivers, E. C., Reuwer, J. F., Jr., and Schaad, L. J., *J. Am. Chem. Soc.,* **80,** 5885 (1958).
18. Reference 13, Chap. 10.
19. Swain, C. G., and Brown, J. F., *J. Am. Chem. Soc.,* **74,** 2534, 2538 (1952).
20. Rony, P. R., *J. Am. Chem. Soc.,* **90,** 2824 (1968).
21. Brouwer, D. M., in *Chemistry and Chemical Engineering of Catalytic Processes,* R. Prins and G. C. A. Schuit, eds., Sijthoff and Nordhoff, Alphen an den Rijn, The Netherlands, 1980, p. 137.
22. Brouwer, D. M., and Hogeveen, H. *Prog. Phys. Org. Chem.,* **9,** 179 (1972).
23. Venuto, P. B., and Habib, E. T., Jr., *Fluid Catalytic Cracking with Zeolite Catalysts,* Marcel Dekker, New York, 1979.
24. Brouwer, D. M., and van Doorn, J. A., *Rec. Trav. Chim.,* **91,** 903 (1972).
25. Bittner, E. W., Arnett, E. M., and Saunders, M., *J. Am. Chem. Soc.,* **98,** 3734 (1976).
26. Saunders, M., and Kates, M. R., *J. Am. Chem. Soc.,* **100,** 7082 (1978).
27. Brouwer, D. M., and Hogeveen, H., *Rec. Trav. Chim.,* **89,** 211 (1970).
28. Brouwer, D. M., and van Doorn, J. A., *Rec. Trav. Chim.,* **90,** 535 (1971).

29. Olah, G. A., *Angew. Chem. Int. Ed. Engl.*, **12**, 173 (1973).

30. Olah, G. A., Halpern, A. Y., Shen, J., and Mo, Y. K., *J. Am. Chem. Soc.*, **93**, 1251 (1971).

31. Simon, A. L., Csiz, L., and Markó, L., in *Mechanisms of Hydrocarbon Reactions—A Symposium*, F. Marta and D. Kallo, eds., Akademiai Kiado, Budapest, 1975, p. 33.

32. Kramer, G. M., *Prepr. ACS Div. Petrol. Chem.*, **22**(2), 301 (1977).

33. Higginson, W. C. E., Rosseinsky, D. R., Stead, J. B., and Sykes, A. G., *Disc. Faraday Soc.*, **29**, 49 (1960).

34. Lyons, J. E., *Aspects Homog. Catal.*, **3**, 1 (1977).

35. Sheldon, R. A., and Kochi, J. K., *Adv. Catal.*, **25**, 272 (1976).

36. Parshall, G. W., *J. Mol. Catal.*, **4**, 243 (1978).

37. Collman, J. P., Hegedus, L. S., Norton, J. R., and Finke, R. G., *Principles and Applications of Organotransition Metal Chemistry,* second edition, University Science Books, Mill Valley, California, 1987.

38. Halpern, J., in *Fundamental Research in Homogeneous Catalysis,* Vol. 3, M. Tsutsui, ed., Plenum Press, New York, 1979, p. 25.

39. James, B. R., *Homogeneous Hydrogenation,* Wiley, New York, 1974.

40. Jardine, F. H., Osborn, J. A., Wilkinson, G., and Young, J. F., *Chem. Ind.* (London), **1965**, 560.

41. Osborn, J. A., Jardine, F. H., Young, J. F., and Wilkinson, G., *J. Chem. Soc.* (A), **1966**, 1711.

42. Halpern, J., *Trans. Am. Crystallogr. Assoc.*, **14**, 59 (1978).

43. Halpern, J., Okamoto, T., and Zakhariev, A., *J. Mol. Catal.*, **2**, 65 (1976).

44. Halpern, J., and Wong, C. S., *J. Chem. Soc. Chem. Commun.*, **1973**, 629.

45. O'Connor, C., and Wilkinson, G., *Tetrahedron Lett.*, **18**, 1375 (1969).

46. Henrici-Olivé, G., and Olivé, S., *Coordination and Catalysis,* Verlag Chemie, Weinheim, 1977.

47. Daniel, C., Koga, N., Han, J., Fu, X. Y., and Morukuma, K., *J. Am. Chem. Soc.*, **110**, 3773 (1988).

48. Knowles, W. S., Sabacky, M. J., and Vineyard, B. D., *Adv. Chem. Ser.*, **132**, 274 (1974).

49. Halpern J., Riley, D. P., Chan, A. S. C., and Pluth, J. J., *J. Am. Chem. Soc.*, **99**, 8055 (1977).

50. Halpern, J., *Science,* **217**, 401 (1982).

51. Olivier, K. L., and Booth, F. B., *Hydrocarbon Process.*, **49** (4), 112 (1970).

52. Orchin, M., and Rupilius, W., *Catal. Rev.*, **6**, 85 (1972).

53. Pruett, R. L., *Adv. Organomet. Chem.*, **19**, 1 (1979).

54. Falbe, J., *Carbon Monoxide in Organic Synthesis,* trans. C. R. Adams, Chapter 1, Springer, New York, 1970.

55. Martin, A. R., *Chem. Ind.* (London), Dec. 11, 1954, p. 1536.

56. Bach, H., Gick, W., Konkol, W., and Wiebus, E., *Proc. 9th Int. Congr. Catal.* (Calgary), 1988, Vol. 1, p. 254.

57. (a) Forster, D., *Adv. Organomet. Chem.*, **17**, 255 (1979);

57. (b) Forster, D., *J. Am. Chem. Soc.*, **98**, 846 (1976).
58. Hjortkjaer, J., and Jensen, V. W., *Ind. Eng. Chem. Prod. Res. Dev.*, **15**, 46 (1976).
59. Webber, K. M., Gates, B. C., and Drenth, W., *J. Catal.*, **47**, 269 (1978).
60. Bogdanovic, B., Henc, B., Karmann, H. G., Nussel, H. G., Walter, D., and Wilke, G., *Ind. Eng. Chem.*, **62**(12), 34 (1970).
61. Keim, W., *Angew. Chem. Int. Ed. Engl.*, **29**, 235 (1990).
62. Ewen, J. A., *J. Am. Chem. Soc.*, **106**, 6355 (1984).
63. Kaminsky, W., Külper, K., Brintzinger, H. H., and Wild, F. R. W. P., *Angew. Chem. Int. Ed. Engl.*, **24**, 507 (1985).
64. Jordan, R. F., *J. Chem. Educ.*, **65**, 285 (1988).
65. Waymouth R., and Pino, P., *J. Am. Chem. Soc.*, **112**, 4911 (1990).
66. (a) Ballard, D. G. H., *Adv. Catal.*, **23**, 263 (1973);
66. (b) Ballard, D. G. H., Dawkins, J. V., Key, J. M., and van Lienden, P. W., unpublished results, 1971, cited in ref. 66a.
67. Eleuterio, H. S., German Patent 1,072,811 (1960); *J. Mol. Catal.*, **65**, 55 (1991).
68. Grubbs, R. H., in G. Wilkinson, F. G. A. Stone, and E. W. Abel, eds., *Comprehensive Organometallic Chemistry*, Vol. 8, Pergamon, Oxford, 1982, p. 499.
69. Ivin, K. J., *Olefin Metathesis*, Academic Press, London, 1983.
70. Kress, J., Wesolek, M., and Osborn, J. A., *J. Chem. Soc. Chem. Commun.*, **1982**, 514.
71. Schrock, R. R., DePue, R. T., Feldman, J., Schaverien, C. J., Dewan, J. C., and Liu, A. H., *J. Am. Chem. Soc.*, **110**, 1423 (1988).
72. Calderon, N., Ofstead, E. A., Ward, J. P., Judy, W. A., and Scott, K. W., *J. Am. Chem. Soc.*, **90**, 4133 (1968).
73. Youinou, M. T., Kress, J., Fischer, J., Aguero, A., and Osborn, J. A., *J. Am. Chem. Soc.*, **110**, 1488 (1988).
74. (a) Birk, J. P., Halpern, J., and Pickard, A. L., *J. Am. Chem. Soc.*, **90**, 4491 (1968);
74. (b) Halpern, J., and Pickard, A. L., *Inorg. Chem.*, **9**, 2798 (1970);
74. (c) Birk, J., Halpern, J., and Pickard, A., *Inorg. Chem.*, **7**, 2672 (1968).
75. Smid, J., Hafner, W., Jira, R., Sedlmeier, J., Seiber, R., Ruttinger, R., and Kojer, H., *Angew. Chem.*, **71**, 176 (1959).
76. Moskovits, M., ed., *Metal Clusters*, Wiley, New York, 1986.
77. Calabrese, J. C., Dahl, L. F., Chini, P., Longoni, G., and Martinengo, S., *J. Am. Chem. Soc.*, **96**, 2614 (1974).
78. Albano, V. G., Cerotti, A., Chini, P., Ciani, G., Martenengo, S., and Anker, M., *J. Chem. Soc. Chem. Commun.*, **1975**, 859.
79. Washecheck, D. M., Wucherer, E. J., Dahl, L. F., Ceriotti, A., Longoni, G., Manassero, M., Sansoni, M., and Chini, P., *J. Am. Chem. Soc.*, **101**, 6110 (1979).
80. Markó, L., and Vizi-Orosz, A., in B. C. Gates, L. Guczi, and H. Knözinger, eds., *Metal Clusters in Catalysis*, Elsevier, Amsterdam, 1986, p. 89.

81. Deeming, A. J., and Hasso, S., *J. Organomet. Chem.*, **114,** 313 (1976).
82. Keister, J. B., and Shapley, J. R., *J. Am. Chem. Soc.*, **98,** 1056 (1976).
83. Muetterties, E. L., *Bull. Soc. Chim. Belg.*, **84,** 959 (1975).
84. Gates, B. C., Guczi, L., and Knözinger, H., eds., *Metal Clusters in Catalysis,* Elsevier, Amsterdam, 1986.
85. Bayer, E., and Schurig, V., *CHEMTECH,* **6,** 212, 1976.
86. Sakurada, I., Sakayuchi, Y., Ono, T., and Ueda, T., *Makromol. Chem.,* **91,** 243 (1966).
87. Tabushi, I., and Kuroda, Y., *Adv. Catal.,* **32,** 417 (1983).
88. Overberger, C. G., Salamone, J. C., and Yaroslavsky, S., *J. Am. Chem. Soc.,* **89,** 6231 (1977).
89. Overberger, C. G., Smith, T. W., and Dixon, K. W., *J. Polym. Sci. Symp.,* **No. 50,** 1 (1975).
90. Challa, G., *J. Mol. Catal.,* **21,** 1 (1983).
91. Weber, W. P., and Gokel, G. W., *Phase Transfer Catalysis in Organic Synthesis,* Springer, Berlin, 1977.
92. Starks, C. M., and Liotta, C., *Phase Transfer Catalysis—Principles and Techniques,* Academic Press, New York, 1978.
93. Alper, H., *Adv. Organomet. Chem.,* **19,** 183 (1981).
94. Bellare, J. R., Kaneko, T., and Evans, D. F., *Langmuir,* **4,** 1066 (1988).
95. Fendler, J. H., and Fendler, E. J., *Catalysis in Micellar and Macromolecular Systems,* Academic Press, New York, 1975.
96. Boudart, M., *Kinetics of Chemical Processes,* Prentice-Hall, New York, 1968, Chapter 7.
97. Hill, C. G., Jr., *An Introduction to Chemical Engineering Kinetics and Reactor Design,* Wiley, New York, 1977.
98. Froment, G. F., and Bischoff, K. B., *Chemical Reactor Analysis and Design,* second edition, Wiley, New York, 1990.
99. Forster, D., Hershman, A., and Morris, D. E., *Catal. Rev.—Sci. Eng.,* **23,** 89 (1981).
100. Wai, J. S. M., Markó, I., Svendsen, J. S., Finn, M. G., Jacobsen, E. N., and Sharpless, K. B., *J. Am. Chem. Soc.,* **111,** 1123 (1989).

FURTHER READING

Acid–base catalysis in solution is reviewed extensively in the book by Bender [14]; a more concise account appears in R. P. Bell's *The Proton in Chemistry* (Cornell University Press, 1973). The background organic chemistry is given in the recent edition of L. P. Hammett's classic *Physical Organic Chemistry* (McGraw-Hill, 1970). Acid-catalyzed hydrocarbon conversion is summarized in Olah's book [12]. The principles of carbenium ion reactivity are set out in reviews by Brouwer [21] and, with more detail, by Brouwer and Hogeveen [22]. The issues of kinetics are summarized by Boudart [96], whose approach has been adopted here to introduce the Brønsted relation and the Hammett acidity function.

A statement of the principles of inorganic chemistry is given by D. F. Shriver, P. W. Atkins, and C. H. Langford, *Inorganic Chemistry* (Oxford University Press, 1990); the readable introduction to organometallic chemistry by J. P. Collman et al. [37] is recommended for understanding organometallic catalysis. G. W. Parshall's book, *Homogeneous Catalysis* (Wiley, 1980) is a concise summary of the chemistry of the important industrial processes involving organometallic catalysis, and C. Masters's paperbound *Homogeneous Transition-Metal Catalysis—a Gentle Art* (Chapman and Hall, London, 1981) provides more detail at an introductory level. Chiral catalysis is the subject of the series *Asymmetric Synthesis*, edited by J. D. Morrison (Academic Press, New York).

PROBLEMS

2-1   A commercial process involves the reaction of isobutylene with methanol to give methyl-*t*-butyl ether (MTBE), a high-octane gasoline blending component. The reaction is acid catalyzed in solution. Predict the catalytic cycle and show the elementary steps in the cycle. What side reactions do you expect to occur?

2-2   Explain why tertiary hexyl cations have a virtually unlimited life in a superacid solution.

2-3   The isomerization reaction

$$\text{cyclohexane} \rightleftharpoons \text{methylcyclopentane} \qquad (2\text{-}205)$$

occurring in a superacid solution has been found to be characterized by an activation energy of 17 kcal/mol, with a rate constant of $9 \times 10^2 \text{ s}^{-1}$ at 100°C. What is the mechanism of this reaction?

2-4   Predict the kinetics of hydride transfer from a tertiary carbon in a paraffin to a secondary alkylcarbenium ion. Illustrate with an energy diagram.

2-5   Reactions between alkylcarbenium ions and olefins can form paraffins and stable allylic polyalkylcyclopentyl cations:

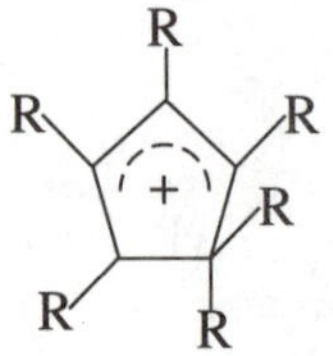

This reaction can lead to deactivation of acid catalysts because it consumes protons. Predict the mechanism of the reaction and compare your prediction with the suggestions of Deno et al. [*J. Am. Chem. Soc.*, **86**, 1745 (1964)].

2-6   Consider the chemistry described in Eqs. 2-79–2-83 and propose a catalytic cycle for the conversion of methane into isobutane in a superacid solution.

2-7 Draw a catalytic cycle for the isobutane–ethylene alkylation.

2-8 Complexes of rhodium(I) catalyze the hydrosilation reaction

$$R_3SiH \ + \ \begin{array}{c} \diagdown \\ \diagup \end{array} C{=}C \begin{array}{c} \diagup \\ \diagdown \end{array} \ \rightarrow \ R_3Si{-}\overset{|}{\underset{|}{C}}{-}\overset{|}{\underset{|}{C}}{-}H \qquad (2\text{-}106)$$

Predict the catalytic cycle and identify each elementary step. Which species are d^6 and which are d^8 in the cycle?

2-9 Consider the low-temperature oxidation of cyclohexane to give cyclohexanone, cyclohexanol, and cyclohexyl hydroperoxide in the presence of a redox initiator, Co^{3+} (present as cobalt acetate). Conceive a process flow diagram and explain the basis for the conception.

2-10 The scheme shown in Fig. 2-56 has been suggested for hydrocarboxylation of olefins and related reactions. Not all the reactions are elementary steps. Designate each elementary step by name (e.g., oxidative addition). What species are coordinatively unsaturated? What catalytic cycles occur? Draw each cycle separately and designate the reactants entering and the products leaving each cycle.

2-11 When rhodium phosphine complexes are applied as catalysts for olefin hydrogenation or hydroformylation, air is rigorously excluded. Explain why. Why are sulfur compounds also excluded?

2-12 Do the electron counting for the rhodium complexes in the Wilkinson hydrogenation cycle (Fig. 2-31). Which are Rh(I) and which are Rh(III) complexes? Which are coordinatively unsaturated?

2-13 Suggest a process flow diagram for methanol carbonylation that uses pentachlorothiophenol instead of an iodide cocatalyst. How are the advantages and disadvantages of this cocatalyst reflected in the process?

2-14 In the presence of a metathesis catalyst and cocatalyst, cyclic olefins can react to give linear high-molecular-weight polymers. Illustrate how for cyclohexene.

2-15 Predict the rate constants for the dehydration of acetaldehyde hydrate catalyzed by (1) acetic acid, (2) chloroacetic acid, and (3) dichloroacetic acid.

2-16 Predict catalytic cycles for (1) the specific-base-catalyzed, and (2) the specific-acid-catalyzed aldol condensation of acetaldehyde.

2-17 What products do you expect in the acid-catalyzed reaction of toluene and propylene? Do not neglect side reactions accompanying the alkylation.

2-18 Laboratory reagents such as cumene, diethyl ether, and tetrahydrofuran are dangerous potential explosives if stored for long after the bottle has been opened and the contents has been exposed to air. Why?

2-19 (a) Predict the approximate distribution of the two products in the chiral hydrogenation depicted in Fig. 2-35 by using the following kinetics and

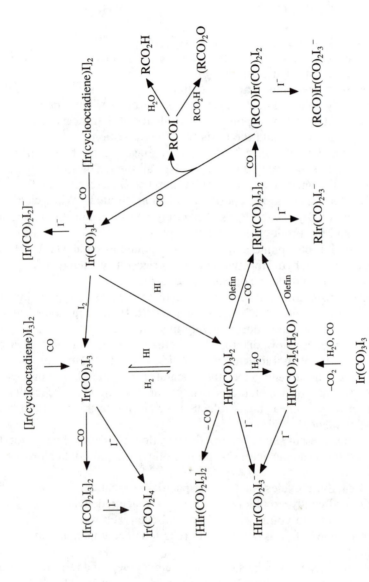

Figure 2-56
Reaction scheme proposed for hydrocarboxylation of olefins and related reactions [99].

equilibrium parameters reported for the reaction in methanol at 25°C [50]: $k' = 5.3 \times 10^3$ L/(mol·s); $k'_{-1} = 0.15$ s^{-1}; $k'_2 = 1.1$ L/(mol·s); $k''_2 = 1.1 \times 10^4$ L/(mol·s); $k''_{-1} = 3.2$ s^{-1}; $k''_{-2} = 6.4 \times 10^2$ L/(mol·s); $K'_1 = k'_1/k_1^{-1} = 3.5 \times 10^4$ L/mol; and $K''_1 = k''_1/k''_1 = 3.3 \times 10^3$ L/mol. (b) Look up the definition of the term "enantiomeric excess" (it is not used elsewhere in this book) and evaluate it for this example.

2-20 The methanol carbonylation reaction catalyzed by a rhodium complex in the presence of CH$_3$I cocatalyst has the kinetics stated in Section 2.4.2. If there is no mass transfer influence, $T = 175°C$, $C_{\text{CH}_3\text{I}} = 1$ mol/L, and $C_{\text{Rh complex}} = 10^{-4}$ mol/L, determine the average residence time in a well-mixed stirred-tank reactor. Assume that conversion is 90%.

2-21 In some liquid phase catalytic reactions the activation energy may be large (tens of kcal/mol) at low temperatures but much smaller (a few kcal/mol) at high temperatures. How does the possible influence of diffusion account for such observations?

2-22 A flexible method for asymmetric catalytic oxidation of substituted olefins (e.g., styrene and numerous natural products) gives diols. Figure 2-57 shows a postulated catalytic sequence involving two cycles, the one on the left giving a high yield of enantioselective diol and the other not [100]. It has been observed that rates of reaction and selectivities for the desired asymmetric hydroxylation are much higher when the olefin is added slowly to the batch reactor rather than being added initially in a high concentration. Explain this observation in terms of rates of steps

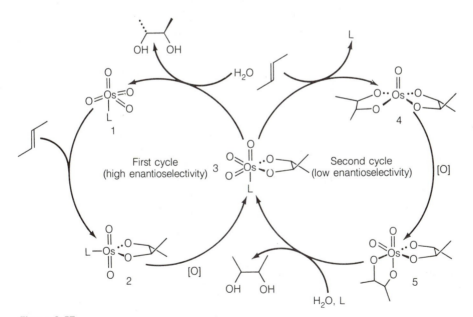

Figure 2-57
Cycles postulated for oxidation of substituted olefins to give diols [100].

shown in the cycles. What type of flow reactor would be most suitable for the asymmetric hydroxylation?

2-23 As this book was being written, researchers were reporting initial successes in the catalytic conversion of paraffins to alcohols. Consult the work of P. E. Ellis and J. E. Lyons [*Catal. Lett.*, **3**, 389 (1989)] and more recent work and identify candidate catalysts and a preliminary process for converting isobutane to *t*-butyl alcohol.

2-24 In his book *Kurze Geschichte der Katalyse in Praxis und Theorie* (Julius Springer, Berlin, 1939), Alwin Mittasch referred to a process predating the sixteenth century and possibly known to Paracelsus. Distillation of "spiritus" with oil of vitriol in a glass retort gave "oleum vitrioli dulce." Frobenius in 1790 reported that the oil of vitriol could be reused six times. What is the process?

3

Catalysis By Enzymes

3.1
INTRODUCTION

The preceding chapter marks the beginning of a transition from catalysis by small, simple molecules to catalysis by large molecules (polymers). This chapter is about catalysts that are large polymer molecules called enzymes. Most enzymes are proteins, which are formed from amino acid monomers. Enzymes regulate virtually all the reactions in biology.

Enzymes are gigantic molecules with typical molecular weights ranging from about 6,000 to more than 600,000. Moreover, enzymes are much more complex structurally than the catalysts considered thus far in this book, and their catalytic sites are far more intricate. Enzymes have evolved by the life process to be extremely specific catalysts; each enzyme is a catalyst for only one biological reaction or, sometimes, one class of reaction. The typical enzyme has a catalytic site that bonds specifically to a single reactant and activates it. The catalytic site is multifunctional, bonding to the reactant at more than one position, almost to the exclusion of all but closely similar molecules (often products), and these similar molecules are reaction inhibitors. Enzymes catalyze conversion of biological molecules with very high selectivities and at high rates, typically in the range of 10 to 10^3 molecules/(enzyme·s).

Each enzyme has a unique three-dimensional structure that usually includes a pocket or cleft presenting an array of functional groups positioned to bond to complementary functional groups of the reactant molecule. The region of the cleft is a three-dimensional microenvironment, and the bonding of the reactant is comparable to a highly specific chemisorption. Nature's process of Darwinian selection has led to the biosynthesis of individual enzyme molecules each of which is in essence a unique multifunctional catalytic site that allows bonding and conversion of a particular reactant or class of reactants. Other chemical properties of enzymes besides those of the catalytic site are important also; the functional groups on the outer surface, for example, determine the

143

enzyme's affinity for particular locations in the biological cell, such as the aqueous environment of the cytoplasm or the hydrophobic environment of a cell wall or other membrane.

This chapter is a brief summary of the structure and chemistry of enzymes, with illustrations of the catalytic sites and mechanisms of catalytic action. The subject of enzyme chemistry and catalysis is too large and complex for a thorough review in this text. The purpose of this chapter is to summarize the essential characteristics of enzyme catalysis with a minimum of biochemical terminology so that the book will provide a perspective for understanding catalysis as a whole. Readers are referred to recent textbooks on enzyme catalysis for more details [1, 2].

The functional groups constituting the catalytic sites of enzymes, many of them acids and bases, are familiar from the preceding chapters. The catalytic action is explained essentially by the same chemistry described above, but the enzymes are deserving of a separate chapter because of their complexity, subtlety, efficiency, and importance in biology.

3.2
COMPOSITION AND STRUCTURE
OF ENZYMES [1, 2]

Enzymes are proteins, macromolecules called polypeptides, which in living organisms are continually being formed in the biosynthetic pathways directed by genes. Relatively small enzymes have also been synthesized in the laboratory by methods that have now become conventional organic chemistry. In the laboratory syntheses pioneered by R. B. Merrifield, the first amino acid in the chain is bonded to a functionalized solid support. Monomers are added one at a time in sequence in automated apparatus, with the growing polypeptide chain remaining anchored to the support until the final product is formed and unlinked from it. The supports are typically functionalized cross-linked polystyrene of the kind to be considered in Chapter 4. The isolation of the growing polypeptide chain in a solid phase ensures a high yield in each step, which is necessary for a good overall yield in a synthesis involving hundreds of steps.

The amino acid monomers from which proteins are formed have the following compositions:

$$
\begin{array}{c}
\text{H} \\
| \\
\text{H}_2\text{N}-\text{C}-\text{COOH} \\
| \\
\text{R}
\end{array}
$$

The amino acids react with each other in processes directed by genes, splitting out water and forming bonds between C and N (peptide bonds):

$$
\cdots-\overset{\text{H}}{\underset{\text{R}}{\text{N}}}-\overset{\text{H}}{\underset{}{\text{C}}}-\overset{\text{O}}{\underset{}{\text{C}}}-\overset{\text{}}{\underset{\text{H}}{\text{N}}}-\overset{\text{R}'}{\underset{\text{H}}{\text{C}}}-\overset{}{\underset{\text{O}}{\text{C}}}-\cdots
$$

There are 20 naturally occurring amino acids, which form virtually all the building blocks of the enzymes (Table 3-1). The R groups on the amino acids, which become the functional groups on the enzymes, include (1) nearly inert hydrocarbon groups such as $-CH_3$ and $-CH_2-\hspace{-0.3em}\text{⟨benzene ring⟩}$; (2) charged groups capable of being proton donors or acceptors, such as $-CH_2CH_2CH_2CH_2NH_3^+$ and $-CH_2COO^-$; (3) uncharged groups that can be proton donors or acceptors, for example, $-CH_2OH$; (4) bifunctional groups capable of being both proton donors and acceptors, for example, $-CH_2CONH_2$; and (5) groups that can form bridged (dimerlike) structures, the latter being $-CH_2SH$ groups, which combine with each other to give disulfide bridges that form intramolecular cross-links in the enzymes:

$$\cdots\!-CHCH_2\underset{|}{S}$$
$$\cdots\!-CHCH_2S$$

This list of functional groups by itself leads to the anticipation of the importance of general and specific acid and base catalysis by enzymes. The concepts of acid–base catalysis are central to enzyme catalysis, but they are only a beginning. The functional groups in enzymes are also ligands for metals; naturally occurring metal ions in enzymes include Mg^{2+}, Zn^{2+}, Ca^{2+}, Ni^{2+}, Fe^{2+}, Fe^{3+}, Co^{3+}, and Mo^{6+}. Metal ions are parts of the catalytic sites in many enzymes.

There are some 1500 known enzymes and more yet to be discovered. Each enzyme molecule is built up from as many as hundreds of amino acids, arranged in a precise order dictated by the biosynthetic process. The number of possible permutations is enormous (can you estimate how many?), and the catalytic nature of the enzyme is highly sensitive to the arrangement of the building blocks. Replacement of even one amino acid unit in the sequence may eliminate the catalytic activity of an enzyme.

The catalytic specificity of an enzyme may be absolute; that is, the enzyme is active for only a single reaction. Enzymes demonstrate almost complete stereochemical specificity; they distinguish between optical or geometrical isomers, and they almost always convert just one form of an enantiomeric pair, unless their biological function is to convert one isomer to another.

The great specificity of enzymes is explained by their structures, which in recent years have yielded to determination by the methods of X-ray diffraction crystallography. Enzymes are stereospecific catalysts because they have asymmetric (chiral) structures formed uniquely of L (levo) amino acid centers.

The structures of enzymes are categorized at several different levels:

1. The so-called *primary structure* is just the sequence of amino acid units in the macromolecule with a specification of the peptide bonds as well as the disulfide bridges, which can link units at various positions in the polymer chain. The primary structure of the enzyme lysozyme [3, 4, 5] is illustrated in Fig. 3-1. Here, abbreviations are used for the amino

Table 3-1
THE COMMON AMINO ACIDS

Amino Acid	Abbreviation	Side Chain R
Glycine	Gly	H
Alanine	Ala	CH_3
Valine	Val	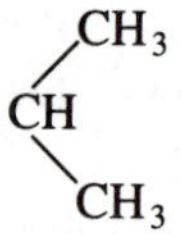 $CH{\scriptstyle<}^{CH_3}_{CH_3}$
Leucine	Leu	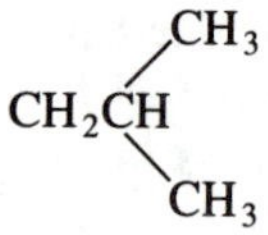 $CH_2CH{\scriptstyle<}^{CH_3}_{CH_3}$
Isoleucine	Ile	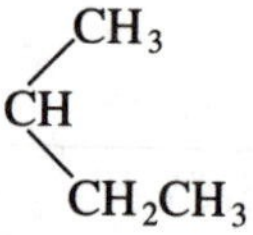 $CH{\scriptstyle<}^{CH_3}_{CH_2CH_3}$
Phenylalanine	Phe	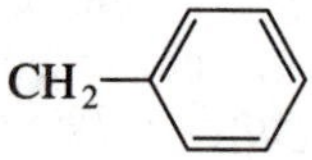 CH_2—
Tyrosine	Tyr	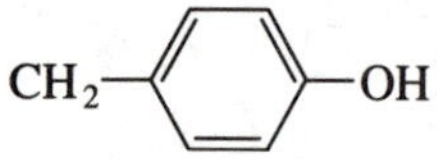 CH_2——OH
Tryptophan	Trp	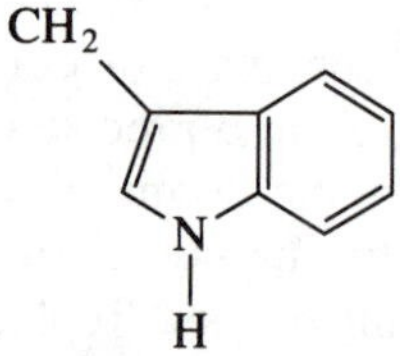 CH_2 (indole)
Serine	Ser	CH_2OH
Threonine	Thr	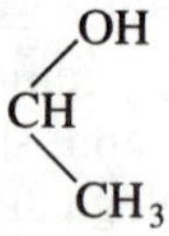 $CH{\scriptstyle<}^{OH}_{CH_3}$

acid units as defined in Table 3-1. For example, the units in the figure labeled Cys refer to the bridging units formed from cysteine.

2. The *secondary structure* refers to segments of a polypeptide chain, specifying the structural elements that result from regular coiling and folding of the chain. Parts of many enzyme molecules assume regular structures including the α-helix, depicted in Fig. 3-2. All the monomer units must have a particular stereochemistry for the α-helix to form.

Table 3-1 (*continued*)
THE COMMON AMINO ACIDS

Amino Acid	Abbreviation	Side Chain R
Cysteine	Cys	CH_2SH
Methionine	Met	$CH_2CH_2SCH_3$
Asparagine	Asn	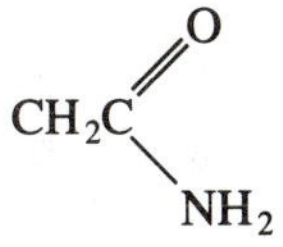 CH_2C with $=O$ and NH_2
Glutamine	Gln	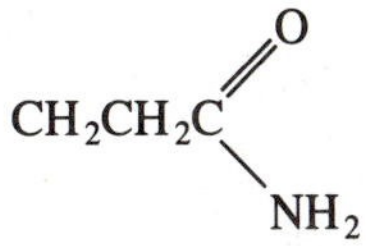 CH_2CH_2C with $=O$ and NH_2
Aspartic acid	Asp	CH_2COOH
Glutamic acid	Glu	CH_2CH_2COOH
Lysine	Lys	$CH_2(CH_2)_3NH_2$
Arginine	Arg	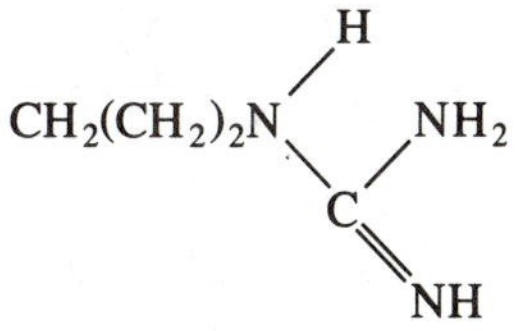 $CH_2(CH_2)_2N$ — with H, NH_2, C, NH
Histidine	His	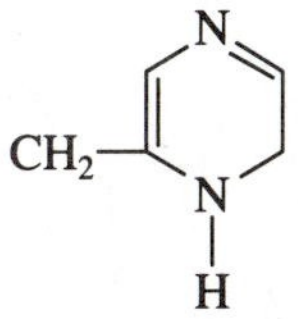 CH_2— imidazole ring with N, N, H
Proline	Pro	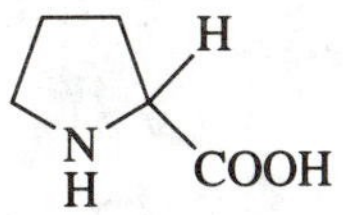 pyrrolidine ring with N, H, H, $COOH$

Not all the R groups (side chains) of the amino acids are small enough to be compatible with the helix structure, and therefore only certain parts of some proteins are α-helices. Another characteristic structure found in the segments of many enzymes is the β-pleated sheet, shown in Fig. 3-3. The organization of these and other components in an enzyme is facilitated by connecting regions and by parts that serve as hinges.

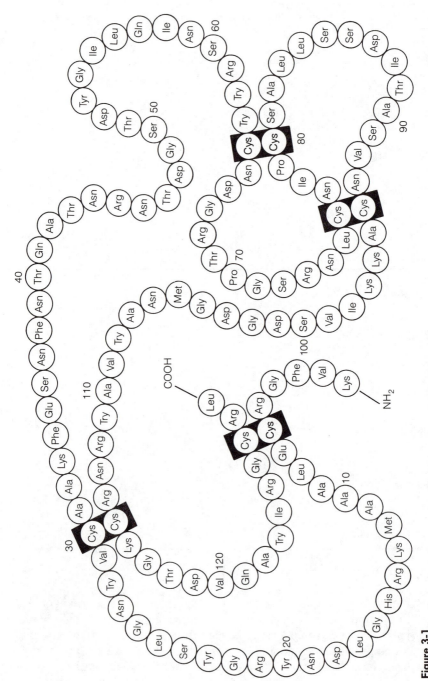

Figure 3-1
Primary structure of the enzyme lysozyme, giving the sequence of amino acids [5] (identified in Table 3-1).

3. The *tertiary structure* is the three-dimensional structure of the enzyme molecule determined by X-ray diffraction crystallography. This structure specifies the arrangement of subunits such as the α-helix into a higher state of organization. The secondary and tertiary structures of an enzyme in its natural state are determined by the genetically encoded sequence of amino acids; they are unique to a given enzyme. If the structure is disrupted (e.g., if the enzyme is denatured as a result of

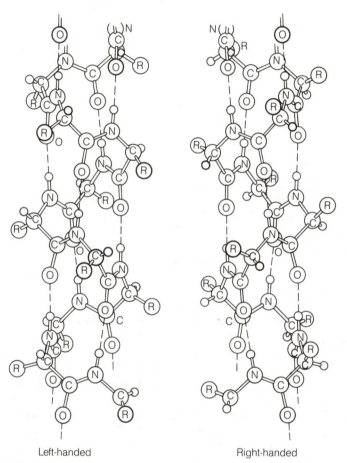

Left-handed   Right-handed

Figure 3-2
The α-helix, a configuration occurring commonly in enzymes. The chain is held in the helical configuration by hydrogen bonds that link the peptide groups:

$$
\begin{array}{ll}
\vdots & \vdots \\
C{=}O\cdots H{-}N \\
N{-}H\cdots O{=}C \\
\vdots & \vdots
\end{array}
$$

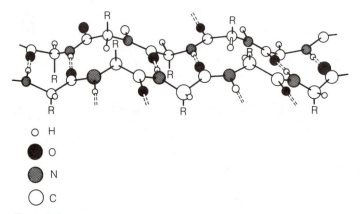

O H
● O
◉ N
○ C

Figure 3-3
The β-pleated sheet, another configuration occurring commonly in enzymes [1].

heating or of coming in contact with a new chemical environment), then the enzyme loses its catalytic activity because it loses the unique geometry of the catalytic site. Enzymes that have been denatured may spontaneously rearrange back into the catalytically active state, which is thermodynamically preferred.

4. The *quaternary structure* refers to the linking of the polypeptide units described above into a multiunit enzyme. Many enzymes consist of more than one unit.

Many enzymes require **cocatalysts**. Some cocatalysts are called *cofactors*, which combine with the enzyme to form the catalytically active site; a cofactor is regenerated with each turnover of reactant (i.e., each occurrence of the catalytic reaction). Other cocatalysts are called *coenzymes*. Typical coenzymes are complex organic molecules. The coenzyme reacts with the reactant to give a new compound as a result of the enzymatic reaction. The coenzyme is therefore a reactant, but it is not exhausted; rather, it is converted in a cycle, being regenerated in separate enzyme-catalyzed reactions that are coupled to the first. Sometimes the function of the coenzyme is to transfer a reactive intermediate from one enzyme-catalyzed reaction to another.

The ideas about enzyme structure can be understood better by preliminary consideration of an example. The lysozyme depicted in Fig. 3-1 is a relatively small, single-unit enzyme that has been investigated in detail, with the tertiary structure determined by X-ray diffraction (Fig. 3-4) [3, 4]. Only a few enzyme structures are known in such detail. The structure contains sections of α-helix constituting about 25 percent of the 129 amino acid units; the schematic diagram of Fig. 3-5 is helpful for visualization of the structure. The molecule is ellipsoidal, with dimensions of about 4.5 by 3 by 3 nm. There is a large cleft where the catalytic site is located in the molecule. Most of the interior regions of the enzyme are composed of amino acid units with hydrophobic R groups, and

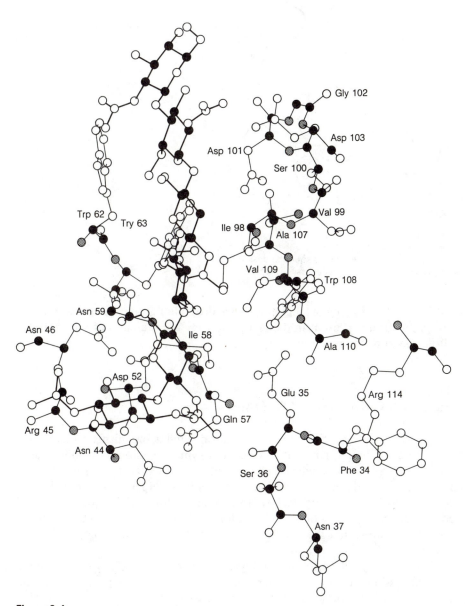

Figure 3-4
Tertiary structure of the enzyme egg white lysozyme [4].

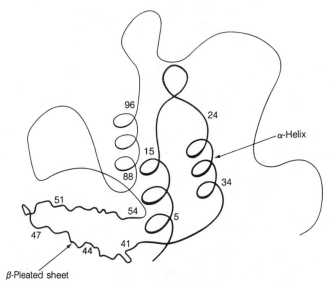

Figure 3-5
Schematic representation of part of the tertiary structure of lysozyme.
The numbers refer to the amino sequence designated in Fig. 3-1. The
α-helix regions and the cleft are easily visualized.

most of the exterior regions have polar (hydrophilic) groups. There are a few
hydrophilic groups present in the cleft, and they are the parts of the catalytic
site that bond to the reactant.

3.3
REACTIONS CATALYZED BY ENZYMES

Enzymes catalyze biological reactions; and since living organisms usually exist
only at relatively mild temperatures, most enzymes function only under these
conditions. Enzymes are delicate and do not retain their structures under conditions much outside the range characteristic of living organisms, although they
may function *in vitro* in atypical environments such as nonaqueous solvents.
Some organisms, such as those living in hot springs, have adapted to relatively
high temperatures, and they could not exist without enzymes that are stable
and active at these temperatures.

There is a great variety of biological reactions, including, for example,
(1) the breakdown of proteins and sugars occurring in the digestive tracts of
animals, (2) the biosynthetic reactions that lead to growth and replacement of
living organisms, (3) photosynthesis, and (4) oxidations that proceed by intricate pathways that convert food (e.g., sugars, proteins, etc.) into CO_2, water,
and energy in a biologically useful form. The reactions are beyond the scope
of this book. The reader is referred to textbooks listed in the Further Reading
section at the end of this chapter.

Many enzyme-catalyzed reactions take place in biological cells (Fig. 3-6) [6]. The contents of cells are organized into well-defined domains. The cytoplasm is an aqueous biological soup with many molecules, including reactants, products, and enzymes in solution. The membranes are much more highly structured. Enzymes are also present in membranes and in cellular subunits called organelles, illustrated by mitochondria (Fig. 3-7). These bound enzymes are sometimes present in arrays that can be described as assembly lines. Evidently, the product of reaction in one enzyme is transported to the next enzyme in line so that a sequence of reactions takes place efficiently. Nature's catalysts function in highly sophisticated reactors.

Enzymes are usually named for the reactions they catalyze. For example, lysozyme catalyzes reactions that lead to the breakdown by hydrolysis (lysing) of polysaccharides; isomerases catalyze isomerization reactions; dehydrogenases catalyze dehydrogenation reactions; and so forth. More specifically, al-

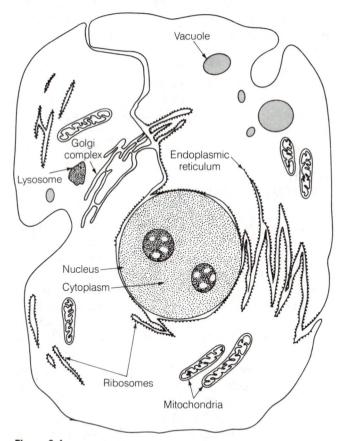

Figure 3-6
Schematic representation of a biological cell; there are wide variations from one cell type to another, but most of the components are common to many cells [6].

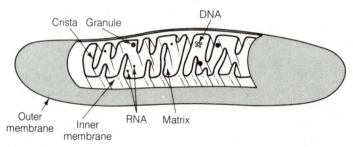

Figure 3-7
Schematic representation of a mitochondrion, an organelle found in the cell [5].

cohol dehydrogenases catalyze alcohol dehydrogenations, and L-lactate dehydrogenase catalyzes the dehydrogenation of L-lactate (to give pyruvate). Because the composition and structure of one kind of enzyme from one plant, animal, or bacterium may be different from those of the same kind of enzyme from another plant, animal, or bacterium, a further specification includes the source, for example, horse liver alcohol dehydrogenase.

In the biochemical literature (and much of the chemical literature), a reactant in a catalytic conversion is referred to as a substrate. This term is avoided here because it has a different meaning in much of the literature related to catalysis; the term "substrate" also refers to a solid surface or a support.

3.4
NATURE OF CATALYTIC SITES:
THE UNIQUENESS OF ENZYMES AS CATALYSTS

The catalytic sites of enzymes are far more complicated in structure than any others considered in this book, and they are also more uniform. The typical enzyme catalytic site consists of an array of functional groups (amino acid side chains and metal-containing groups) positioned in a three-dimensional region that is usually the inside of a cleft. Most of the amino acid side chains defining this microscopic phase are hydrophobic so that they interact (weakly) with the hydrophobic groups of reactants or other biological molecules that come in contact with the enzyme. The more reactive functional groups positioned within the largely hydrophobic cleft include charged groups, proton donors and acceptors, metal ions coordinated to the ligands offered by the amino acid side chains, and more complicated metal-containing structures such as the clusters of iron, molybdenum, and sulfur found in nitrogenase [7].

A relatively well-characterized enzyme cleft is shown schematically in Fig. 3-8 [8]. This figure illustrates the bonding of NAD^+ (a coenzyme) to a dehydrogenase. The figure depicts the bonding of the NAD^+ ion through hydrogen bonds at seven positions; this two-dimensional depiction does less than full justice to the preciseness of the interaction, but it shows clearly the geometric complementarity between the NAD$^+$ and the dehydrogenase. The importance of a complementarity between enzymes and their reactants was recognized long ago as generally important in enzyme catalysis; in 1894, E. Fischer referred to it with a lock-and-key analogy [9].

The cleft is a three-dimensional surface with precisely positioned functional groups. The reactant bonds to these groups through its own complementary groups. Molecules similar to the reactant, for example, a fragment of a reactant molecule which may be a product of the catalytic reaction, obviously can also bond to the catalytic site, or to part of it. In this way such compounds act as competitive inhibitors.

At first sight, what sets the enzymes apart from the catalysts considered earlier in this book is the complexity and preciseness of the structure of the catalytic site. The idea of a microscopic three-dimensional cavity has already been illustrated by the cyclodextrin catalyst and the coiled polymers of Chapter 2, but the cyclodextrin pocket is simply a hydrophobic region with a chemical

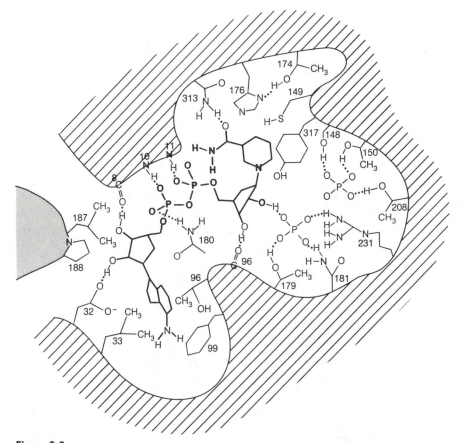

Figure 3-8
Schematic representation of the binding of the coenzyme NAD$^+$ to glyceraldehyde 3-phosphate dehydrogenase from *Bacillus stearothermophilus* [8].

affinity and a geometric fit for a phenyl ring. The cyclodextrin is viewed as a primitive analogue of an enzyme, and research with cyclodextrin and a host of other organic catalysts (and even micelles) has been motivated primarily by the goal of mimicking enzymes and understanding them better.

The idea of an array of complementary functional groups on a surface was illustrated by multifunctional polymers mentioned in Chapter 2, but these groups are arranged randomly on the polymer backbone, which itself is usually randomly arranged in solution. The groups in these synthetic polymers are therefore unable to bond specifically to a highly structured reactant. The complexity and subtlety of structure of the enzyme catalytic site are still far beyond the synthetic chemist's reach, except when the chemist imitates nature's synthetic methods.

Other distinctive characteristics of enzymes as catalysts are the following:

1. Enzymes have flexible structures, and the flexibility is important in facilitating catalytic reactions. For example, bonding of a reactant to an enzyme may induce a small but significant change in the geometry of the catalytic site, such as results from narrowing of the cleft (like a partial closing of a clam shell), so that the complementarity of the fit between the enzyme and the catalytic intermediate (transition state) is better than the complementarity of the fit between the enzyme and the reactant. This idea was articulated by J. B. S. Haldane [10] and Linus Pauling [11].

   The term *induced fit* is applied to describe the action of the reactant on the enzyme: The reactant, often a relatively rigid molecule, bonds to the catalytic site; consequently, the flexible enzyme is slightly distorted, with the result being a facilitation of the catalytic conversion as the transition state structure [a catalyst–reactant (or enzyme–substrate) complex] is optimized. Evidence of this principle has been obtained in experiments in which the enzyme was allowed to bond to each of a series of model reactants chosen to mimic the reactant and postulated transition state structures. Evidence of a shift in the tertiary structure of an enzyme (hexokinase) upon bonding to a reactant (glucose) was obtained by X-ray diffraction; results are illustrated in Fig. 3-9 [12].

   The term *strain* is applied to describe the reverse of induced fit, that is, the action of the enzyme on the reactant to distort it and facilitate its reaction. Strain seems to be less important than induced fit.

2. Enzymes are subject to subtle processes of regulation resulting from structural changes induced by bonding of compounds called *effectors*. Effectors can be inhibitors or activators. The usual kind of inhibitor of an enzyme (or any catalyst) is structurally similar to a reactant and bonds to the catalytic site in competition with the reactant. But *allosteric* effectors, which are generally structurally different from the reactants, bond to the enzyme at their own separate site (or sites) remote from the catalytic site. As a result of the bonding, the structure of the

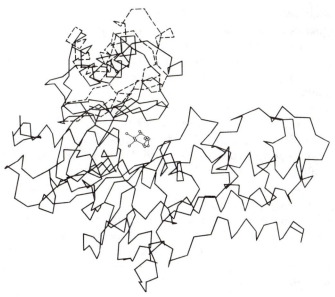

Figure 3-9
Conformational changes in hexokinase resulting from the bonding of
the reactant glucose. Solid lines indicate the carbon backbone of the
enzyme crystallized in the presence of glucose. Dashed lines indicate
a different structure, that of the enzyme crystallized in the absence of
the glucose [12].

catalytic site may change just enough that it becomes markedly more
active in the case of the activator or less active in the case of the
inhibitor.

Not many structural results demonstrate these ideas simply and clearly
for enzymes, but there is a wealth of experimental evidence for hemoglobin.
(Hemoglobin is not an enzyme, but a protein component of blood, which binds
to oxygen and is responsible for its transport.) Binding of oxygen to one of the
iron-containing heme groups of hemoglobin leads to changes in the quaternary
structure of this multiunit protein (and also to changes in the tertiary structure),
with the result that the affinity of the remaining heme groups for oxygen is
increased.

The actions of effectors are important in the control of biological processes. Living organisms regulate the rates of reactions by a variety of subtle
processes, including (1) regulating the activities of enzymes with effectors; (2)
regulating the concentrations of the enzymes, for example, by induction and
repression of their synthesis; and (3) compartmentalizing of enzymes in organs
and organelles. Furthermore, some enzymes are synthesized in inactive (pro-
enzyme) forms called *zymogens*. The zymogen conversion (which is itself enzyme catalyzed) leads to formation of the enzyme and onset of the reaction
that it catalyzes. For example, the zymogen chymotrypsinogen is synthesized

by cells of the mammalian pancreas and is secreted in that form to the intestine. There, the single polypeptide chain undergoes a limited proteolysis catalyzed by the endopeptidase trypsin, forming chymotrypsin. Thus, the enzyme becomes activated only in the extracellular environment of the intestine, where it catalyzes degradation of ingested proteins, but it does not destroy the contents of the living organism's own cells.

Another kind of control is demonstrated in biosynthesis. Enzymes are catalysts in biosynthetic processes that require specificities beyond theoretical thermodynamic limits for simple enzymes [1]. The required specificity is possible because some enzymes perform an *editing* or *proofreading* function: some enzymes involved in polymerization have, in addition to the catalytic site for synthesis, an additional hydrolytic active site where incorrect intermediates or products are destroyed as they are formed.

Still another kind of control of biological reactions is exerted by *suicide inhibitors*, also known as *Trojan-horse inhibitors*. These inhibitors are unreactive in the absence of the target enzyme and are activated by the target enzyme itself and then bind to it almost irreversibly. The enzyme appears to commit suicide under these conditions.

EXAMPLE 3-1   Michaelis–Menten Kinetics

Problem

Derive the form of rate equation that results if an enzyme-catalyzed reaction proceeds by a sequence involving bonding of a reactant to the enzyme followed by the slow reaction of the resulting intermediate (the enzyme–reactant) complex.

Solution

The sequence, in its simplest form, is

$$R + E \underset{k_{-1}}{\overset{k_1}{\rightleftharpoons}} RE \tag{3-1}$$

$$RE \overset{k_2}{\longrightarrow} P + E \tag{3-2}$$

To simplify matters, assume that reaction 3-1 achieves a virtual equilibrium and, therefore, that reaction 3-2 is rate determining. In the simplest case, the enzyme is present in only two forms, E and RE. Therefore, the total concentration $(C_E)_t$ is $C_E + C_{RE}$. At equilibrium,

$$k_1 C_R C_E = k_{-1} C_{RE} = k_1 C_R [(C_E)_t - C_{RE}] \tag{3-3}$$

Solving for C_{RE} gives

$$C_{RE} = \frac{C_R(C_E)_t}{\dfrac{k_{-1}}{k_1} + C_R} \tag{3-4}$$

The rate is $k_2 C_{RE}$, or, with k_{-1}/k_1 defined as K_M, the Michaelis constant,

$$r = \frac{k_2 C_R (C_E)_t}{K_M + C_R} \tag{3-5}$$

The rate constant k_2 is sometimes called k_{cat}.

This is an expression of saturation kinetics, similar to Eq. 1-24 of Chapter 1; it is called Michaelis–Menten kinetics and accounts for many examples in enzyme catalysis.　　　　　　　　　　　　　　　　■

EXAMPLE 3-2　Enzyme Activities

Problem

Kinetics of enzyme-catalyzed reactions may often be reported as a *turn-over number*, k_{cat}, or as an apparent second-order rate constant k_{cat}/K_M. Explain these terms.

Solution

The interpretations follow from the two limiting cases of Eq. 3-5. Consider the limiting case of a high reactant concentration which is so high that the catalytic sites are saturated and $C_R \gg K_M$. Then, the rate equation reduces to

$$r = k_{cat}(C_E)_t \tag{3-6}$$

and k_{cat} is recognized as a first-order rate constant. If the rate were written per enzyme molecule rather than per unit volume, then the reaction would be of zero order, and k_{cat} would be the rate at saturation (the maximum number of reactant molecules converted per catalytic site per unit time); this is the definition of the turnover number.

In the other limiting case, $C_R \ll K_M$, and the kinetics reduces to

$$r = \left(\frac{k_{cat}}{K_M}\right) C_R (C_E)_t \tag{3-7}$$

In this limiting case of low coverage of the catalytic sites by reactant, the apparent second-order rate constant is k_{cat}/K_M. If the rate were written per catalytic site, an apparent first-order rate constant would result.　■

3.5
EXAMPLES OF ENZYME STRUCTURE AND CATALYSIS

3.5.1 Organic Catalysis

Catalysis by the many enzymes that do not contain metals is explained by the patterns of organic chemistry. The catalytic sites in these enzymes are made up of the functional groups provided by the R groups of the amino acids that constitute the enzymes. The sites are typically intricate combinations of proton donors and acceptors, and the catalysis is often, in essence, general acid and base catalysis. Specific acid and base catalysis (catalysis by H_3O^+ and OH^-) is not common because most enzymes function in media with pH values of approximately 7, where the concentrations of H_3O^+ and OH^- are very low. Because the catalytic sites are usually complex combinations of the organic functional groups, they bind to reactant molecules in highly specific ways. Often the reactants are large molecules (e.g., polymers to be degraded) that interact with the enzymes through a number of functional groups that are complementary to the functional groups of the catalytic site.

The generalizations of the preceding sections are borne out by many experimental results; but, because of the complexity of enzyme structures, the understanding of most individual enzymes is fragmentary. The next section is a summary of structural and mechanistic results for several of the most thoroughly characterized enzymes. These results by themselves only begin to justify the generalizations of the preceding sections, but they illustrate the methods of investigation and many of the properties that make enzymes unique catalysts.

LYSOZYME [1, 2, 13]

Hen egg white lysozyme was the first enzyme to be characterized by X-ray diffraction (Fig. 3-4), and NMR data indicate that the structure in solution is the same as the crystal structure. The availability of the structural information inspired many experiments with the goal of understanding the catalytic site and reaction mechanism. The cleft in the lysozyme structure was inferred to contain the catalytic site, consisting of a few polar amino acid side chains among many hydrophobic groups. Results of experiments providing evidence of the structure of this site are summarized in the following paragraphs.

Lysozyme catalyzes the hydrolysis of bacterial cell wall polysaccharides. The basic repeating unit in these polymers is a disaccharide, with two types of modified glucose residues in a β-1,4 linkage:

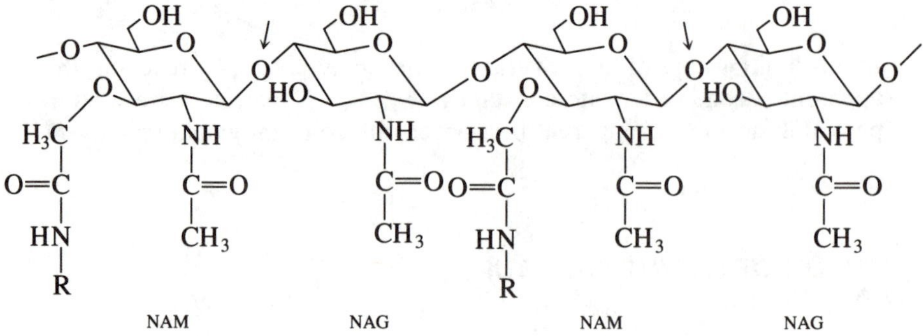

The monomers here are N-acetylmuramic acid (NAM), alternating with N-acetylglucosamine (NAG). In the cell wall, the lactic ether carboxyl group is in an amide linkage to peptides, denoted above by NHR [2].

When lysozyme acts on this reactant, it acts specifically as an N-acetylmuramidase, transferring the NAM group to water and cleaving the polymer at the positions indicated above by the arrows.

The polymeric reactants are difficult to investigate, but simpler compounds with similar structures (model compounds) react in the same way and have facilitated characterization of the activity. Results for various NAG oligomers are summarized in Table 3-2 [14]. The smallest NAG oligosaccharide that is hydrolyzed rapidly is the hexamer $(NAG)_6$, which is cleaved between the fourth and fifth monomer units. Longer-chain oligomers are converted at about the same rate as the hexamer, whereas shorter-chain oligomers are converted at much lower rates and have more than one cleavage site. These results suggest that the catalytic site incorporates at least six amino acid side chains for binding to reactants and that the specific conversion requires that they all be occupied by the reactant.

Further evidence of the nature of the catalytic site was obtained in experiments with a dye called Biebrich scarlet, which has the following structure:

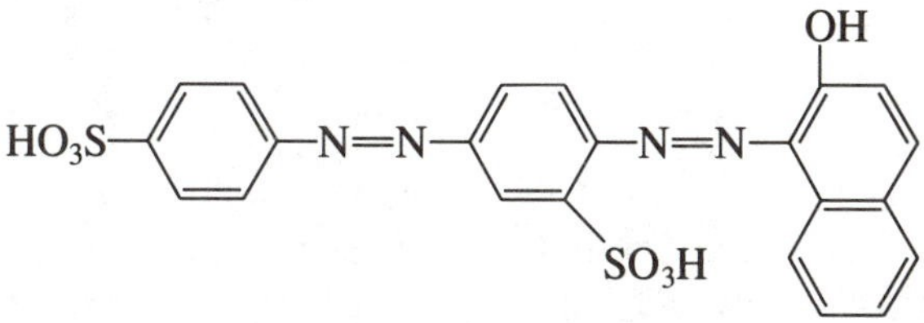

This compound is a competitive inhibitor, which binds to one specific functional group in lysozyme, found to be at the end of the sequence of amino acid side

Table 3-2
HYDROLYSIS RATES AND PATTERNS OF CLEAVAGE OF NAG
OLIGOMERS [14]

Reactant	Cleavage Position	Relative Rate
Dimer	$X_1 \overset{\downarrow}{-} X_2$	0.003
Trimer	$X_1 \overset{\downarrow}{-} X_2 \overset{\downarrow}{-} X_3$	1
Tetramer	$X_1 - X_2 - X_3 \overset{\downarrow}{-} X_4$	8
Pentamer	$X_1 - X_2 - X_3 \overset{\downarrow}{-} X_4 \overset{\downarrow}{-} X_5$	4,000
Hexamer	$X_1 - X_2 - X_3 - X_4 \overset{\downarrow}{-} X_5 - X_6$	30,000

$^a k_{cat} = 0.25 \ s^{-1}$.

chains. Experiments characterizing competitive binding of the hexameric reactant and the dye allowed researchers to distinguish the various modes of productive binding (that which leads to catalysis) and unproductive binding of the reactant. Some of the possibilities are shown schematically in Fig. 3-10 [15], where the six amino acid side chains of the catalytic site are depicted as A, B, C, D, E, and F. According to this representation, the specific interaction that is productive involves bonding of reactive groups of six different monomer units with six enzyme functional groups; when this is blocked by the dye, the binding of (NAG)₆ is unproductive. Apparently (NAG)₆ is just long enough to fill the catalytic site with proper positioning for rapid hydrolysis.

To identify some of the amino acid side chains involved in the catalytic site of lysozyme, reagents have been used that specifically bond with particular groups [16]. For example, oxidation of the six tryptophans with N-bromosuccinamide deactivates the enzyme; this result is consistent with the inference that at least one tryptophan is part of the catalytic site. When all the carboxyl groups are esterified, the enzyme is deactivated. When all the carboxyl groups except aspartate-52 and glutamic acid-35 are modified (which is possible because binding of inhibitors to lysozyme prevents modification of these two groups in the cleft), the enzyme is still active; and when the aspartate-52 is esterified, the enzyme is inactive. These results imply that aspartate-52 is part of the catalytic site and are consistent with the suggestion that glutamic acid-35 is part of the catalytic site.

The foregoing results and related information, combined with knowledge of the lysozyme structure, have led to the schematic representation of the catalytic site shown in Fig. 3-11 [17]. Here, the monomer units of the hexasaccharide are shown positioned next to the amino acid groups in the groove of the enzyme. A more detailed representation of this enzyme–reactant complex is shown in Fig. 3-12 [17].

According to this structural model (Fig. 3-12), the aspartate-52 and the glutamic acid-35 are on two sides of the reactive glycosyl group of the hexameric NAG reactant. To elucidate the role of these groups, measurements

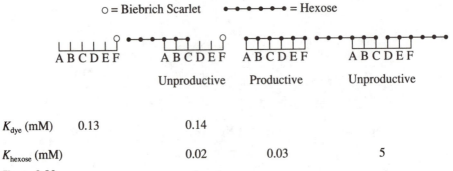

Figure 3-10
Schematic representation indicating productive and unproductive binding of a hexameric reactant to lysozyme [15].

have been made of the enzyme in solution at various pH values. Two pK_a values have been observed, 4.7 and 6.1, and attributed to ionization of the aspartic acid-52 and glutamic acid-35, respectively [17]. The latter value is not representative of a carboxylic acid in an aqueous solution, but the surrounding of this group in the enzyme is hydrophobic (and the measured pK_a value characterizes the surrounding solution and must be adjusted to represent the microenvironment of the group). If these pK_a assignments are correct, then at a pH of 5 (the value at which the lysozyme is most active), the aspartate-52 would be in the form of —COO⁻, whereas the glutamic acid-35 would be in the form of —COOH. The former is a good candidate for electrostatic stabilization of an incipient glycosyl oxocarbenium ion, and the latter is a good candidate proton donor for general acid catalysis to protonate the leaving group [2]:

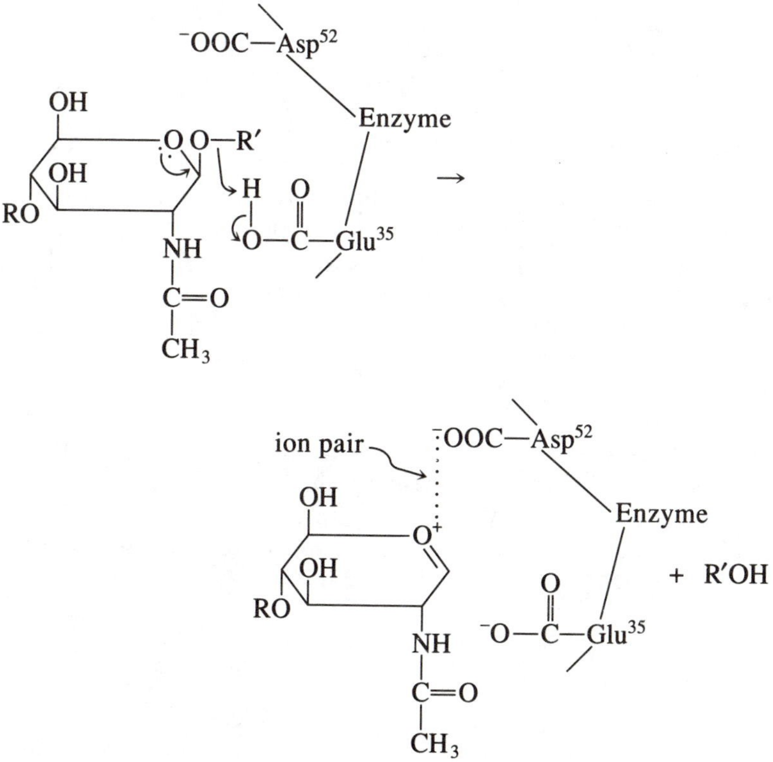

Measurements of isotope effects (D vs. H) on the catalytic kinetics support the suggestion of the oxocarbenium ion intermediate [2]. Further details, such as the conformation of the ring in the above depiction and the possibility of an induced fit, are still debated.

In summary, the catalytic action of lysozyme is relatively well understood, and there is good consistency between the structural results obtained by X-ray diffraction crystallography and the reactivity data determined in experiments with the enzyme in solution, such as competitive inhibition of ca-

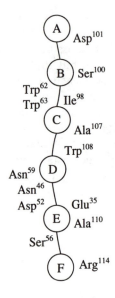

Figure 3-11
Schematic representation of the catalytic site of lysozyme. A–F represent the glycosyl units of a hexasaccharide reactant. Some of the amino acids in the cleft near the active site are identified [17].

talysis, reaction of selected amino acid groups to determine whether they are part of the catalytic site, and determination of whether some groups in the catalytic site are ionized. A charged group in the enzyme apparently stabilizes a carbenium ion intermediate. This is atypical; carbanion intermediates are much more common in enzyme catalysis.

EXAMPLE 3-3   Dependence of Rates on pH
Problem
Determination of rates of enzyme-catalyzed reactions as a function of the solution pH provides information about the involvement of ionizable groups in the catalytic site. Show how to interpret such data for the following simple scheme representing the enzyme and the enzyme–reactant complex each in an ionized and an unionized form:

$$E^- \underset{-R}{\overset{R}{\rightleftarrows}} ER^- \xrightarrow{\text{slow}} P + E^-$$

$$K_E H^+ \Big\updownarrow -H^+ \qquad H^+ \Big\updownarrow -H^+ K_{ER} \qquad (3\text{-}9)$$

$$HE \underset{-R}{\overset{R}{\rightleftarrows}} HER$$

Solution
Define equilibrium constants for each of the four reactions on the left in Eq. 3-9. These cannot all be independent; they are related by the equation:

$$K_E K'_R = K_{ER} K_R \qquad (3\text{-}10)$$

Using the Michaelis–Menten equation of Example 3-1 with the equilibrium information and expressing the rate in terms of the concentration of the enzyme in all forms, $(C_E)_t$, gives

$$r = \frac{k_{cat} C_R (C_E)_t}{K_R + C_R \left(1 + \dfrac{C_{H^+}}{K_{ER}}\right) + \dfrac{K_R C_{H^+}}{K_E}} \tag{3-11}$$

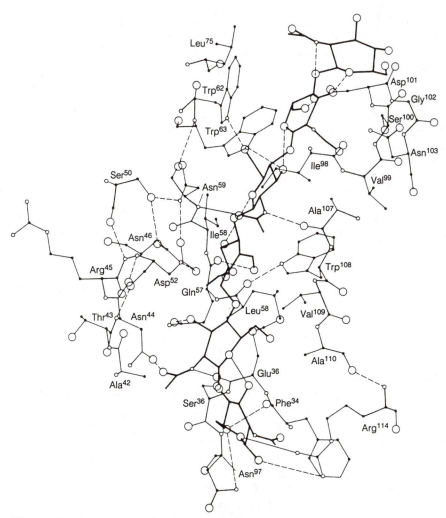

Figure 3-12
Bonding of the reactant hexa-N-acetylglucosamine to the catalytic site of lysozyme. The three sugar residues in the upper part of the figure are bound in the manner observed for tri-N-acetylglucosamine; the binding of the other three was inferred from models. The positions of the atoms in the enzyme correspond to the structure determined in the absence of reactant [17].

In the limiting case for which $C_R \gg K_R$,

$$r_{max} = \frac{k_{cat}(C_E)_t\, K_{ER}}{K_{ER} + C_{H^+}} \tag{3-12}$$

This equation represents the dependence of rate on pH.   ■

RIBONUCLEASE [1, 2]

Another of the best-understood enzymes is bovine pancreatic ribonuclease A. This was one of the first enzymes to be purified, and it was the first to be made in the laboratory by methods of organic chemistry.

Ribonuclease is a small protein composed of a single chain with 124 amino acid monomer units. The structure has been determined by X-ray diffraction (Fig. 3-13) [18].

Ribonuclease catalyzes the hydrolysis of ribonucleic acid (RNA) by a two-step process in which a chain cleavage first occurs as a cyclic phosphate intermediate is formed:

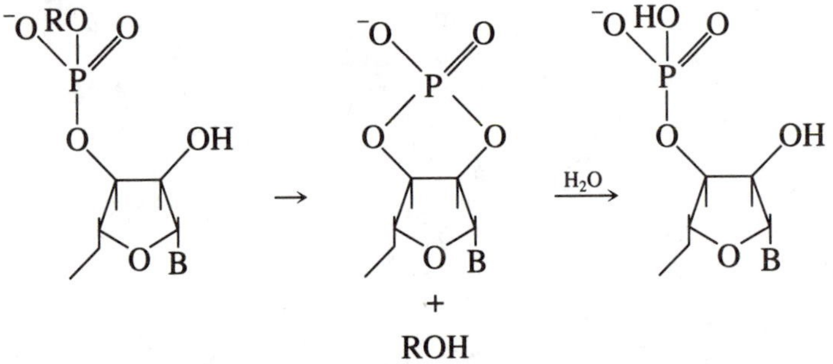

$$\tag{3-13}$$

The enzyme is specific for the monomer units cytidine and uridine but not adenosine and guanosine units in the polymeric RNA (these have different B groups in the structure shown in Eq. 3-13).

The enzyme has been the subject of numerous investigations involving the specific chemical modification of amino acid side chains in attempts to determine which functional groups are part of the catalytic site [19]. For example, dinitrofluorobenzene reacts rapidly with lysine-41 at its ε-amino group, leading to complete deactivation of the enzyme. And the alkylating agent iodoacetamide deactivates ribonuclease maximally at pH 5.5, whereby histidines are the only amino acid units that are modified. Structural investigations implicate histidine-12 and histidine-119, which are therefore suggested to be part of the catalytic site. All three of the amino acid units mentioned in this paragraph have been shown by X-ray diffraction to be located in the cleft of the enzyme.

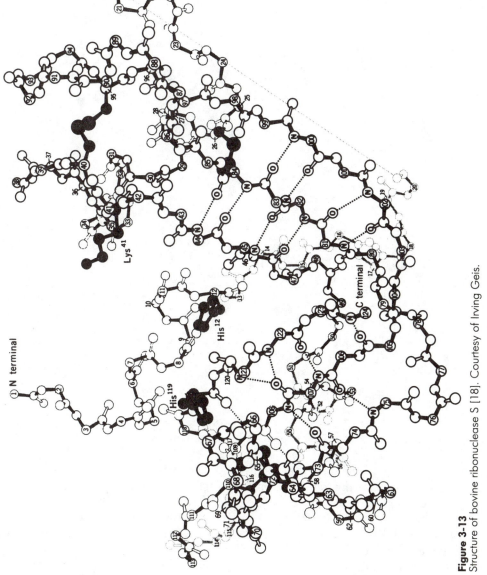

Figure 3-13
Structure of bovine ribonuclease S [18]. Courtesy of Irving Geis.

The dependence of catalytic activity on pH is described by a bell-shaped curve, with maximum rates at about pH 7. These results are consistent with the implication that groups with pK_a values of 6.3 and 8.1 are involved in the enzyme–reactant complex, and it was therefore proposed [20] that the catalysis takes place by a concerted general acid–general base (push-pull) reaction involving two histidine units, later identified as histidine-12 and histidine-119 [1]:

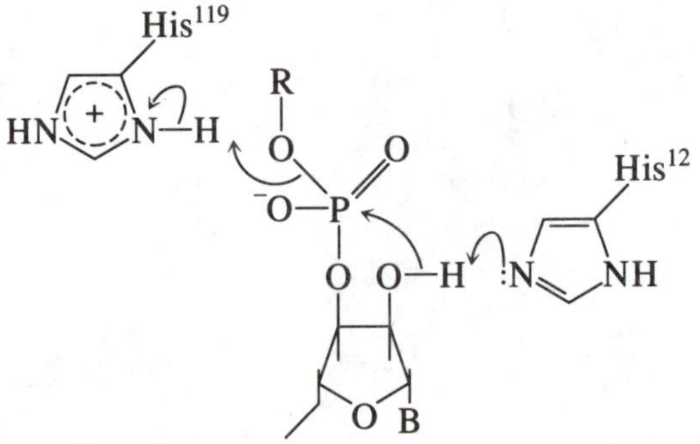

$$(3\text{-}14)$$

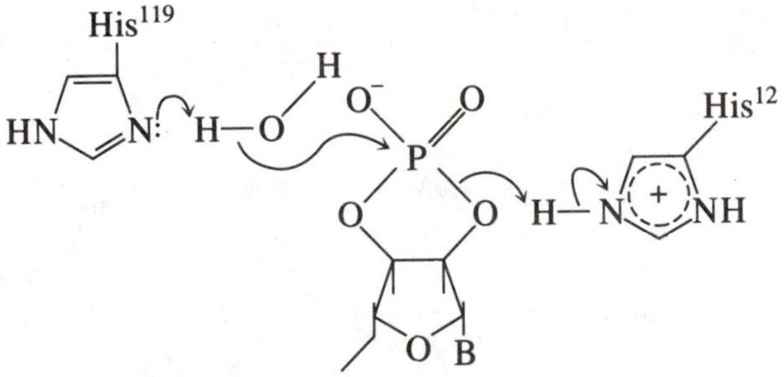

$$(3\text{-}15)$$

A detailed picture of the catalytic site has been developed from X-ray data for the enzyme and enzyme–reactant complexes. The structure of the enzyme–reactant complex for the cyclization step was inferred from the crystal structure of the enzyme–UpcA complex, where UpcA is a close analogue of the reactant, differing only in having a CH_2 group in place of an oxygen atom. The structure shown at the top of page 169 is therefore inferred to be closely similar to that of a productively bound enzyme–reactant complex [1]. This structure shows that histidine-119 is within hydrogen-bonding distance of the leaving group, and histidine-12 is within hydrogen-bonding distance of the 2′ —OH group of the pyrimidine ribose, consistent with the possibility of general acid–general base catalysis involving these groups.

Additional structural data have been obtained by X-ray diffraction for enzyme–inhibitor complexes and used as a basis for locating other amino acid side chains in the catalytic site [1]. The lysine units 7, 41, and 66 are important, and threonine-45 is positioned to form hydrogen bonds responsible for the enzyme's specificity for pyrimidines, as shown schematically below [21]:

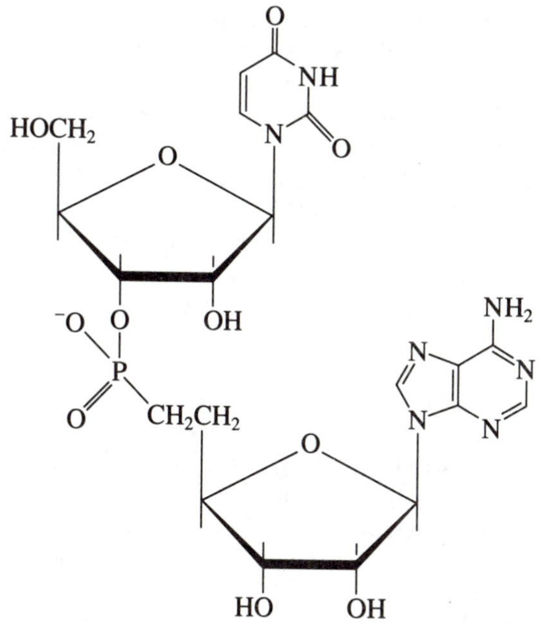

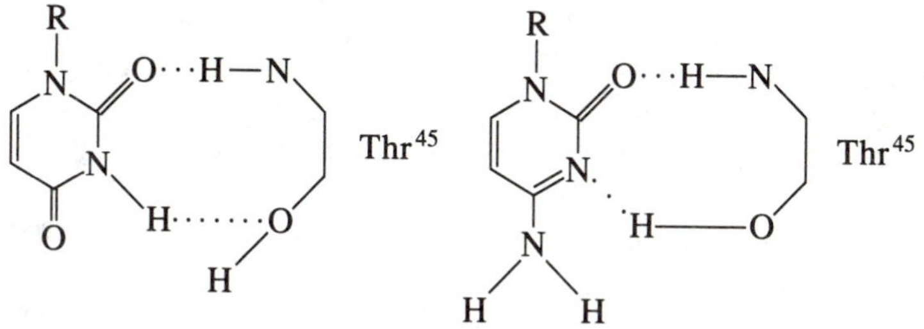

Some of the details of this model are not well established, but the depth of understanding places ribonuclease among the best-understood enzymes. This example illustrates the multifunctional character of the enzyme and the preciseness of the interaction between catalyst and reactant. The nature of the individual catalyst–reactant interactions and the elementary steps are just what might be expected on the basis of the summary of acid–base catalysis given in Chapter 2. The enzyme is doing the same chemistry, but in a much more intricate and precise manner.

3.5.2 Metalloenzyme Catalysis [22]

Many enzymes contain metal ions as essential parts of the catalytic sites. The ligands are the R groups of the enzymes. The enzyme molecule is in essence a single, large, multidentate ligand that bonds to the metal to occupy three or more coordination sites. The individual ligands that coordinate to the metal ion are almost all different from the ligands mentioned in Chapter 2, such as the phosphines. The biological metals are also for the most part different from the metals of man-made catalysts. The biological metals must be reliably and consistently available to living organisms; they include copper, iron, and zinc, but not rhodium and platinum.

The metals in catalytic sites of enzymes play a variety of roles, ranging from simple Lewis acids, typified by Zn^{2+}, to redox centers involved in one-electron transfers, typified by Fe^{2+} and Fe^{3+}. For the most part, the catalytic behavior of these metal complexes is not what would at first be expected from what is known of the chemistry of similar metal complexes in solution. Rather, the reactivities of the metal centers are uniquely determined by the combination of ligands provided by the enzyme; it is not just the identities of the ligands that are important, but also the geometry, which is determined by the three-dimensional structure of the enzyme. The enzyme provides an enforced geometry for the metal–ligand combination. The unique structure and bonding of the metalloenzyme control both the electronic state and the geometry for bonding of ligands, including reactants and inhibitors. Both the electronic state and the stereochemistry of a metal in an enzyme are unusual by the standards of classical inorganic chemistry.

Metals are bonded near the surfaces of some enzyme molecules. These can be replaced readily by other metals as a result of ion exchange. Consequently, metals in such positions are not stable enough to meet catalytic demands; the typical catalytic site is in the interior (often hydrophobic) region of an enzyme, where it is virtually inaccessible to cations in a surrounding aqueous medium. The nature and geometric arrangement of the ligands determine which metal has an affinity for the particular site. The sites that accommodate one metal in an enzyme exclude other metals with an amazing specificity.

The ligand surroundings of metals in many enzymes have been determined by X-ray diffraction crystallography. Other powerful structural tools are extended X-ray absorption fine-structure (EXAFS) spectroscopy (mentioned again in Chapters 5 and 6) and NMR spectroscopy. These latter techniques have provided precise characterizations of the local structure, even when the full three-dimensional structure of the enzyme is not known [23].

For example, the structure shown below has been determined from X-ray data for the zinc-containing enzyme carboxypeptidase A [24]:

The coordination environment of Zn, in the absence of a reactant or inhibitor, includes five ligands, with nitrogens from two histidines, two oxygens from a single glutamate, and a water molecule; the geometry is that of a distorted pseudotetrahedron.

The structure of the iron-containing site in cytochrome P-450 has been determined by X-ray crystallography [25]:

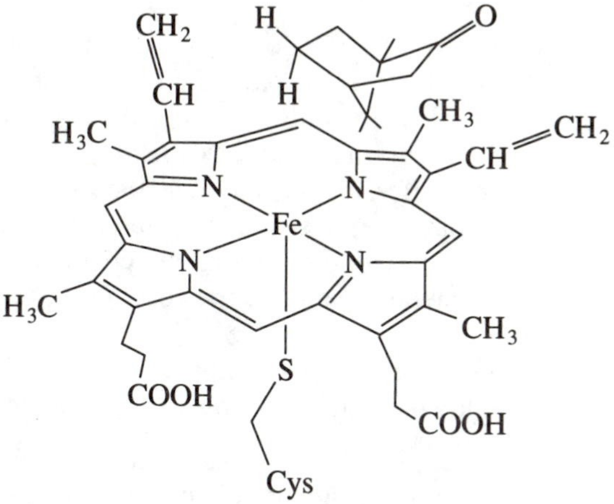

This structure has been inferred for the catalytic site with a bound reactant molecule, camphor, which is shown at the upper right. Prior to binding of camphor, the Fe^{3+} ion is coordinated by the nitrogens of four pyrroles, a sulfur from cysteine, and a water molecule or hydroxide ion. Camphor displaces the water or hydroxide ion. The enzyme catalyzes the stereoselective oxidation of camphor. The exo side of C5 of the camphor is positioned adjacent to the Fe, allowing regio- and stereoselective hydroxylation at that site.

Iron is often bonded to sulfur, as exemplified by the following structure for rubredoxin [26]:

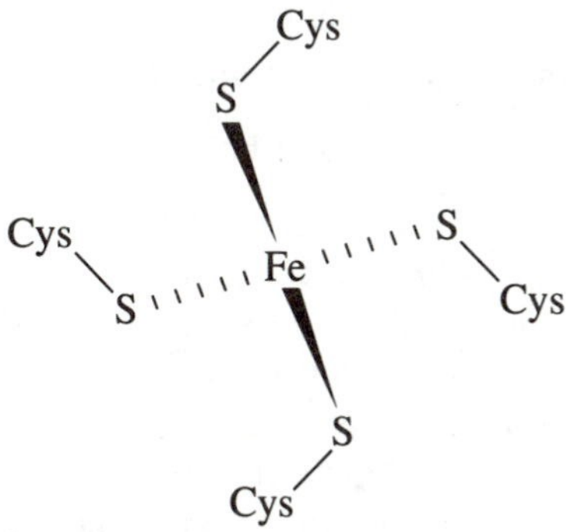

This is a distorted tetrahedral arrangement of sulfur atoms from cysteine at nearly equal distances from the iron.

The metals in enzymes are often not mononuclear complexes, like those shown above, but clusters. Iron–sulfur clusters are found in enzymes, such as the following cubanelike structure in ferredoxin [23, 27]:

One of the striking characteristics of metalloenzymes as catalysts is that specific metal ions are used for specific reactions. Often these are reactions of very small molecules and ions. (In contrast, most reactions of large molecules in biology are catalyzed by enzymes that lack metal centers.) What opportunities do the metal centers offer in catalysis that the organic groups do not? Williams [22] concluded that whenever a good attacking group is required in an enzyme, it is of necessity a metal or an organic coenzyme center. Without the unique reactivities of metals, many of the reactions of biology could not occur.

Examples of the roles of metal ions in biological catalysis are summarized in Table 3-3 [22]. The following generalizations have been suggested [22]:

1. When the pH is near 7, almost all acid–base catalysis by metal ions involves Zn. When the medium is acidic, Mn and Fe ions play this role instead.

2. Free-radical rearrangements are catalyzed by Co^{2+} in the low-spin state. Alternatively, they may be initiated by free radicals generated by metals. There are parallels to the roles of metals as redox initiators mentioned in Chapter 2.

3. Two-electron (oxygen atom) reactions involve Mo when it occurs at a low redox potential and Fe when the redox potential is high. The reactions of N_2 and NO_3^- require Mo.

4. The reactions of H_2 and CH_4 require Fe or Ni.

5. Oxygen utilization is associated with Fe or Cu.

6. Oxygen evolution is associated with Mn.

The following explanations have been offered for these generalizations [22]:

1. Zn^{2+} is the only good Lewis acid available that does not undergo redox reactions. In acidic media (outside cells), the Zn complexes are not stable, and Zn^{2+} is replaced by Mn^{2+} or Fe^{3+}.

2. Co functions in a stable low-spin state, Co^{2+}, which has an exposed unpaired electron, d^7. The stable low-spin states of the other metal ions, Mn, Fe, and Ni, do not provide such a stable, simple, directed radical.

3. Mo is stable in high oxidation states and undergoes oxygen atom transfer reactions at low redox potentials; Mo has the greatest range of oxidation states at low potentials.

TABLE 3.3
EXAMPLES OF CATALYSIS BY METALLOENZYMES [22]

Small-Molecule Reactant	Metal Ion	Examples
Glycols, ribose	Co in B_{12}	Rearrangements
CO_2, H_2O	Zn	Carbonic anhydrase
Phosphate esters	Zn	Alkaline phosphatase
RNA	Fe, Mn	Acid phosphatase
N_2	Mo (Fe)	Nitrogenase
NO_3^-	Mo	Nitrate reductase
SO_4^{2-}	Mo	Sulfate reductase
CH_4, H_2	Ni (Fe)	Methanogenesis
$O_2 \rightarrow H_2O$	Fe	Cytochrome oxidase
	Cu	Lactase
O insertion (high redox potential)	Fe	Cytochrome P-450
SO_3^{2-}, NO_2^-	Fe	Reductase
$H_2O \rightarrow O_2$	Mn	Oxygen-generating system of plants
H_2O_2/Cl^-, I^-	Fe (Se)	Catalase, peroxidase
H_2O/urea	Ni	Urease

4. Fe and Cu readily undergo one-electron redox reactions in the range of biological redox potentials. Both can bind O_2 and assist in its activation.

CARBONIC ANHYDRASE [1, 22, 28, 29]

Some of the best-understood metalloenzymes are the carbonic anhydrases, which catalyze the hydration of CO_2 and its reverse, the dehydration of bicarbonate:

$$CO_2 + H_2O \rightleftharpoons HCO_3^- + H^+ \qquad (3\text{-}16)$$

There are several well-characterized human carbonic anhydrases, denoted A, B, and C. A and B each have 260 amino acid units; C has 259 [30, 31]. Each has one tightly bound Zn^{2+} ion. The three enzymes have similar sequences of amino acids and similar tertiary structures. The ligands of the Zn^{2+} ion include three imidazole rings of three histidines at the bottom of the cleft, about 1.2 nm from the surface; the fourth ligand is believed to be H_2O or OH^-, depending on the pH [32].

The C enzyme is a highly active catalyst. With saturation kinetics of the commonly observed form (Michaelis–Menten kinetics),

$$r = \frac{kC_{CO_2}C_{enzyme}}{C_{CO_2} + K_M} \tag{3-17}$$

k is 10^6 s^{-1} and K_M is 8.3×10^{-3} mol/L; for the reverse reaction,

$$r = \frac{kC_{HCO_3^-}C_{enzyme}}{C_{HCO_3^-} + K_M} \tag{3-18}$$

k is 6×10^5 s^{-1} and K_M is 3.2×10^{-2} mol/L [33].

The activity depends on the pH; a group with a pK_a of 7 needs to be in the dissociated form for the forward (hydration) reaction and in the undissociated form for the reverse (dehydration) reaction [34]. The only groups at the active site that can ionize with this pK_a are the histidines and water. It has been suggested that water bound to Zn^{2+} is the key group, being in the form of OH$^-$ at high values of pH.

Infrared spectra of CO_2 bound to the enzyme show that it is not polarized and thus is not strongly bonded to the Zn^{2+} ion [34]. NMR and other experiments with the enzyme containing other cations in place of Zn^{2+} (which have a reduced catalytic activity) are consistent with this interpretation, although it is still debated.

A postulated mechanism consistent with this picture is shown schematically in Fig. 3-14 [22]. It is thought that the Zn^{2+} may also bond loosely to the CO_2, possibly orienting it and weakly polarizing it, in addition to providing a nucleophilic water molecule [1, 35]:

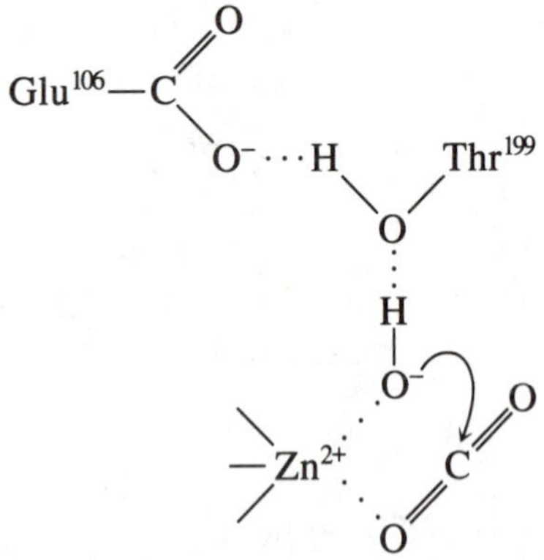

$$\tag{3-19}$$

Here, the bound water molecule is postulated to be hydrogen bonded to the hydroxyl group of threonine-199, which in turn is postulated to be hydrogen bonded to the carboxylate of glutamic acid-106. These details are still uncertain.

The metal center opens the way to the reaction. The role of the Zn^{2+} is that of a Lewis acid; it does not change oxidation state during the catalytic cycle. Only a few other metal ions are believed to function the way Zn^{2+} does

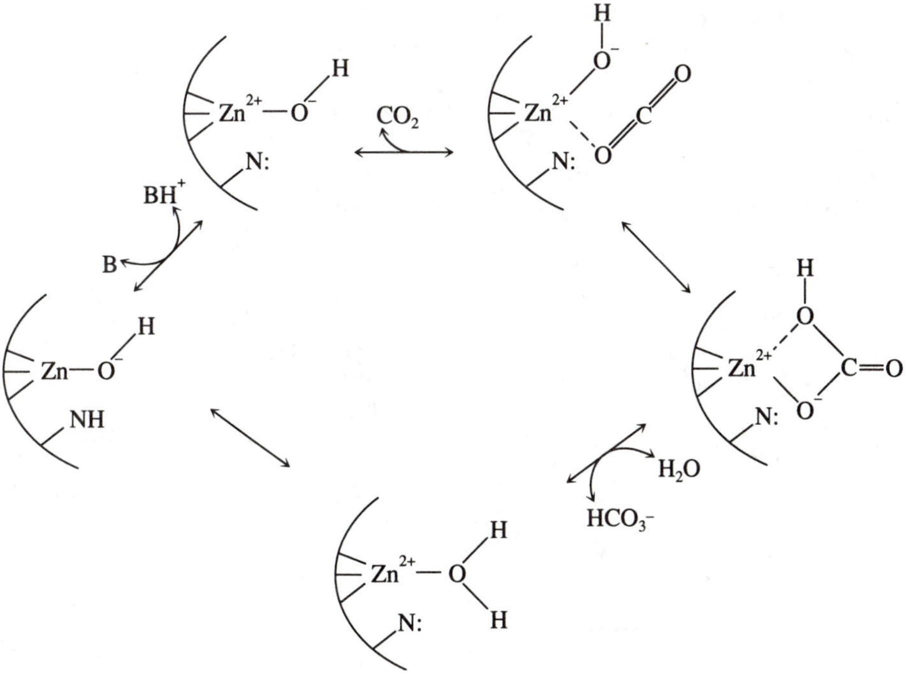

Figure 3-14
Schematic representation of the catalytic cycle for carbonic anhydrase [22].

in the catalysis: Co^{2+} and Cd^{2+}. These, like Zn^{2+} in carbonic anhydrase, are Lewis acids and meet the following requirements [22]: The metal ion is present with a low coordination number so that two sites are present between which reactants can undergo rapid exchange. The Lewis acids are strong, and the strength is greater with the neutral nitrogen-containing ligands than it would be with anionic ligands. Other metal ions lack these characteristics; for example, Cu and Ni do not provide two ligand positions for rapid exchange, and Mn^{2+} and Fe^{2+} are too weakly acidic [22].

The cycle shown in Fig. 3-14 illustrates that the crucial role of the Zn^{2+} center is to activate (dissociate) water. The organic groups in amino acid side chains lack this reactivity; only a metalloenzyme could be active for this reaction.

In summary, carbonic anhydrase is an enzyme that very rapidly catalyzes a difficult reaction of small molecules. The anchoring ligands are part of histidine. The heart of the catalytic site is the Zn^{2+} ion acting as a neutral acid to dissociate water.

A FREE-RADICAL REACTION CATALYZED BY VITAMIN B₁₂

Numerous reactions catalyzed by metalloenzymes involve the metals in redox roles. Sometimes the intermediates include free radicals. A relatively well-understood example is the dehydration of a diol catalyzed by vitamin B_{12} [36]. A postulated catalytic cycle is shown in Fig. 3-15.

The metal is Co^{3+}, which is rare in biology. The role of the metal is associated with the requirement for a free-radical catalyst with a restricted reactivity [22]. The radical is sometimes referred to as a bound radical [36] to emphasize the restriction. Indiscriminate free-radical reactions must be avoided in biology; they are associated with mutagenesis and cancer.

3.5.3 Summary

Understanding of the nature of enzyme action springs primarily from the structural evidence of enzymes and complexes of enzymes with inhibitors and analogues of reactants, as determined by X-ray diffraction. It would be desirable

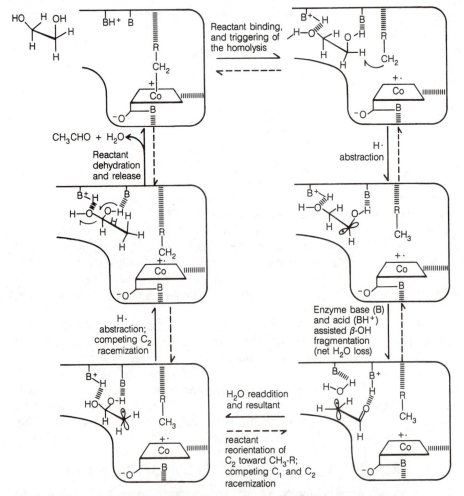

Figure 3-15
Postulated catalytic cycle for diol dehydration catalyzed by vitamin B_{12} [36].

to have more direct evidence from the structures of enzyme–reactant complexes, but this is difficult to obtain because of the high reactivities of these intermediates. Complementary evidence emerges from probing the catalytic sites with reagents that are selective for the various amino acid side chains to determine which ones are related to catalytic activity. Complementary evidence also emerges from the pH dependence of catalytic activity and from experiments identifying the acidic and basic groups that are involved in the catalysis and in determining whether they are ionized. Further information comes from physical methods such as EXAFS and NMR spectroscopy.

The results of these experiments provide the basis for models of catalytic sites and reaction mechanisms. In a few instances, these models are firmly established, although details remain to be elucidated. But only relatively simple and thoroughly investigated enzymes have been chosen as examples here. Most enzymes are so complicated that understanding of structure and mechanism is far from complete.

A number of common patterns emerge: Enzymes function by combinations of general acid and base catalysis, electrostatic effects, metal ion catalysis, and nucleophilic catalysis. The unique structures of enzyme catalytic sites, multifunctional arrays of catalytic groups precisely placed in hydrophobic pockets, bring to bear a subtle complementarity of the above effects combined with phenomena such as induced fit (distortion of the enzyme), strain (distortion of the reactant), and neighboring group interactions (precise orientation of reactants), and reaction steps that are often concerted, involving networks of functional groups.

Enzymes epitomize the control of chemical reactivity. Researchers will have to work for generations to design and produce catalysts that can do nearly as well. With so much opportunity for improvement, catalysis research is assured a long future.

3.6
SUPPORTED ENZYMES

Enzymes were the first catalysts used in technology.[1] The processes have not changed much; they are fermentations, and the major products are still wine and other alcoholic beverages. Today fermentation is used on a large scale to manufacture many biological products. Many new applications are expected, and the revolutionary techniques of gene splicing are leading to tailored organisms and economical processes for the large-scale production of complex compounds that had previously been available only in minute quantities. Enzyme catalysis will be exploited on an enormous scale as the new biotechnology develops.

Fermentations are carried out with complex organisms containing large numbers of enzymes. But there are also applications with individual enzymes

[1] Fermentation was practiced by the ancient Egyptians some 5,000 years ago.

that are used to produce biological products on a commercial scale. The enzymes used in these processes are often bound to the surface of a solid, such as porous glass, or are entrapped in a solid polymer [37]. The motivation for isolating the enzyme in a separate phase is the ease of separation from the products. Many other examples of supported catalysts are considered in the chapters that follow. The most important technological example of catalysis by a supported enzyme is the conversion of glucose (from corn starch) into fructose, which is used widely as a sweetener. The catalyst is glucose isomerase ionically bonded to DEAE-cellulose.

There are also applications of whole biological cells anchored to solid supports. Examples include immobilized microbial cells for production of L-aspartic acid, L-malic acid, and L-alanine.

REFERENCES

1. Fersht, A., *Enzyme Structure and Mechanism,* 2nd edition, Freeman, New York, 1985.
2. Walsh, C., *Enzymatic Reaction Mechanisms,* Freeman, San Francisco, 1979.
3. Phillips, D. C., *Proc. Nat. Acad. Sci. U.S.,* **57,** 484 (1967); Blake, C. C. F., Mair, G. A., North, A. C. T., Phillips, D. C., and Sarma, V. R., *Proc. R. Soc.* (London), **B167,** 378 (1967).
4. Pincus, M. R., and Scheraga, H. A., *Macromolecules,* **12,** 633 (1979).
5. Blake, C. C. F., Koenig, D. F., Mair, G. A., North, A. C. T., Phillips, D. C., and Sarma, V. R., *Nature* (London), **206,** 757 (1965).
6. Bailey, J. E., and Ollis, D. F., *Biochemical Engineering Fundamentals,* McGraw-Hill, New York, 1977.
7. Holm, R. H., and Simbon, E. D., in T. G. Spiro, ed., *Molybdenum Enzymes,* Wiley, New York, 1985, p. 1.
8. Biesecker, G., Harris J. I., Thierry, J. C. Walker, J. E., and Wonacott, A. J., *Nature* (London), **266,** 328 (1977).
9. Fischer, E., *Ber. Deut. Chem. Ges.,* **27,** 2985 (1894).
10. Haldane, J. B. S., *Enzymes,* Longmans, London, 1930, reprinted by MIT Press, Cambridge, MA, 1965.
11. Pauling, L., *Am. Scientist,* **36,** 51 (1948).
12. Bennett, W. S., and Steitz, T. A., *Proc. Nat. Acad. Sci. U.S.,* **75,** 4848 (1978).
13. Hammes, G. G., *Enzyme Catalysis and Regulation,* Academic Press, New York, 1982.
14. Rupley, J. A., and Gates, V., *Proc. Nat. Acad. Sci. U.S.,* **57,** 496 (1967).
15. Holler, E., Rupley, J. A., and Hess, G. P., *Biochemistry,* **14,** 1088 (1975).
16. Imoto, T., Johnson, L. N., North, A. C. T., Phillips, D. C., and Rupley, J. A., *The Enzymes,* **7,** 665 (1972).
17. Blackburn, S., *Enzyme Structure and Function,* Marcel Dekker, New York, 1976.

18. Cantor, C., and Schimmel, P. R., *Biophysical Chemistry,* Freeman, San Francisco, 1980.
19. Means, G., and Feeny, R., *Chemical Modification of Proteins,* Holden-Day, San Francisco, 1971.
20. Findlay, D., Herries, D. G., Mathias, A. P., Rabin, B. R., and Ross, C. A., *Nature* (London), **190,** 781 (1961).
21. Blackburn, P., and Moore, S., *The Enzymes,* **15,** 317 (1982).
22. Williams, R. J. P., *J. Mol. Catal.,* **30,** 1 (1985).
23. Armstrong, W. H., in *Metal Clusters in Proteins,* L. Que, Jr., ed., American Chemical Society, Washington, D.C., 1988, p. 1.
24. Rees, D. C., Lewis, M., and Lipscomb, W. N., *J. Mol. Biol.,* **168,** 367 (1983).
25. Poulos, T. L., Finzel, B. C., Gusalus, I. C., Wagner, G. C., and Kraut, J., *J. Biol. Chem.,* **260,** 16,122 (1985).
26. Watenpaugh, K. D., Sieker, L. C., and Jensen, L. H., *J. Mol. Biol.,* **138,** 615 (1980).
27. Berg, J. R., and Holm, R. H., *Met. Ions Biol.,* **4,** 1 (1982).
28. Lindskog, S., Henderson, L. E., Kannan, K. K., Liljas, A., Nyman, P. O., and Strandberg, B., *The Enzymes,* **5,** 587 (1971).
29. Coleman, J. E., *Progr. Bioorg. Chem.,* **1,** 159 (1971).
30. Andersson, B., Nyman, P. O., and Strid, L., *Biochem. Biophys. Res. Commun.,* **48,** 670 (1972).
31. Henderson, L. E., Henriksson, D., and Nyman, P. O., *J. Biol. Chem.,* **251,** 5457 (1976).
32. Kannan, K. K., Notstrand, B., Fridborg, K., Lövgren, S., Ohlsson, A., and Petef, M., *Proc. Nat. Acad. Sci., U.S.,* **72,** 51 (1975).
33. Steiner, H., Jonsson, B. H., and Lindskog, S., *Eur. J. Biochem.,* **59,** 253 (1975).
34. Riepe, M. E., and Wang, J. H., *J. Biol. Chem.,* **243,** 2779 (1968).
35. Kannan, K. K., Petef, M., Fridborg, K., Cid-Dresdener, H., and Lövgren, S., *FEBS Lett.,* **73,** 115 (1977).
36. Finke, R. G., Schiraldi, D. A., and Mayer, B. J., *Coord. Chem. Rev.,* **54,** 1 (1984).
37. Chibata, I., Tosa, T., and Sato, T., *J. Mol. Catal.,* **37,** 1 (1986).

FURTHER READING

Background biochemistry is presented in textbooks such as L. Stryer's *Biochemistry*, 3rd edition, Freeman, New York, 1988. Details of enzyme structure and mechanism are given in Fersht's introductory text [1], Hammes's monograph [13], and Walsh's thorough advanced text [2]. Excellent representations of enzyme structures are presented in the book by R. E. Dickerson and I. Geis, *The Structure and Action of Proteins,* Harper & Row, New York, 1969. The physical chemistry of enzymes is summarized in books by Cantor and Schimmel [18] and T. E. Creighton, *Proteins: Structures and Molecular Principles,* Freeman, New York, 1984. Williams's review [22] is a good entry into the literature of metalloenzymes.

PROBLEMS

3-1   Enzymes are often denatured when brought in contact with detergent solutions. Use chemical reasoning to explain how.

3-2   Consider the following sequence of steps for an enzyme-catalyzed reaction:

$$R + E \underset{k_{-1}}{\overset{k_1}{\rightleftharpoons}} RE \tag{3-20}$$

$$RE \xrightarrow{k_2} E + P \tag{3-21}$$

Assume that the steady-state approximation can be applied to describe the concentration of RE. Determine the form of the kinetics. What is the interpretation of the Michaelis constant in this equation and how does it differ from the interpretation in Example 3-1?

3-3   For simple Michaelis–Menten kinetics, show that the Michaelis constant is simply the reactant concentration at which the rate is half the maximum.

3-4   Use the results of Example 3-3 to show how the apparent value of the Michaelis constant depends on the pH.

3-5   Derive the steady-state rate equation for the following sequence:

$$E + R \rightleftharpoons X_1 \rightleftharpoons X_2 \rightarrow E + P \tag{3-22}$$

3-6   The steady-state kinetics data shown below were measured for the hydration of CO_2 catalyzed by bovine carbonic anhydrase (reaction 3-16) at 0.5°C and pH 7.1 with a solution of 2×10^{-3} mol/L of phosphate buffer and an enzyme concentration of 2.8×10^{-9} mol/L [De Voe, H., and Kistiakowsky, G. B., *J. Am. Chem. Soc.*, **83**, 274 (1961).]

Forward Reaction (Hydration)		Reverse Reaction (Dehydration)	
$10^3 \times C_{CO_2}$, mol/L	$10^5 \times$ Rate, mol/(L·s)	$10^3 \times C_{HCO_3^-}$, mol/L	Rate, mol/(L·s)
1.25	2.8	2	1.1
2.5	5.0	5	2.2
5	8.3	10	3.4
20	16.7	15	4.0

Estimate rate equations for the forward and reverse reactions; what are the turnover numbers for each?

3-7 Consider an inhibitor (I) and reactant (R) which both bind to the active site of an enzyme and are mutually exclusive. This may be represented as

$$E + R \rightleftharpoons ER \xrightarrow{k_{cat}} E + P$$
$$+$$
$$I$$
$$\Big\Updownarrow K_I \qquad\qquad (3\text{-}23)$$
$$EI$$

For the case of Michaelis–Menten kinetics, obtain an expression relating the rate to the reactant and inhibitor concentrations, the disassociation constant for the ER complex, and the disassociation constant for the EI complex. Show how the parameter values can be estimated graphically from the rate as a function of the reactant concentration at various inhibitor concentrations; assume that a control experiment is run with no inhibitor present.

3-8 An example of Michaelis–Menten kinetics in enzyme catalysis is the reduction of V^{5+} to V^{4+} by the enzyme gluathione reductase in the presence of the coenzyme NADPH. The reaction sequence can be represented as

$$E + R \rightarrow (ER) \rightarrow E + P \qquad\qquad (3\text{-}24)$$
$$E = \text{enzyme} \qquad\qquad (3\text{-}25)$$
$$R = V^{5+}, \text{NADPH} \qquad\qquad (3\text{-}26)$$
$$P = V^{4+}, \text{NADP}^{+} \qquad\qquad (3\text{-}27)$$

Reaction kinetics was monitored by measuring the concentration change of NADPH by optical absorption of NADPH at 340 nm, with the following data:

r_i^{-1}, $(mM)^{-1}$ min·mg enzyme	C_R^{-1}, $(mM)^{-1}$
41.5	4
31.8	2
22.6	1
18.4	0.5

Calculate the Michaelis constant K_M and the maximum rate. (Courtesy of Professor N. Dalal.)

3-9 Why are heavy metals toxic?

3-10 Consult ref. 22 and summarize the Lewis acid strengths of the metal ions in enzymes.

3-11 Why is CO a physiological poison?

3-12 What are catalytic antibodies? Consult the recent literature by P. Schultz to find out.

4

Catalysis by Polymers

4.1
THE NATURE OF POLYMERS

The preceding chapters introduced molecular catalysis and included catalytic cycles with well-characterized intermediates. Identification of intermediates becomes much more difficult when catalysis involves structures other than indentifiable molecular species. Many catalysts are solids. In this chapter, the solids are organic polymers. They are considered before the more common inorganic solid catalysts because they provide an ideal transition: One of the important characteristics of the polymers is that groups bonded to them may be nearly uniform in character and function catalytically just as their molecular analogues do in solution. The subject is developed in terms of catalysis by functional groups bonded to polymers, with emphasis initially on the parallels to solution catalysis. Then, more complex examples of catalytic groups immobilized in a rigid solid phase are introduced in a development of themes that carry over into catalysis by surfaces of inorganic solids.

An organic polymer molecule can be constructed by stringing together a set of building blocks called monomers. The simplest kind of polymer is made up of only one kind of monomer unit, such as linear polystyrene, an example from Chapter 2. The physical properties of solid polymers can be varied widely. They are influenced by the average molecular weight (which may be as much as millions), by the chemical nature of the monomer or the combination of monomers, and by the conditions of polymerization, which affect the arrangement of the polymer molecules and their interactions with each other.

A simple kind of solid polymer is a gel, a phase in which strands of polymer molecules are intertwined randomly, as sketched in Fig. 4-1. A gel of linear polystyrene is much like a hydrocarbon liquid. The tangled molecules are in constant, but restricted, random motion. They accommodate high concentrations of molecules of similar chemical character, for example, benzene and other hydrocarbons, although they lack an affinity for water and other polar molecules. Benzene is said to swell the polymer. Small molecules that swell the gel diffuse through it rapidly, encountering almost as little resistance to passage between the chains as they would diffusing through a solution of long hydrocarbon molecules. The interactions of the polymer chains are similar to the interactions of hydrocarbon molecules in solution, and the polymers dissolve in some solvents.

Polymers can also be constructed to form robust solids with physical properties that make them suitable as materials of construction. The wide range of physical properties is illustrated by *copolymers* of styrene and a closely related bifunctional monomer, divinylbenzene:

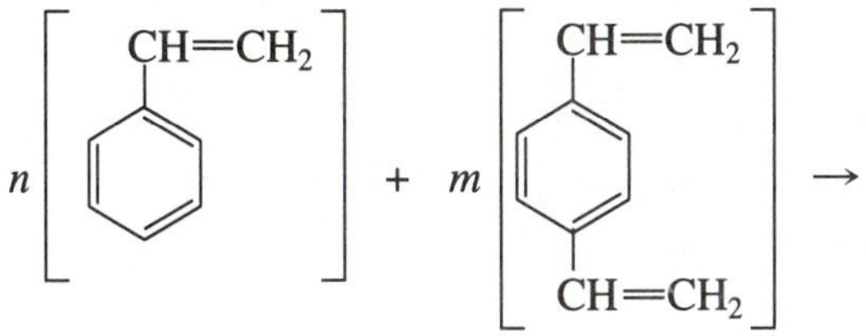

styrene   divinylbenzene

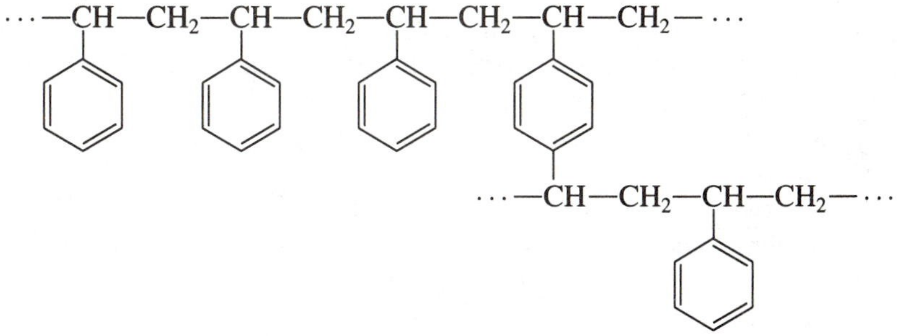

poly(styrene-divinylbenzene)

$$(4\text{-}1)$$

The divinylbenzene is a cross-linking agent, and the monomer unit that ties two chains together is a cross-link (recall the cross-links in enzymes). The polymer molecules have a chemical character almost identical to that of the linear polystyrene, but they now are more rigidly structured three-dimensional molecules.

The divinylbenzene-to-styrene ratio in the monomer mixture has a strong influence on the physical properties of the resultant polymer. If this ratio is only 0.01, the polymer can be produced as solid gelular particles, typically prepared as spherical beads. The hydrocarbon chains in a bead are so loosely packed that a particle can swell to about five times its original (shrunken) volume if immersed in a good swelling agent like benzene.

Figure 4-1
Schematic depiction of a polymer chain in a gel. The chain is disordered, mobile, and flexible.

If the divinylbenzene-to-styrene ratio is increased to 0.1, the resultant polymer beads are still flexible and swellable, but the maximum increase in volume is only about 30 percent. More and more added cross-links hold the polymer network together more and more tightly. As the ratio approaches values of 1 or more, the polymer becomes a highly structured solid, nearly impermeable even to molecules having a strong chemical affinity for it.

Such polymers, including cross-linked polystyrene, are available with a wide variety of cross-link densities and physical properties. These properties can be varied widely, and tailor-made polymers can be designed to hold catalytic groups. The variations in physical properties of the polymers are associated with wide variations in the catalyst performance, a topic developed in the discussion that follows.

4.2
ATTACHMENT OF CATALYTIC
GROUPS TO POLYMER SUPPORTS

The catalysis by acidic, basic, and transition metal complex groups described in Chapter 2 can be translated into a new phase as the catalytic groups are bonded to a polymer. The polymer isolates the groups; it is a *support*. If a stream containing reactants flows through a tube packed with beads of the polymer, the familiar catalytic reactions take place, and the reaction products flow on through the tube. The catalyst is isolated in the tubular reactor, and the practical problems of corrosion by catalysts in solution and of separation of soluble catalysts from reaction products are alleviated. It is primarily for these reasons that most industrial catalysts are solids. (Nature also relies on supported catalysts: many enzymes are held within membranes.)

Well-known organic synthesis routes allow incorporation of catalytic groups into solid polymers. Sulfonic acid groups are incorporated by direct sulfonation of cross-linked polystyrene:

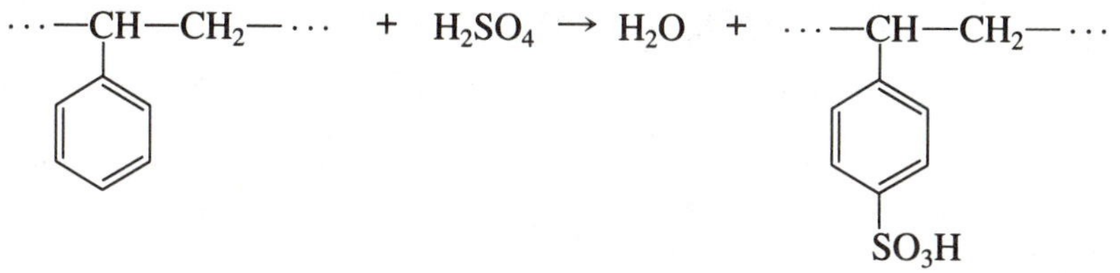

$$\cdots -CH-CH_2- \cdots \quad + \quad H_2SO_4 \quad \rightarrow \quad H_2O \quad + \quad \cdots -CH-CH_2- \cdots$$

$$\text{SO}_3\text{H}$$

$$(4\text{-}2)$$

Amine groups are incorporated by reaction of ammonia or amines with $-CR_2Cl$ groups [formed by chloromethylation of the polymer with dichloroethyl methyl ether (a carcinogen)], giving, for example, a primary amine:

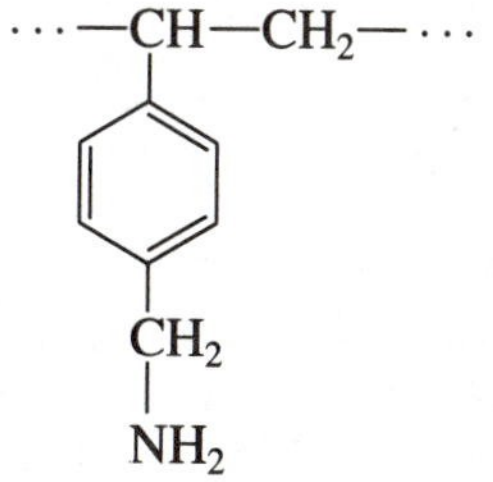

$$\cdots -CH-CH_2- \cdots$$

$$\text{CH}_2$$

$$\text{NH}_2$$

Polymers containing redox groups can be made by the incorporation of monomer units such as quinone:

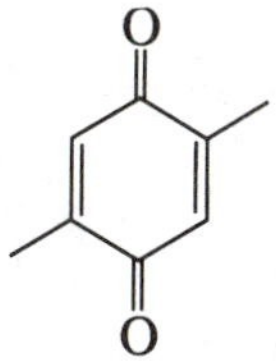

Polymers containing groups similar to the phosphine ligands used in transition metal complex catalysts are prepared by routes known from organometallic synthesis. For example, phosphine-functionalized poly(styrene-divinylbenzene),

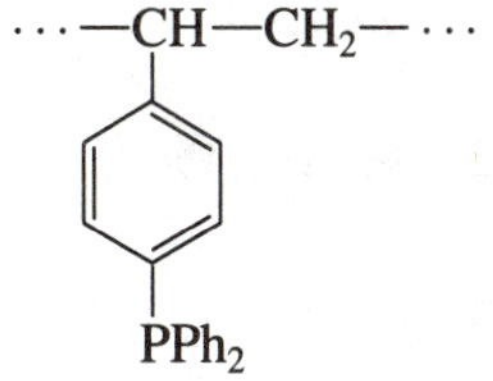

$$\cdots -CH-CH_2- \cdots$$

$$\text{PPh}_2$$

is an often-used polymer support since, when it is brought in contact with metal complexes like Rh(PPh)$_3$Cl, it undergoes phosphine–phosphine ligand exchange and forms metal-containing polymers, such as

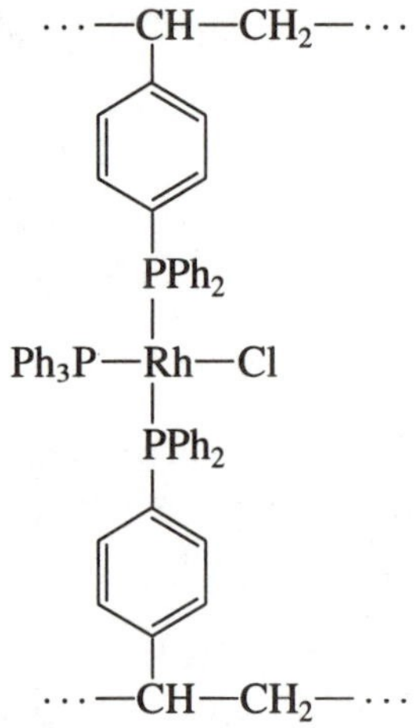

$$\cdots -CH-CH_2- \cdots$$

$$PPh_2$$
$$Ph_3P-Rh-Cl$$
$$PPh_2$$

$$\cdots -CH-CH_2- \cdots$$

These polymers have chemical and catalytic properties similar to those of their soluble analogues. Many acidic, basic, and metal-containing polymers have been prepared; some syntheses are shown in Fig. 4-2.

In the following pages the similarities between catalysis in solution and catalysis in gelular polymers are illustrated. Then, some new points are developed, with an emphasis on the characteristics distinguishing catalytic groups in solution from the same groups bonded to rigid polymer supports. This discussion leads to a recognition of the importance of the physical properties of the support in regulating the catalytic properties of the attached groups.

4.3
CATALYSIS IN POLYMER GELS

Polymer gels, being similar to solutions, are among the simplest catalysts. Many of the reactions mentioned in Chapter 2 take place in gels that incorporate the appropriate catalytic groups, much as they take place in solutions. Consider, for example, catalysis by —SO$_3$H groups, the strong proton donors mentioned previously to illustrate solution catalysis by p-toluenesulfonic acid and linear poly(styrenesulfonic acid). The acid groups in the poly(styrene-divinylbenzene) gel are so strongly polar that the gel, when suspended in water, becomes strongly hydrated (or "swelled"), having perhaps five or more water molecules associated with each acid group; the interactions involve hydrogen bonds and resemble solvation. The acid groups in the hydrated gel are dissociated, and specific acid catalysis takes place readily; reaction rates are often nearly the same as those in an aqueous solution of a strong acid at the same concentration as in the gel.

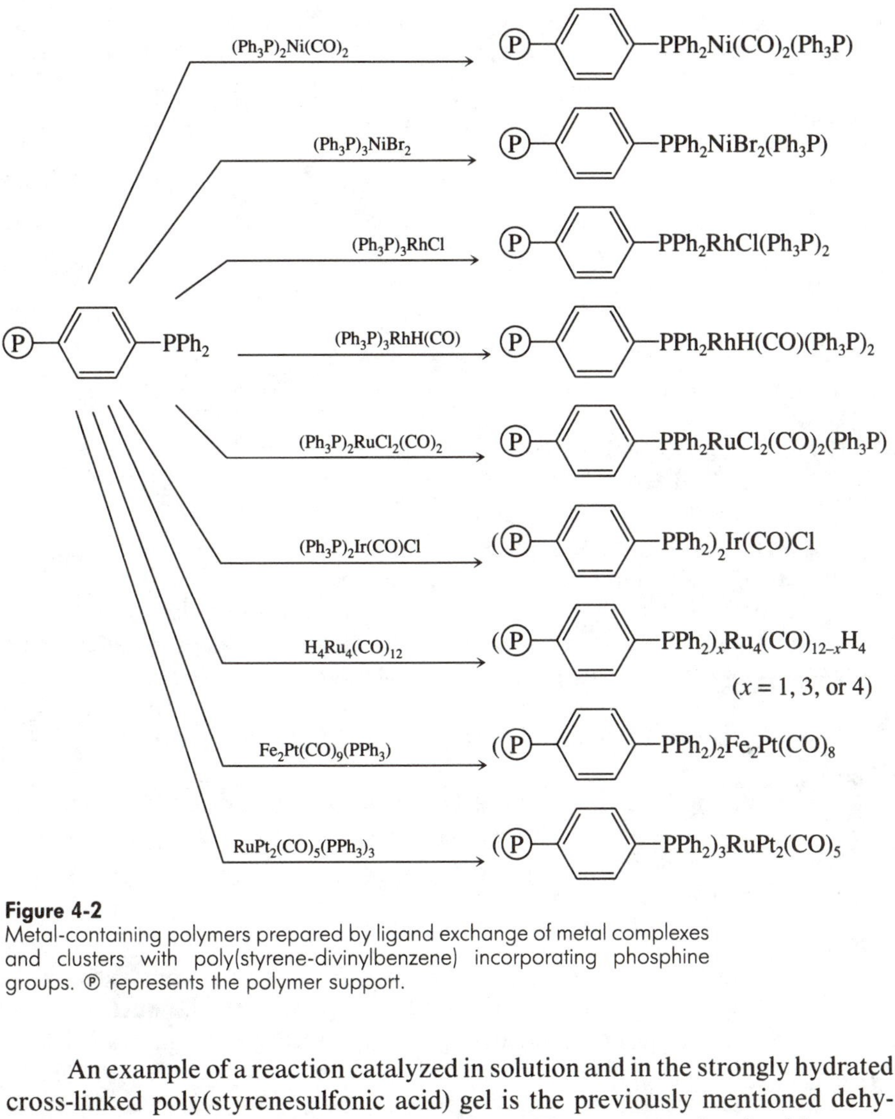

Figure 4-2
Metal-containing polymers prepared by ligand exchange of metal complexes and clusters with poly(styrene-divinylbenzene) incorporating phosphine groups. Ⓟ represents the polymer support.

An example of a reaction catalyzed in solution and in the strongly hydrated cross-linked poly(styrenesulfonic acid) gel is the previously mentioned dehydration of t-butyl alcohol to give isobutylene. The rate equation in either case is of the form:

$$r = kC_A \tag{4-3}$$

where the rate constant k is proportional to the number of $-SO_3H$ groups, and the subscript A refers to alcohol. There are many other acid-catalyzed reactions that take place with nearly the same kinetics in an aqueous strong acid solution and in the hydrated sulfonic acid-containing gel.

Other illustrations of the parallels between solution and gel phase catalysis include many reactions catalyzed by basic groups (such as aldol condensation catalyzed by amines), olefin hydrogenation and hydroformylation catalyzed by rhodium complexes that are analogous to the soluble complexes mentioned in Chapter 2, and redox reactions (such as dehydrogenation catalyzed by polyquinone):

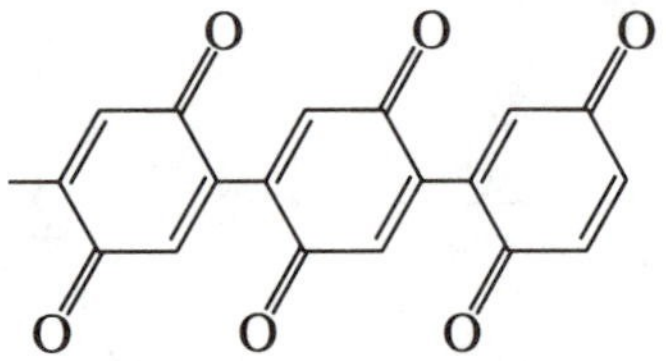

4.4
ADSORPTION AND THE KINETICS
OF POLYMER-CATALYZED REACTIONS

For a catalytic reaction to take place in a polymer, the reactants must first be transported into the polymer and combine chemically with the catalytic groups. This combination is referred to as **adsorption**, specifically **chemisorption**. In the next section the ties between adsorption and catalysis are demonstrated. The lessons are valid not just for polymer catalysts but for solids generally.

EXAMPLE 4-1   Measurement of Kinetics of a Solid-Catalyzed Reaction

Problem

Devise experiments for the measurement of the rates of *t*-butyl alcohol dehydration catalyzed by beads of cross-linked poly(styrenesulfonic acid). Assume that the reactant is in the liquid phase; alternatively, assume that it is in the gas phase.

Solution

Measurement of the rate of the reaction is especially easy for this reaction at the normal boiling point of the alcohol, since one of the products (isobutylene) is a gas, whereas the other product (water) and the reactant are liquids [1, 2]. If the alcohol is brought to reflux in a flask and the particles of catalyst are added, the rate of reaction is simply the flow rate of the isobutylene gas evolved through the condenser, which can be measured with a wet-test flow meter or a rising soap film flow meter. In a thorough determination of kinetics, one would vary the mass of catalyst (with the mass being so little that the rate per unit mass of catalyst would be constant) and the composition of the reactant solution. The paraffin methylcyclohexane is expected to be nearly inert (a diluent) and conveniently forms an azeotrope with the alcohol, boiling at a temperature within a few degrees Celsius of the boiling point of the pure alcohol; the product water also forms an azeotrope that boils at approximately the same temperature. Therefore, the reactant composition can be varied systematically with reactant mixtures refluxing at nearly the same temperature.

Rate data determined in this way are shown in Fig. 4-3. The data were extrapolated to time zero and the initial reactant compositions [2].

In a kinetics experiment one usually varies the temperature. If the pressure is to be maintained at 1 atm, then the experiment can be carried out conveniently at temperatures less than the boiling point of the alcohol (82°C), provided that the reactor is held in a thermostat; but measurement of the rate is no longer so simple, because the isobutylene initially accumulates in the liquid and is not all evolved as a gas. An alternative is to maintain all the products in the liquid phase in the reactor, drawing liquid samples periodically with a syringe and analyzing them, for example, by gas chromatography. For short times, the conversion is linearly dependent on time, and differentiation of the conversion-vs.-time data determines the reaction rate.

Variations on these themes allow the measurement of reaction rates when the reactant is in the vapor phase. A simple device for this purpose is a Schwab reactor, named after the inventor G.-M. Schwab, a pioneer

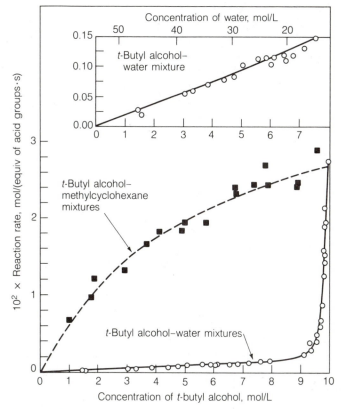

Figure 4-3
Kinetics of the dehydration of *t*-butyl alcohol at 80 ± 2°C catalyzed by particles of cross-linked poly(styrenesulfonic acid) [2].

in catalysis, who more than 50 years ago helped to place the subject on the firm quantitative foundation of reaction kinetics [3]. The reactor (Fig. 4-4) consists of a thermostated tube holding the catalyst particles in place (a *fixed-bed reactor*). As the reactant vapor flows through the bed of particles, it undergoes a conversion that depends on the time of contact with the catalyst. The product flows through a condenser, and the volatile product is taken off, its flow rate determining the rate of reaction. The condensed reactant may be revaporized to recirculate through the catalyst bed.

Again, with a gas chromatograph or another instrument for analysis of the product stream, much more flexibility is afforded. With a modern laboratory flow reactor system (Fig. 4-5), the restriction to a noncondensable product is removed, and all the products are maintained in the gas phase, typically flowing from the reactor through a heated sampling valve (where the stream is intercepted periodically and injected into the

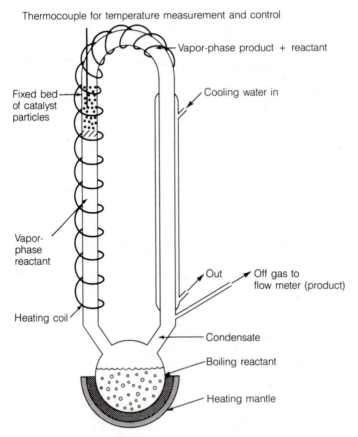

Figure 4-4
Schwab reactor for measurement of catalytic reaction rates when one of the products is noncondensable.

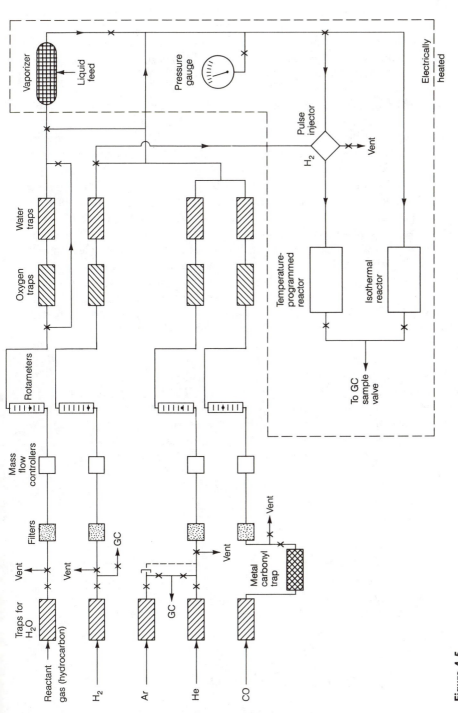

Figure 4-5
Schematic representation of a laboratory fixed-bed flow reactor system for measurement of kinetics of solid-catalyzed reactions [4].

gas chromatograph). Usually, the experiment is carried out at steady state, with the reactor temperature and the feed flow rates carefully controlled. In recent years, great strides have been made in automating laboratory flow reactors and the attendant flow meters, temperature and pressure controllers, and analytical devices. Interpretation of data from such a reactor is considered in Example 4-3.   ■

The example of *t*-butyl alcohol dehydration illustrates some more similarities between gel and solution catalysis, as well as some fundamental differences. Suppose the liquid medium that surrounds the particles of gel containing —SO₃H groups is made to contain smaller and smaller concentrations of water in solution with the reactant alcohol. Reaction rate data (inset, Fig. 4-3) show that as the water concentration is reduced, the rate of the catalytic reaction increases in proportion to the alcohol concentration, consistent with Eq. 4-3. This pattern, however, holds true only over a limited range of water concentration, as shown in Fig. 4-3. Complications set in as the water concentration is reduced to low values; in this range, the rate increases much more than linearly with increasing alcohol concentration, rising sharply as the last traces of water are removed. The data show that water in low concentrations depresses the rate of the catalytic reaction much more than the paraffin methylcyclohexane at the same concentration. In other words, water present in low concentrations *inhibits* the reaction, whereas water present in high concentrations simply dilutes the reactant.

When only little water is present in the polymer, the chemistry of catalysis is different from that in the hydrated polymer. Without the water, the —SO₃H groups are not dissociated, and therefore general acid catalysis prevails. When excess water is present, the acid groups are dissociated and specific acid catalysis prevails. The data therefore illustrate a transition from general to specific acid catalysis as water is added to the catalyst. The data show that the —SO₃H groups are more active catalytically than the hydrated protons. In the case of general acid catalysis, water is a reaction inhibitor because it competes favorably with the alcohol for the —SO₃H groups, which are the proton donors and sites of catalysis in the polymer.

The rate equation describing the data of Fig. 4-3 is

$$r = kC_A + \frac{k' K_A C_A}{1 + K_A C_A + (K_W C_W)^2} \tag{4-4}$$

where the C's refer to the liquid phase concentrations of A (alcohol) and W (water) [2]. A strong parallel exists between Eq. 4-4 and Eq. 2-14. There are two limiting cases: If the water concentration C_W is high, then the second term on the right-hand side becomes negligible, and specific acid catalysis prevails (Eq. 4-3). In the other limit, as C_W approaches zero, the first term on the right-hand side of Eq. 4-4 becomes negligible, and general acid catalysis prevails:

$$r = \frac{k' K_A C_A}{1 + K_A C_A + (K_W C_W)^2} \tag{4-5}$$

This equation accounts for the inhibition of reaction by water and shows the competition between water and alcohol for the catalytic groups; the competition is evident from the term $K_A C_A + (K_W C_W)^2$ in the denominator.

Consider the limit as C_W approaches zero:

$$r = \frac{k' K_A C_A}{1 + K_A C_A} \qquad (4\text{-}6)$$

The rate increases linearly with increasing C_A at low values of C_A, so that the rate-vs.-C_A curve bends over to approach a horizontal line. The shape of the curve is just that of Fig. 1-4 (cf. Fig. 4-3). This is an illustration of saturation kinetics: At low alcohol concentrations in paraffin, adding more alcohol to the reactant solution proportionately increases the rate of reaction, presumably by proportionately increasing the concentration of alcohol associated with the catalyic groups. At the highest alcohol concentrations in the liquid phase, further increases scarcely affect the rate, since the catalytic groups are almost saturated with the reactant alcohol.

In summary, the first special case (large C_W) corresponds to solutionlike catalysis, with the specific acid catalysis described by an equation (Eq. 4-3) typical of specific acid catalysis in solution. In the other limiting case (small C_W), catalysis by the $-SO_3H$ groups is described by saturation kinetics and a competition between the reactant and a product for the catalytic groups. Kinetics of the latter form is typical for catalysis by solids.

A simple model helps to explain Eq. 4-5. Consider the limit as C_W approaches zero and postulate the following sequence of steps:

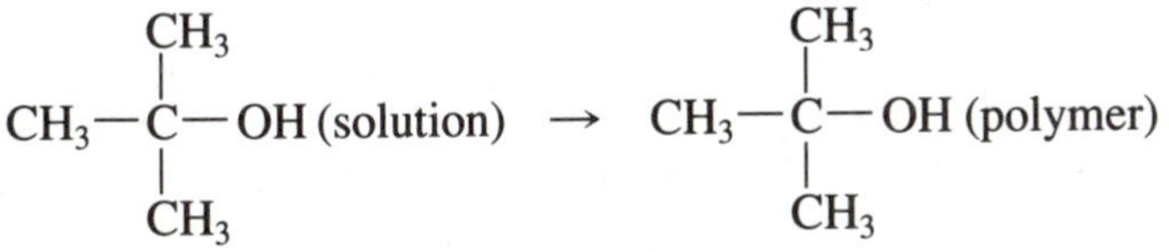

$$\text{(4-7)}$$

(This is a transfer of alcohol between phases.)

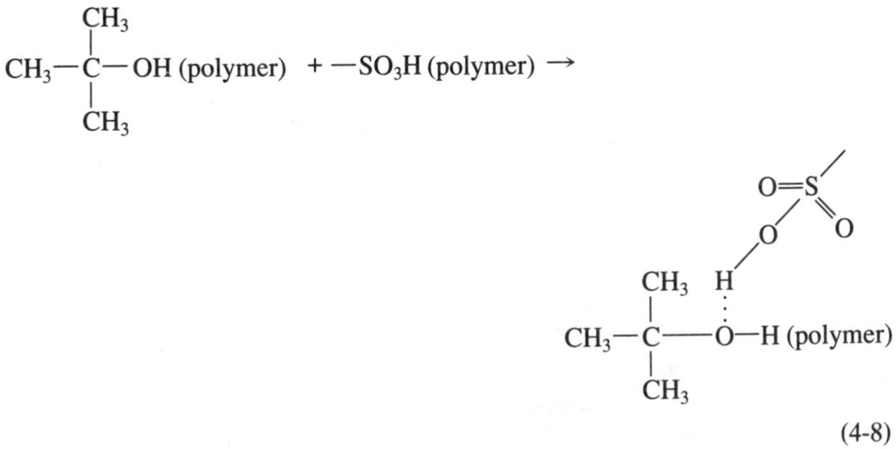

$$\text{(4-8)}$$

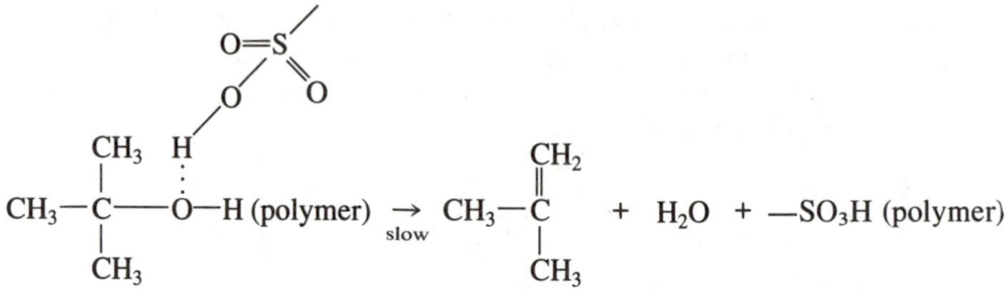

$$O = S \diagdown$$

$$CH_3 - \underset{\underset{CH_3}{|}}{\overset{\overset{CH_3}{|}}{C}} - O - H \ (\text{polymer}) \ \underset{\text{slow}}{\rightarrow} \ CH_3 - \underset{\underset{CH_3}{|}}{\overset{\overset{CH_2}{\|}}{C}} \ + \ H_2O \ + \ -SO_3H \ (\text{polymer})$$

$$(4\text{-}9)$$

Consider the last step to be rate determining.

The first two (fast) steps constitute an equilibration of the alcohol between phases. If the equilibrium is described by the ideal (Langmuir) isotherm (introduced in Example 4-2, below)

$$\theta_A = \frac{K_A C_A}{1 + K_A C_A} \tag{4-10}$$

where θ_A is the fraction of $-SO_3H$ groups bonded to the alcohol. Since the rate-determining step is assumed to be reaction 4-9.

$$r = k'\theta_A = \frac{k' K_A C_A}{1 + K_A C_A} \tag{4-11}$$

which has the form of Eq. 4-6.

Now consider nonzero concentrations of water and the competition between alcohol (A) and water (W) for the acid groups. Assume again that the equilibrium is described by the Langmuir isotherm:

$$\theta_A = \frac{K_A C_A}{1 + K_A C_A + K_W C_W} \tag{4-12}$$

This assumption leads to the rate equation:

$$r = \frac{k' K_A C_A}{1 + K_A C_A + K_W C_W} \tag{4-13}$$

This result is similar to the observed rate equation (Eq. 4-5) for the special case of small C_W, but it is not quite the same; the model is an oversimplification.

Related examples are the dehydration reactions of ethanol and methanol in the presence of the sulfonic acid resin. Ethanol reacts to give water, diethyl ether and, under some conditions, ethylene; with methanol, the only possible dehydration product is dimethyl ether. Kinetics data determined for the ethanol dehydration reaction

$$2C_2H_5OH \rightarrow H_2O + C_2H_5-O-C_2H_5 \tag{4-14}$$

are analyzed in Example 4-4.

EXAMPLE 4-2 The Langmuir Adsorption Isotherm

Problem

Derive Eq. 4-10, which describes ideal (Langmuir) adsorption, and determine whether the data of Fig. 4-6 for ethanol adsorbed on the sulfonic acid resin are consistent with the equation.

Solution

The assumptions underlying the Langmuir adsorption model are the following:

1. All the adsorption sites are equivalent.

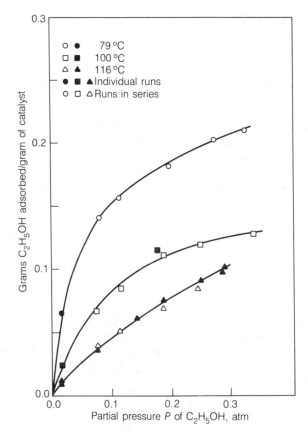

Figure 4-6
Equilibrium adsorption data for ethanol on sulfonated poly-(styrene-divinylbenzene) [5].

2. Interactions between the molecules bonded to these sites are negligible.

3. Only one adsorbing molecule (the **adsorbate**) can be bonded to each site on the solid (the **adsorbent**).

When adsorption equilibrium prevails, the rate at which molecules from the surrounding gas or liquid phase are adsorbed is equal to the rate at which molecules are desorbed from the sites. The rate of adsorption of A is then expected to be proportional to the concentration of A in the surrounding fluid and to the fraction of adsorption sites that are vacant:

$$r_{ads} = kC_A\theta_v \tag{4-15}$$

where θ_v is the fraction of sites vacant.

The rate of desorption of A is expected to be proportional to the fraction of sites occupied by A:

$$r_{des} = k'\theta_A \tag{4-16}$$

where θ_A is the fraction of sites occupied by A.

These rates are equal at equilibrium, and $\theta_A + \theta_v = 1$. Therefore,

$$k'\theta_A = kC_A(1 - \theta_A) \tag{4-17}$$

$$\theta_A = \frac{K_A C_A}{1 + K_A C_A} \tag{4-18}$$

where the ratio of rate constants k/k' has been defined as K_A, the adsorption equilibrium constant of A. It is easy to show that if adsorption of two species, A and B, occurs simultaneously on the sites, then

$$\theta_A = \frac{K_A C_A}{1 + K_A C_A + K_B C_B} \tag{4-19}$$

and

$$\theta_B = \frac{K_B C_B}{1 + K_A C_A + K_B C_B} \tag{4-20}$$

The denominator of these equations shows the competition between A and B for the sites. These equations are the simplest that show a saturation effect: θ_A increases from zero toward the limit of 1 as C_A increases, and the shape of the curve is like that of Fig. 1-4, which is associated with saturation kinetics of a catalytic reaction.

A linearized form of Eq. 4-18 (where the partial pressure P is used instead of the concentration),

$$\frac{P_A}{\theta_A} = \frac{1 + K_A P_A}{K_A} \qquad (4\text{-}21)$$

indicates that a plot of P_A/θ_A vs. P_A will give a straight line if the data conform to the equation. Since θ_A is not measured directly, the amount adsorbed per unit mass of adsorbent is used instead in the plot of Fig. 4-7. The straight lines imply that the data are adequately represented by the Langmuir equation. The adsorption equilibrium constants determined by the data are summarized in Table 4-1. (What values of θ_A are determined by these data? Are these values consistent with the assumption of adsorption of one alcohol molecule per —SO_3H group? There are 5.2 meq of —SO_3H groups/g of polymer.) ■

The Langmuir adsorption isotherm has only restricted applicability, since the sites for adsorption on most solids are not equivalent (they are nonuniform or heterogeneous) and since interactions between adsorbed molecules may not be negligible.

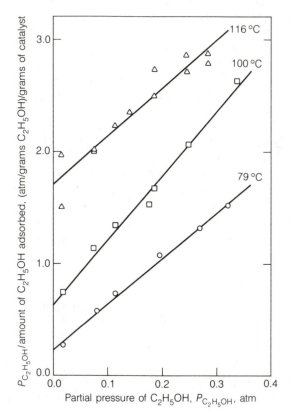

Figure 4-7
Adsorption equilibrium data of Fig. 4-6 for ethanol on sulfonated poly(styrene-divinylbenzene) plotted to test the appropriateness of the Langmuir isotherm [5].

Table 4-1
LANGMUIR ADSORPTION EQUILIBRIUM CONSTANTS FOR ETHANOL AND
WATER ON CROSS-LINKED POLY(STYRENESULFONIC ACID) [5]

Temperature, °C	$K_{C_2H_5OH}$, atm^{-1}	K_{H_2O}, atm^{-1}	$K_{C_2H_5-O-C_2H_5}$, atm^{-1}
100	8.9	13.6	0
116	2.5	7.6	0

EXAMPLE 4-3 Measurement of Catalytic Kinetics with a Fixed-Bed Flow Reactor
Problem

Devise experiments for determination of kinetics of the ethanol dehydration reaction catalyzed by beads of cross-linked poly(styrenesulfonic
acid). Assume that the reactants are in the vapor phase at atmospheric
pressure.

Solution

A simple experiment involves flow of the reactant gases at steady state
through a bed of catalyst particles held at constant temperature, with
analysis of the product stream by an on-line gas chromatograph; apparatus
like that depicted in Fig. 4-5 is suitable.

First, it would be appropriate to determine the dependence of conversion on time of contact of the reactant with the catalyst; to determine
initial rates of the catalytic reaction, low enough conversions are measured so that the conversion increases linearly with contact time. Data
obtained in such an experiment are shown in Fig. 4-8.

Here, conversion is plotted as a function of the *inverse space velocity*, which has units of number of moles of catalyst —SO₃H groups
divided by the feed flow rate in mol/min. If the reactant gas passes through
the reactor in *plug flow* (piston flow) then, by definition, all the fluid
elements have the same residence time in the reactor, and the inverse
space velocity is proportional to the contact time. Since the conversion
has units of mol of feed converted/mol of feed, the slope of the curve in
Fig. 4-8 has units of mol of feed converted/(mol of —SO₃H groups·min);
this is the rate of the catalytic reaction. The data provide a precise determination of the rate in this example. Typically, at higher conversions,
the conversion-vs.-inverse space velocity curve bends over. When the
curvature is negligible within experimental error, the slope of the line
determines the rate, and the conversion is referred to as differential; the
reactor is said to be a *differential reactor*. When the rate is not virtually
constant over the length of the reactor (i.e., when there is curvature in
the plot), the reactor is said to be an *integral reactor*.

To determine kinetics fully, initial rates can be measured as a function of temperature and feed composition. Such data are analyzed in the
following example. ∎

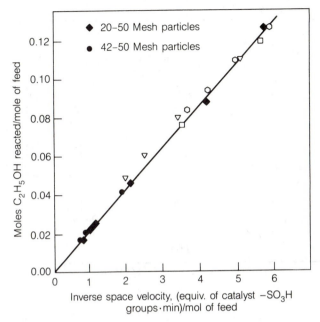

Figure 4-8
Conversion of ethanol into water and diethyl ether catalyzed by
cross-linked poly(styrenesulfonic acid) at 120°C and 1 atm [5, 6].
The slope of the line is the initial rate of the reaction.

EXAMPLE 4-4 Analysis of Kinetics of Ethanol Dehydration Catalyzed by Sulfonated Poly(styrene-divinylbenzene)

Problem
Use the data of Table 4-2 to determine an empirical rate equation for the
ethanol dehydration reaction. The rate of the reverse reaction is negligible.

Solution
A good start can be made by examining the data qualitatively and drawing
preliminary conclusions about the form of the equation. The data are
plotted in a convenient form in Fig. 4-9. There are subsets of data for
each of three pairs of reactant components: ethanol + argon (a diluent),
ethanol + diethyl ether (a product), and ethanol + water (the other product). The comparison of the first two subsets shows that the effect of the
ether is almost the same as that of the diluent argon. Therefore, the rate
equation should reflect the lack of competitive inhibition of reaction by
ether, since it is evidently weakly adsorbed. The curve for the ethanol–
water data falls below that for the ethanol–argon data, and this comparison demonstrates that the water is an inhibitor. The equation therefore
should have a denominator term indicating the competitive adsorption of
ethanol and water.

Table 4-2

KINETICS OF DEHYDRATION OF ETHANOL CATALYZED BY SULFONATED
POLY(STYRENE-DIVINYLBENZENE) AT 120 ± 1°C[a]

$10^4 \times r,$ mol/(min·g of Catalyst)	$P_{C_2H_5OH}$, atm	P_{H_2O}, atm	$P_{C_2H_5-O-C_2H_5}$, atm	P_{argon}, atm	Ref.
1.19	1.00	0.00	0.00	0.00	[7]
0.81	0.65	0.00	0.35	0.00	[7]
0.47	0.42	0.00	0.58	0.00	[7]
0.30	0.59	0.41	0.00	0.00	[7]
0.12	0.40	0.60	0.00	0.00	[7]
0.65	0.58	0.00	0.00	0.42	[8]
0.60	0.47	0.00	0.00	0.53	[8]
0.41	0.33	0.00	0.00	0.67	[8]
0.25	0.20	0.00	0.00	0.80	[8]
0.57	0.75	0.25	0.00	0.00	[5]
0.24	0.55	0.45	0.00	0.00	[5]
0.08	0.38	0.62	0.00	0.00	[5]

[a] The catalyst was 8% cross-linked gel-form resin having 5.2 meq of $-SO_3H$ groups per gram.

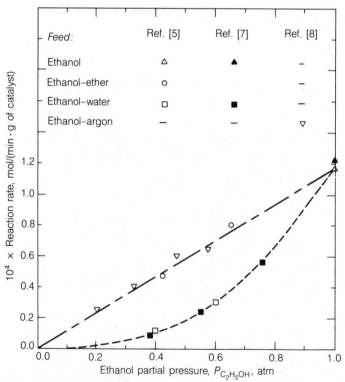

Figure 4-9
Kinetics of ethanol dehydration catalyzed by sulfonic acid resin at 120°C.
The curves are from Eq. 4-24 with the parameter values of Table 4-3.

Next, a group of plausible equations is selected having forms that meet the qualitative criteria stated above (Table 4-3). These have the Langmuir forms just encountered. Exponents other than 1 and 2 in these equations can be tested, but it is prudent to begin with the simple forms to determine how well they represent the data and to introduce complications only if they are needed.

The goodness of fit of the data provided by the four equations of Table 4-3 is indicated in the right-hand column of the table. Estimation of parameters for these nonlinear equations is straightforward with nonlinear regression programs that are available even for personal computers.

Equation C represents the data better than the others. Equation B can be rejected immediately because the negative values of the parameters are physically meaningless. Equations A and D can be rejected because of the large uncertainties associated with the values of K_A.

Equation C fits the data well, as shown in Fig. 4-9. But is the equation more than just an empirical fitting equation? Does it rest on a firm theoretical foundation and provide insight into the mechanism of the catalytic reaction? To answer these questions, consider an interpretation of the kinetics based on the assumptions that (1) the rate-determining step is the reaction of two adsorbed ethanol molecules and that (2) the adsorption equilibria of reactants and products are described by the Langmuir isotherm. The results of Example 4-2 have already confirmed the second assumption.

If the kinetics is

$$r = k\theta_A^2 \tag{4-22}$$

with the adsorption equilibrium described by

$$\theta_A = \frac{K_A P_A}{1 + K_A P_A + K_W P_W} \tag{4-23}$$

then the rate equation agrees with the form determined by fitting the kinetics data:

$$r = \frac{kK_A^2 P_A^2}{(1 + K_A P_A + K_W P_W)^2} \tag{4-24}$$

In terms of the aforementioned model, k' from Table 4-2 is identified as kK_A^2.

Infrared spectra confirm the bonding of about one alcohol molecule per sulfonic acid group, consistent with the assumed rate-determining step [9]. The values of $K_{C_2H_5OH}$, K_{H_2O}, and $K_{C_2H_5-O-C_2H_5}$ determined from the adsorption equilibrium experiments (Table 4-1) and the values determined from the catalytic kinetics (Table 4-3) are in good agreement. The reaction model therefore does rest on a relatively firm foundation. A suggested transition state is shown in Fig. 4-10. The proton transfers are

Table 4-3
COMPARISON OF SUGGESTED FORMS OF RATE EQUATIONS WITH THE DATA OF TABLE 4-2

	Equation[a]	k'[b]	K_A, atm^{-1}	K_W, atm^{-1}	Sum of Squares of Deviations of Predicted and Observed Values of Rate, [mol/(min·g of catalyst)]^2
A	$r = \dfrac{k'P_A}{1 + K_A P_A + K_W P_W}$	$(1.20 \pm 0.09) \times 10^{-4}$	$(3.8 \pm 100) \times 10^{-3}$	3.28 ± 0.52	1.9×10^{-10}
B	$r = \dfrac{k'P_A^2}{1 + K_A P_A + K_W P_W}$	-0.0178 ± 0.056	-150 ± 470	-290 ± 910	7.4×10^{-11}
C	$r = \dfrac{k'P_A^2}{(1 + K_A P_A + K_W P_W)^2}$	$(8.21 \pm 1.1) \times 10^{-4}$	1.65 ± 0.20	2.70 ± 0.33	1.4×10^{-10}
D	$r = \dfrac{k'P_A}{(1 + K_A P_A + K_W P_W)^2}$	$(1.20 \pm 0.08) \times 10^{-4}$	$(7.9 \pm 44) \times 10^{-4}$	1.32 ± 0.14	1.4×10^{-10}

[a] Parameter values estimated by a nonlinear least squares fitting technique; r has dimensions of mol/(min·g of catalyst); A = ethanol; W = water.
[b] k' has units of mol/(min·g of catalyst·atm) in Eqs. A and D and units of mol/(min·g of catalyst·atm^2) in Eqs. B and C.

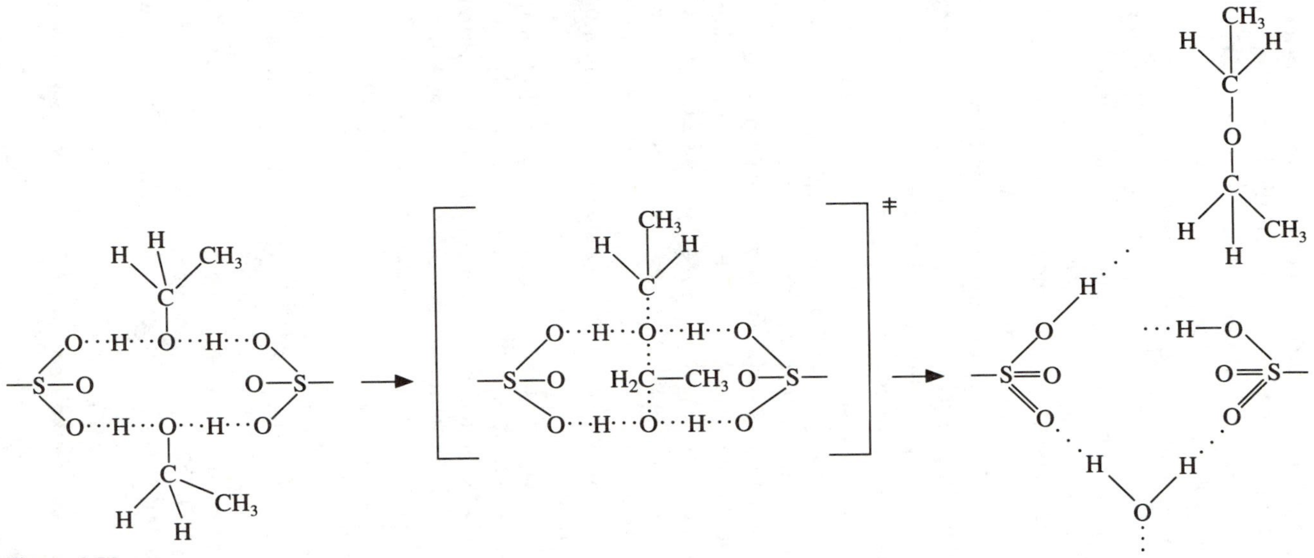

Figure 4-10
Suggested mechanism of ethanol dehydration catalyzed by sulfonated poly(styrene-divinylbenzene) [9].

suggested to be concerted, with the alcohol hydrogen bonded between —SO₃H groups.   ■

In summary, the alcohol dehydration of the preceding example provides an illustration of a catalytic reaction for which the form of the rate equation determined from the kinetics data corresponds to a simple reaction model. The combination of two alcohol molecules adsorbed on (hydrogen bonded to) —SO₃H groups is rate determining, and the adsorption is approximated well by the Langmuir model. This is one of only a few examples for which a simple model of the catalytic mechanism based on Langmuir adsorption is in agreement with experimental results. More typically, simple models are inadequate, often because the Langmuir adsorption model is inappropriate. The gel-form polymer catalyst is different from most solids since it is solutionlike, incorporating an array of catalytic sites that are nearly uniform.

The alcohol dehydration reaction proceeds by the rate-determining combination of two adsorbed reactant molecules, namely, by a *Langmuir–Hinshelwood* mechanism; this term is applied generally when the rate-determining step involves only adsorbed reactants. Langmuir, who was awarded a Nobel prize for his investigations of adsorption and surfaces, recognized that many surface reactions proceed via adsorbed intermediates. Hinshelwood systematically treated the kinetics of surface-catalyzed reactions, using Langmuir's isotherm to account for the adsorption. The example above follows the Hinshelwood treatment; the rate equation (Eq. 4-24) is an example of Langmuir–Hinshelwood kinetics, corresponding to the Langmuir–Hinshelwood mechanism. This kind of treatment of kinetics of solid-catalyzed reactions is used again in the following chapters, although the underlying assumptions are often of questionable validity.

The model presented here is satisfying in its accounting for the available experimental evidence, but it is only superficial in comparison with some of the models (e.g., the Wilkinson olefin hydrogenation cycle presented in Chapter 2). Explicit quantitative information about isolated elementary steps in catalytic cycles is lacking for virtually all solid-catalyzed reactions; even the postulated intermediates are almost always speculative, as it is virtually impossible to isolate and identify them.

Reactions catalyzed by the sulfonic acid resin include many besides alcohol dehydration that are known from Chapter 2, including olefin isomerization and oligomerization, esterification, alkylation of aromatic compounds, and so on. A number have found industrial application, including esterification, bisphenol A synthesis from phenol and acetone, and the conversion of methanol and isobutylene into methyl *t*-butyl ether, a high-octane gasoline blending component.

EXAMPLE 4-5 Langmuir–Hinshelwood and Michaelis–Menten Kinetics
Problem

The form of kinetics describing the typical enzyme-catalyzed reaction is the following (the Michaelis–Menten equation):

$$r = \frac{k_{cat}C_R(C_E)_t}{K_M + C_R} \tag{4-25}$$

where R refers to reactant and E to enzyme, and $(C_E)_t$ is the concentration of enzyme added to the reactor (the total concentration of enzyme in all forms). Interpret the terms k_{cat} and K_M (the Michaelis constant) and show that the form of the kinetics is equivalent to Langmuir–Hinshelwood kinetics.

Solution

To write the equation in a more familiar form, numerator and denominator are divided by K_M, and $1/K_M$ is defined as K_R:

$$r = \frac{k_{cat}K_R C_R(C_E)_t}{1 + K_R C_R} \tag{4-26}$$

This equation now has the form of a Langmuir–Hinshelwood equation, except that the rate is shown to be proportional to the concentration of enzyme. If the rate were written per enzyme molecule, the form would be identical to the Langmuir–Hinshelwood form, with k_{cat} equal to the rate constant and K_R ($= 1/K_M$) equal to the equilibrium constant for the binding of the reactant to the enzyme. The sequence of steps is then

$$R + E \rightleftharpoons RE \tag{4-27}$$

$$RE \rightarrow P + E \tag{4-28}$$

where RE is the reactant–enzyme complex and P is the product, and the second step is rate determining. K_M is identified as the equilibrium constant for dissociation of the reactant–enzyme complex.

It is no surprise that the Langmuir–Hinshelwood and Michaelis–Menten equations are equivalent; the assumptions underlying them are equivalent. Each is based on the assumption that reactant first bonds to uniform catalytic sites (which can be saturated) and then reacts. ■

4.5
INTERACTIONS OF CATALYTIC GROUPS: THE ROLE OF THE SUPPORT

For a number of these reactions that take place in the near absence of water, the dependence of reaction rate on the concentration of catalytically active acid groups in the polymer has been determined. The catalysts are convenient to investigate, since the catalytic groups are identified ($-SO_3H$ groups) and can be replaced to any desired degree by simple ion exchange, for example, by bringing the catalyst in contact with aqueous NaOH to neutralize a fraction of the acid groups and to form $-SO_3Na$ groups in their place.

Often it has been observed that the rate of a catalytic reaction is strongly dependent on the concentration of —SO₃H groups. The data of Fig. 4-11 demonstrate the effect for the dehydration reaction of *t*-butyl alcohol. The —SO₃M groups (where M is Na, K, or Rb) all have the same effect; they dilute the catalytically active acid groups. The reaction is roughly first order in acid group concentration at low acid group concentrations and is roughly fourth order at high concentrations. The results suggest that the catalytic sites are single —SO₃H groups when the population of groups in the polymer is sparse and that the catalytic sites are clusters of —SO₃H groups when their population is dense. Infrared spectroscopy shows that the acid groups in close proximity to each other form a hydrogen-bonded network; an explanation for the high order of reaction in acid group concentration is suggested by the structures of Fig. 4-12. According to this model, the reaction involves concerted proton transfers, and both the reactant alcohol and the inhibitor water can be hydrogen bonded

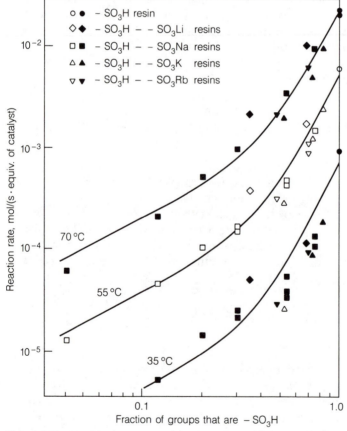

Figure 4-11
Dependence of *t*-butyl alcohol dehydration rate on the composition of sulfonated poly(styrene-divinylbenzene) catalyst [10].

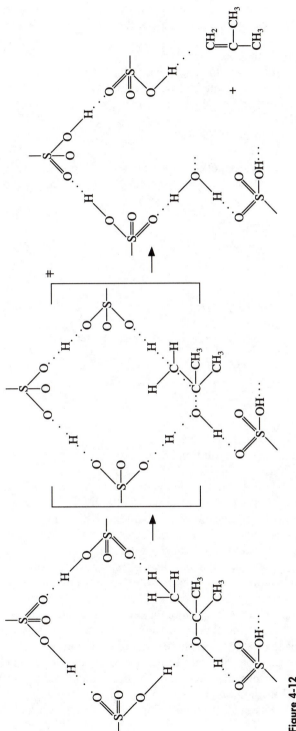

Figure 4-12
Suggested mechanism of *t*-butyl alcohol dehydration involving reactant hydrogen bonded to polymer-bound —SO₃H groups [10].

in the three-dimensional network. This picture illustrates a cooperative action of the proton donor/acceptor groups, as in the push–pull catalysis mentioned on p. 39.

Similarly, high reaction orders in acid group concentration have been observed for general acid catalysis in nonpolar solvents. There is an analogy between catalysis in the flexible hydrocarbon polymer matrix (with its strongly polar catalytic groups) and catalysis in a concentrated solution of a strong acid

such as p-toluenesulfonic acid H_3C—⟨◯⟩—SO_3H in a hydrocarbon sol-

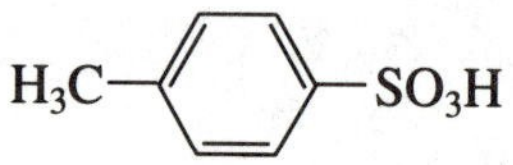

vent such as n-hexane. The polymer appears to be able to solvate and stabilize carbenium ions, and structures like the following have been suggested [11]:

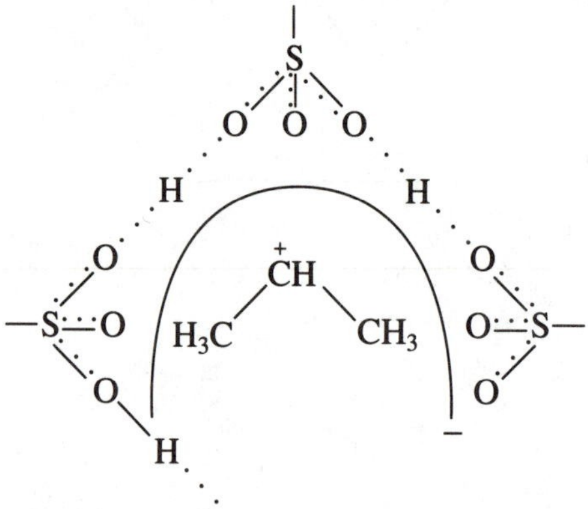

There is, however, an important distinction between the two kinds of catalysis. The acid groups in solution are unrestricted, having a strong tendency to dimerize and otherwise bond with themselves or other polar molecules, as depicted in Fig. 2-3. But the groups anchored to the polymer matrix are restricted in their motions and cannot bond so effectively with each other; consequently, they may be more available for bonding with strongly polar reactants such as alcohols or with weakly polar reactants such as olefins. Groups bonded to relatively rigid polymers (e.g., those having high degrees of cross-linking) may be efficiently held apart from each other and accessible to reactants, whereas groups in more flexible (more solutionlike) polymer gels may be inefficient catalysts because they accommodate reactants poorly.

These points lead to a recognition of some possibilities for catalyst design that are related to the physical properties of the support and not just the chemical properties of the catalytic groups. Anchoring a catalyst to a support can sometimes make possible a catalytic reaction that otherwise could scarcely even take place. This point is illustrated by the results obtained with polymer-bound complexes of titanium [12]. Catalysts were formed by functionalization of a rigid (highly cross-linked) poly(styrene-divinylbenzene), as follows:

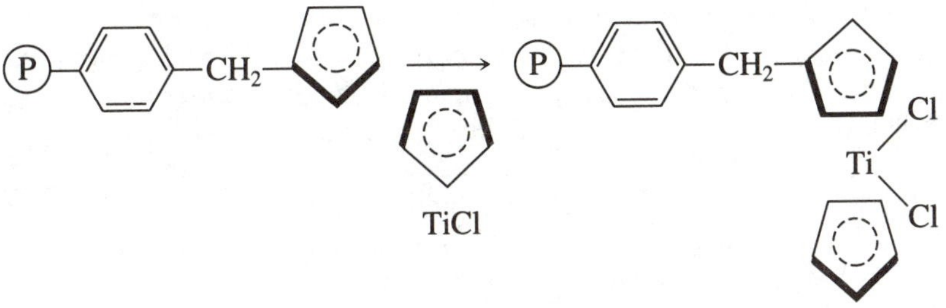

(4-29)

where Ⓟ represents the polymer backbone. Upon reduction with butyllithium, a catalyst was formed, presumably having the following structure:

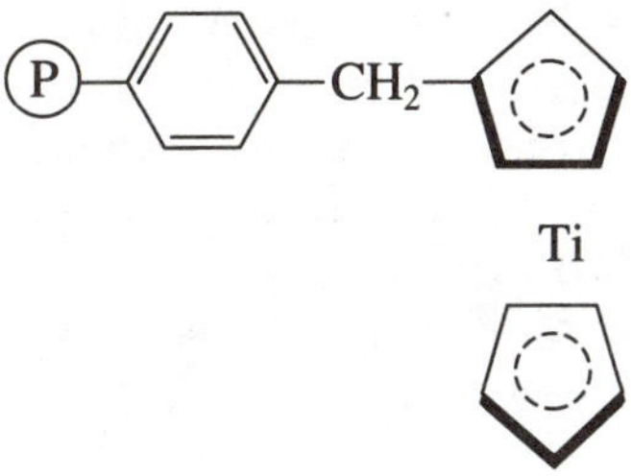

The sandwich complex of Ti has the coordinative unsaturation to allow bonding of reactant hydrogen and olefin (or acetylene), and it is catalytically active for hydrogenation. The molecular analogue of this species

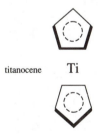

titanocene

is much less active as a catalyst in solution because it is not stable; Ti—Ti bonds form, and the complex oligomerizes or polymerizes to form species lacking the necessary coordinative unsaturation. This is an example of self-inhibition of a catalyst, similar to that mentioned for the sulfonic acid resin. It is also suggestive of the phenomena involved with cobalt and rhodium hydroformylation catalysts in solution (p. 94). In hydroformylation, CO ligands bonded to the Co (or phosphine ligands bonded to the Rh) hinder self-inhibition that would result from metal–metal bond formation. In the present example, rigid polymeric ligands bonded to the Ti isolate the metal centers, preventing self-inhibition by metal–metal bond formation. The physical properties of the polymer support are critical. If the support is too flexible and solutionlike, the catalyst cannot function well; if it is rigid and holds the titanocene groups in their own territories, they cannot nullify each other's catalytic activities.

The polymer-bound titanocene catalyst is a model of the inorganic solid catalysts to be described in Chapters 5 and 6: these metals, metal oxides, and metal sulfides have rigid surfaces presenting metal sites with coordinative unsaturation for bonding and activation of reactants.

4.6
BIFUNCTIONAL AND
MULTIFUNCTIONAL CATALYSIS

The illustrations of catalysis by —SO_3H groups demonstrate the importance of catalytic groups working in combination (Figs. 4-10 and 4-12). There are many opportunities for carrying out intricate catalytic reactions and reaction sequences by incorporating just the right combinations of groups in the polymers. Some of the simplest ideas related to *bifunctional catalysis* by a solid are illustrated by poly(styrene-divinylbenzene) supports containing not just —SO_3H groups, but also —SO_3Ag or —SO_3K groups. The catalytic reactions are

$$H_2O + CH_3-C\overset{O}{\underset{OC_3H_7}{\big\langle}} \rightarrow CH_3COOH + C_3H_7OH \qquad (4\text{-}30)$$

$$H_2O + CH_3-C\overset{O}{\underset{\underset{\text{allyl acetate}}{OCH_2CH=CH_2}}{\big\langle}} \rightarrow CH_3COOH + H_2C=CHCH_2OH$$

$$(4\text{-}31)$$

The dependence of reaction rate on —SO_3H group concentration in the polymer beads suspended in reactant solutions is shown in Fig. 4-13. When the acid groups are diluted with —SO_3K groups, the rate is directly proportional to the —SO_3H group concentration. The catalytic groups are formed simply by dissociation of hydrated —SO_3H groups. But when —SO_3Ag groups rather than —SO_3K groups are present with the —SO_3H groups, the catalytic activity is markedly higher for allyl acetate but not for propyl acetate. Evidently the —SO_3Ag groups are simply inactive ("diluent") groups for the one reaction but exert a strong influence on the other.

The role of the —SO_3Ag groups is attributed to their ability to coordinate

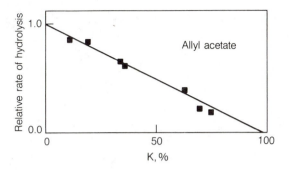

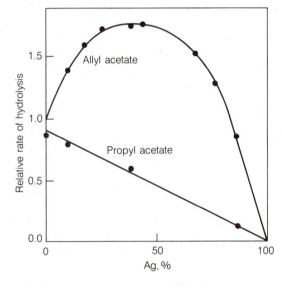

Figure 4-13
Kinetics of hydrolysis of *n*-propyl acetate and allyl acetate at 27.5°C catalyzed by sulfonic acid resin modified by incorporation of —SO₃K and —SO₃Ag groups. The results show the role of the —SO₃Ag groups in concentrating the allyl acetate in the catalytic phase [13].

olefins:

$$\text{C}=\text{C} \quad \underset{\text{Ag}^+}{\cdots}$$

When the —SO₃Ag groups are present, the concentration of reactant allyl acetate in the polymer is greater than when —SO₃K groups are present instead. The silver-containing groups tend to concentrate reactant molecules in the phase containing the catalytically active acid groups. They affect the distribution of reactants between phases; they influence the phase equilibrium and not the catalytic cycle. Polymeric catalysts can be designed to incorporate groups to control the swelling, that is, the phase equilibrium.

EXAMPLE 4-6 Polymer Catalysts Comparable to Phase Transfer Catalysts

Problem

Conceive a means to use a polymer to accelerate the reaction of two compounds present in mutually insoluble liquids.

Solution

The phase transfer catalysts described on pp. 127–129 accelerate reactions of compounds soluble in separate liquid phases. They do this by forming a complex with one of the reactants which is then soluble in the other phase; the phase transfer agent shuttles the reactant across the phase boundary.

Something akin to phase transfer catalysis can be effected with a swellable polymer that incorporates groups chosen to have affinities for the different reactants [14]. For example, the hydrolysis of 1-bromoadamantane in the presence of toluene and water has been shown to be accelerated by a lightly cross-linked polystyrene incorporating catalytically active $—CH_2OCH_2CH_2OCH_3$ groups; these groups have an affinity for the water, and the hydrocarbon matrix has an affinity for the 1-bromoadamantane. When the reactants are brought together in the microenvironment of the swelled polymer, they react much more rapidly than when the polymer is absent. ∎

A more complex example illustrates the opportunities for catalyst design by the selection of combinations of groups in a polymer matrix. In the aldox process of Example 2-8, a cobalt or rhodium complex in solution is used for hydroformylation of propylene to give butyraldehyde; a solution of KOH is used for aldol condensation of the aldehyde; and a metal such as Ni is used for hydrogenation, finally giving 2-ethylhexanol. A polymer incorporating rhodium complexes is active for olefin hydroformylation, and a polymer incorporating amine groups is active for aldol condensation. A polymer incorporating both kinds of groups (Fig. 4-14) catalyzes the sequence of three reactions shown in Fig. 4-15. The reactions are the aldox sequence, except that the final hydrogenation of the olefinic double bond in the aldol condensation product is not accompanied by reduction of the —CHO group to give —CH₂OH. The rhodium complex might be sufficiently active for this reaction if it were coordinated to the amine instead of the phosphine ligands.

The aldox catalyst is an example of a multifunctional catalyst designed for a sequence of reactions; the catalytic groups were identified as those appropriate for the individual reactions. One characteristic of this multifunctional catalyst is its high activity when it is compared with a combination of monofunctional catalysts, that is, a mixture of particles incorporating just Rh complexes with particles containing just amine groups [15]. The multifunctional catalyst gives higher rates of reaction under comparable conditions, presumably because the presence of the polar amine groups in the polymer with the Rh complexes gives a higher concentration of reactants in the polymer phase, near the catalytic Rh groups. This effect of the amine groups is comparable to that

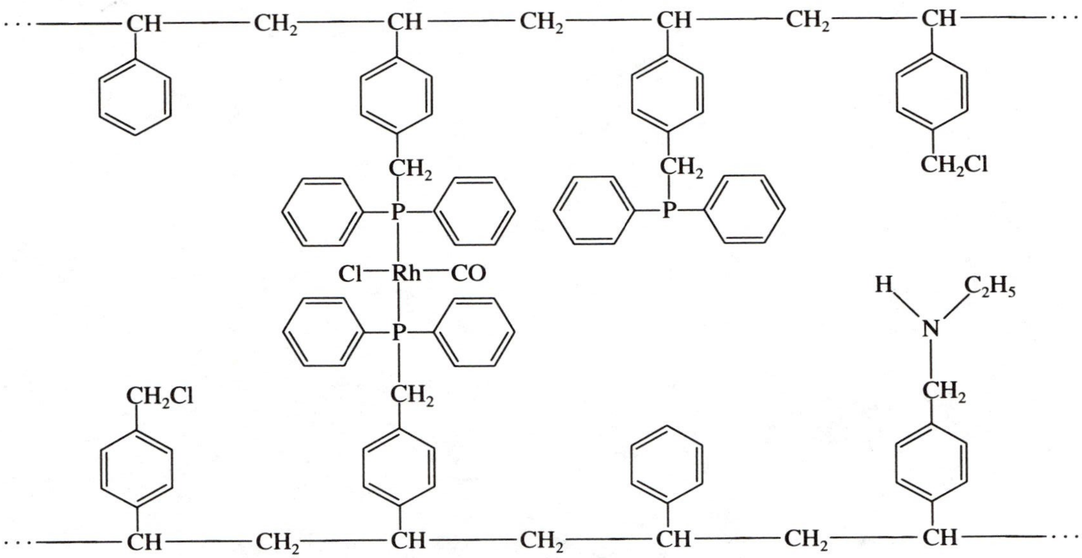

Figure 4-14
A representative structural element of a multifunctional aldox catalyst [15].

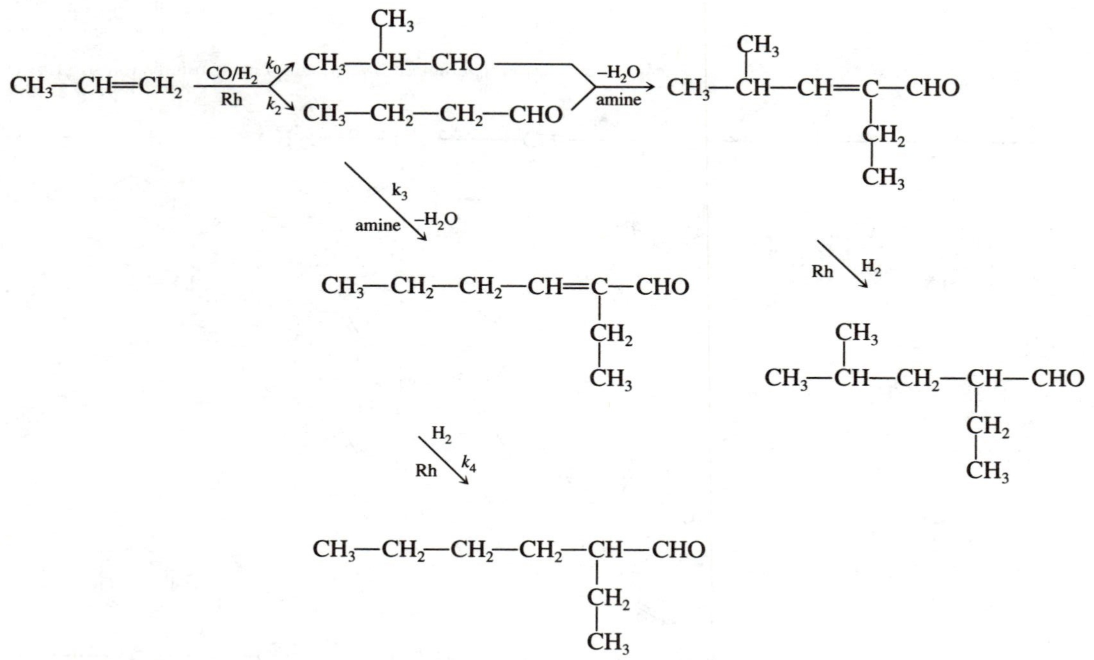

Figure 4-15
The aldox reaction network catalyzed by a styrene-divinylbenzene copolymer incorporating Rh complex and amine groups.

of the —SO₃Ag groups in the catalyst for allyl acetate esterification discussed above.

EXAMPLE 4-7 Design of a Solid Catalyst for Bisphenol A Synthesis

Problem

Use the information of Example 2-1 as a basis for designing a solid polymeric catalyst for the synthesis of bisphenol A from phenol and acetone.

Solution

The reaction is catalyzed by strongly acidic solutions, and the sulfonic acid resin is expected to be a good catalyst. Experimental results confirm the expectation [16]. The polymeric catalyst is used in industrial processes for bisphenol A synthesis, having replaced the cheaper mineral acids, which are corrosive and present a spent acid disposal problem.

A more challenging catalyst design problem is to take advantage of the promotion of the reaction by mercaptans. Incorporation of sulfur-containing groups in the polymer might be expected to produce a bifunctional catalyst having a higher activity than the polymer containing only —SO₃H groups. One method for incorporation of sulfur-containing groups involves neutralization of a fraction of the —SO₃H groups with dimethylthiazoladine:

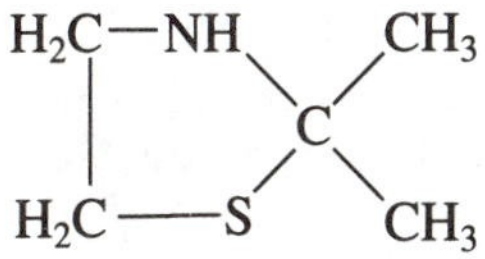

$$\begin{array}{c} H_2C\!-\!NH\quad\ \ CH_3 \\ |\qquad\quad \diagdown C \diagup \\ H_2C\!-\!\!-\!S\quad\ \ CH_3 \end{array}$$

The groups produced are

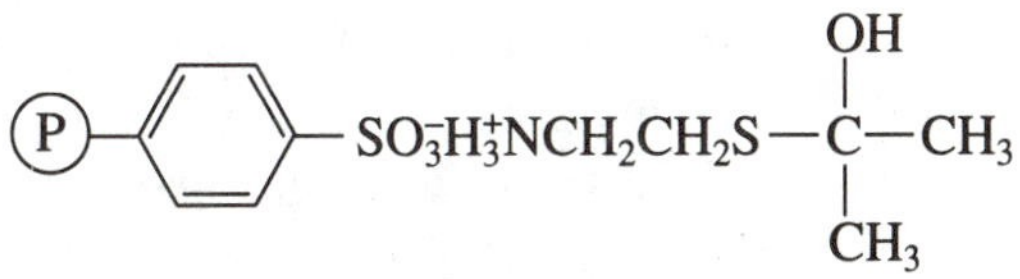

$$(P)\!-\!\!\!\bigcirc\!\!\!-\!SO_3^-H_3^+NCH_2CH_2S\!-\!\underset{\underset{CH_3}{|}}{\overset{\overset{OH}{|}}{C}}\!-\!CH_3$$

This suggestion appears in a patent [17], in which it is claimed that incorporation of the sulfur-containing groups (with neutralization of a fraction of the acid groups) increases the activity of the catalyst. [What catalytic cycle might take place in the promoted (bifunctional) catalyst that does not take place in the simple acid catalyst?] ■

EXAMPLE 4-8 Design of a Solid Catalyst for the Wacker Reaction

Problem

Use the information in Chapter 2 to design a solid Wacker catalyst.

Solution

Palladium complexes are best for the ethylene oxidation, and a second function is needed to reoxidize the palladium in place. A catalyst design is suggested with monomer units incorporating palladium sulfonate groups

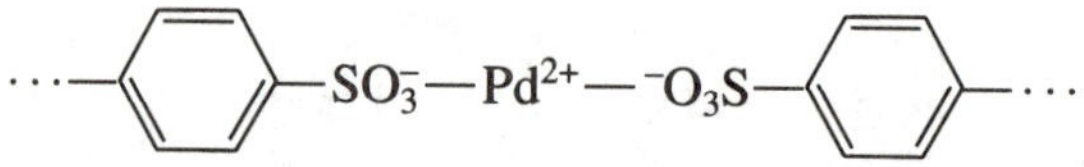

$$\cdots\!-\!\!\left\langle\bigcirc\right\rangle\!\!-\!SO_3^-\!-\!Pd^{2+}\!-\,^-O_3S\!-\!\left\langle\bigcirc\right\rangle\!\!-\!\cdots$$

and quinone groups:

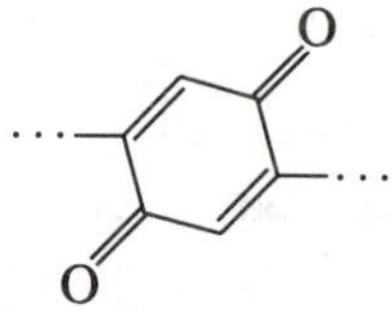

Several catalysts of this type have been reported [18], including the following copolymer:

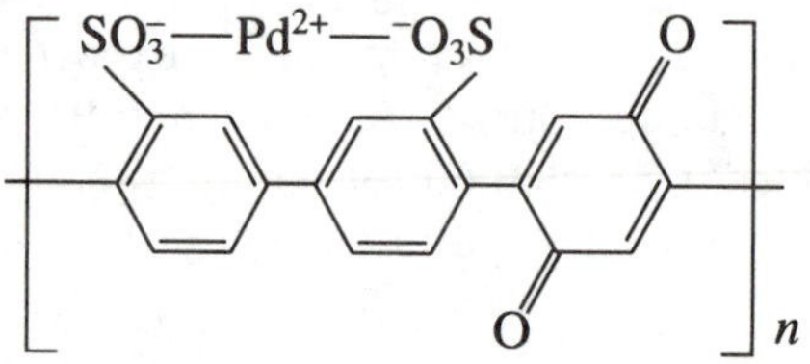

The solid polymer functions in much the same way as Pd^{2+} and quinone together in solution. The redox cycles are inferred to include the Pd^{2+}–Pd^0 conversion and the quinone–hydroquinone conversion. The nature of the electron transfer between the groups remains unclear, but it is known that similar polymers are semiconductors. ∎

The bisphenol A synthesis reaction of Examples 2-1 and 4-7 provides an illustration of a more subtle kind of combination of functions in a polymer catalyst and a good opportunity to use the Langmuir–Hinshelwood formalism in analysis of the kinetics. The kinetics of the reaction of phenol and acetone in the presence of sulfonated poly(styrene-divinylbenzene) to give bisphenol A and water has been measured and represented with the following equation [16]:

$$r = \frac{kK_AC_AK_P^2C_P^2}{(1 + K_AC_A + K_WC_W)(1 + K_PC_P + K_HC_H)^2} \tag{4-32}$$

where A is acetone, P is phenol, W is water, and H is the solvent n-heptane. The role of heptane indicated by Eq. 4-32 is that of an inhibitor, not that of a simple diluent (which would fail to appear in the equation). From the form of the equation it is evident that heptane does not play the role of an inhibitor that competes with water and acetone for the polar catalytic groups. Rather, it appears to compete with phenol for a different kind of site.

A simplified interpretation of the kinetics can be developed by casting

Eq. 4-32 in a form representing the rate in terms of the polymer phase concentrations of reactants. This is an extension of the development illustrated by the simpler examples of ethanol and t-butyl alcohol dehydration in the presence of the same catalyst.

If acetone and water compete for the —SO$_3$H groups, and if the Langmuir model describes the adsorption at equilibrium, then

$$\theta_A = \frac{K_A C_A}{1 + K_A C_A + K_W C_W} \tag{4-33}$$

Any polar molecule can compete with acetone and water for the acid groups, but the data and the form of the rate equation suggest [16] that the term $K_P C_P$ is negligible in comparison with $1 + K_A C_A + K_W C_W$.

If the phenol interacts predominantly with a different kind of site in competition with n-heptane, then

$$\theta'_P = \frac{K'_P C_P}{1 + K'_P C_P + K'_H C_H} \tag{4-34}$$

where the prime denotes a separate phase within the polymer. Now, Eq. 4-32 can be represented as

$$r = k \theta_A \theta_P'^2 \tag{4-35}$$

(Compare this with the solution kinetics of Example 2-1.)

If the model is meaningful, there must be a plausible answer to the following question: What are the sites where phenol and n-heptane bond and the more polar reactants do not? An answer is provided by the hydrophobic or apolar regions of the polymer gel; the paraffin has an affinity for the hyrocarbon backbone and aromatic rings of the polymer, and the aromatic ring of phenol has an affinity for the same groups. There is an analogy to the hydrophobic bonding illustrated in catalysis by macromolecules and by cyclodextrin in solution; the bifunctional character of the catalysis parallels that illustrated for ester hydrolysis with poly(styrenesulfonic acid) and with cyclodextrin in solution. The polymer gel is represented as a bifunctional catalyst that concentrates the reactant acetone at one kind of site and the phenol at another. The reaction is facilitated by the juxtaposition of these two kinds of sites in the polymer.

The catalytic sequence can be approximated as follows (with a parallel to the development of Example 2-1):

$$A_{abs} + \text{—SO}_3H \underset{}{\overset{K_{36}}{\rightleftharpoons}} A_{ads} \tag{4-36}$$

$$A_{ads} + P_{abs} \underset{}{\overset{K_{37}}{\rightleftharpoons}} I_{ads} \tag{4-37}$$

$$I_{ads} + P_{abs} \underset{}{\overset{k_{38}}{\rightleftharpoons}} H_2O_{ads} + BPA_{abs} \tag{4-38}$$

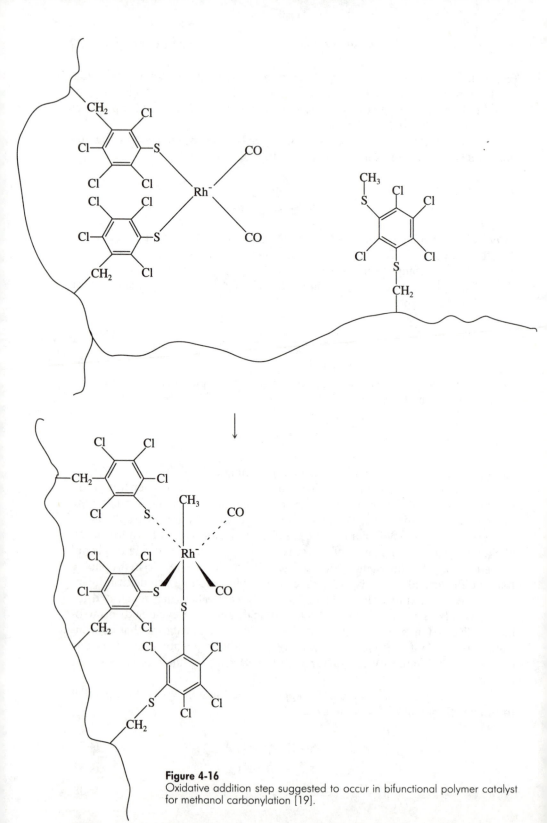

Figure 4-16
Oxidative addition step suggested to occur in bifunctional polymer catalyst for methanol carbonylation [19].

where A_{abs} represents absorbed acetone (swelling the nonpolar part of the polymer matrix) and A_{ads} represents adsorbed (chemically bonded) species (hydrogen bonded to acid groups).

If step 4-38 is assumed to be rate determining, substitution in the equation leads to the kinetics of Eqs. 4-32 and 4-35. (Is the analogous step rate determining in the solution catalysis?)

EXAMPLE 4-9 Design of a Bifunctional Catalyst for Methanol Carbonylation

Problem

Use the information in the cycle of Fig. 2-40 to design a polymer-supported catalyst for methanol carbonylation.

Solution

Both a catalyst (Rh complex) and a cocatalyst (CH_3I) are required for the solution reaction. The solid catalyst must therefore be bifunctional. Stable attachment of iodide to poly(styrene-divinylbenzene) is not possible, but the pseudohalide described in Example 2-9 can be supported through the phenyl ring. The supported pseudohalide is a ligand for rhodium. The bifunctional solid containing rhodium has been shown to work without any added cocatalyst [19]. The roles of the two catalyst functions are believed (by inference from the cycle of Fig. 2-40) to involve the oxidative addition step of Fig. 4-16. (Why would a highly cross-linked polymer be inappropriate as a support?) ∎

Table 4-4
CLASSES OF SOLUTION CATALYSIS AND THE CORRESPONDING CLASSES
OF MULTIFUNCTIONAL CATALYSIS BY POLYMERS

Class of Solution Catalysis	Example of Solution-Catalyzed Reaction	Multifunctional Polymer Catalyst for the Same Reaction
Sequence of reactions, each with its own catalyst	Aldox sequence (olefin hydroformylation, aldol condensation, hydrogenation)	Poly(styrene-divinylbenzene) incorporating a combination of catalytic groups (e.g., Rh complex and amine groups)
Reaction requiring a catalyst plus a cocatalyst	Bisphenol A synthesis in the presence of H_2SO_4 (catalyst) and a mercaptan (cocatalyst)	Poly(styrene-divinylbenzene) incorporating —SO_3H groups and sulfur-containing groups
Reaction in two-phase liquid system	Alkylation of benzene with propylene catalyzed by aqueous H_2SO_4	Poly(styrene-divinylbenzene) incorporating —SO_3H groups; the organic polymer matrix constitutes one function (holding the hydrocarbon reactants) and the acid groups constitute the other function

To summarize, the examples of multifunctional catalysis by polymers have parallels in solution catalysis. Table 4-4 is a generalized list of categories of solution catalysis and their counterparts in multifunctional catalysis by solid polymers. The examples illustrate how one can approach the design of solid catalysts by identifying the appropriate functions. This theme recurs in subsequent chapters: most successful solid catalysts are multifunctional. But there is an important difference between the polymers and the inorganic solids that are more typical of industrial catalysts. The polymer catalysts are structurally rather simple, whereas the surfaces of inorganic solids are complex. The polymers provide a transition from the well-understood soluble catalysts to the more complex surface catalysts.

4.7
POROUS POLYMERS AND SURFACE CATALYSIS

A polymer catalyst offers sites for bonding and activation of a reactant. Often this is not sufficient for catalysis; a second reactant may need to be accommodated in the polymer so that it can find access to the first and undergo conversion. This point is demonstrated with an example of an acid-catalyzed reaction, and it leads to the recognition of another important opportunity in catalyst design.

The reaction is the alkylation of benzene with propylene to give isopropylbenzene. When a solution of benzene and propylene is brought in contact with particles of the 8% cross-linked poly(styrenesulfonic acid) gel, no observable catalytic reaction occurs [11]. The difficulty is that the hydrocarbon reactants have so little affinity for the acid groups of the polymer that these groups are tightly linked to each other through hydrogen bonds:

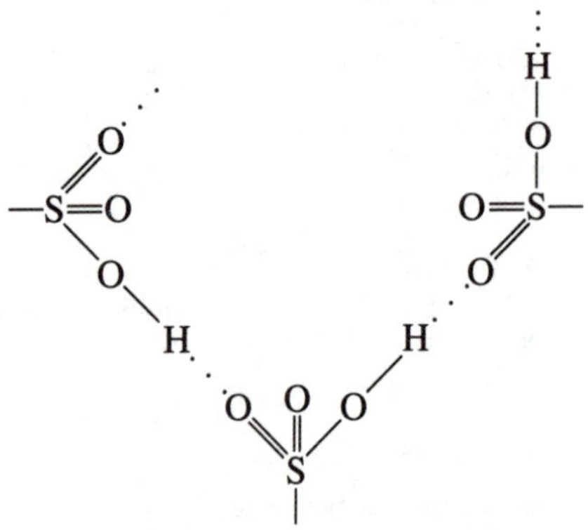

The polymer remains unswelled; the network is collapsed and almost free of reactants.

To make the acid groups effective catalytically, they must be made accessible to the reactants. A good swelling solvent might be added to the reactant solution, but hydrocarbons like the reactant are insufficient, and good polar solvents like water have such a great affinity for the —SO_3H groups that they exclude the reactants and suppress the catalytic reaction almost completely. The practical alternative is to modify the physical properties of the polymer.

An active catalyst results when the polymer is synthesized to have a large fraction of its catalytic groups exposed on a surface rather than buried in a sometimes nearly impenetrable tangle of hydrocarbon chains. The surface groups function effectively because the reactants come in contact with them directly; the groups are present at, or very near, an interface. This goal is met by making the solid porous; a polymer bead now incorporates a labyrinth of channels, and the surface is largely internal. The porous polymer bead may have 10^3 or 10^4 times as much surface area as the gelular bead, which has only the peripheral area. The porous polymer may have a surface area of hundreds of square meters per gram.

Examined under an electron microscope (Fig. 4-17), the porous polymer appears as a jumble of interconnected spheres having typical dimensions of 5 to 10 nm. The labyrinth of void spaces between these microspheres is the pore structure. This space can be filled with the reactant molecules, which readily find access to the —SO_3H groups on the microsphere surfaces.

The porous polymers can be made with wide ranges of physical properties, including the pore dimensions and surface area. The porous polymers are prepared by polymerizing the styrene and divinylbenzene monomers in the presence of a solvent like n-heptane, which is a good solvent for the monomers but not a good swelling agent for the polymer. The polymer as it forms becomes a separate solid phase (which becomes the microspheres), and the spaces originally filled with solvent containing the reservoir of monomers remain as pores after the solvent is removed. The surfaces can be functionalized with any of the acid, base, redox, and metal complex groups mentioned earlier. The porous polymers prepared in this way are usually highly cross-linked and more robust

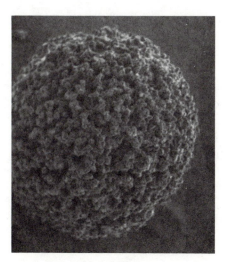

Figure 4-17
Electron micrograph showing the microspheres and void spaces (pores) of macroporous cross-linked poly(styrenesulfonic acid). (Courtesy of Rohm and Haas Co.)

and stable than the typical gels. For these reasons, and others to be mentioned shortly, the porous polymers are the ones applied as industrial catalysts.

EXAMPLE 4-10 Kinetics of Isobutylene Oligomerization Catalyzed by Macroporous Sulfonated Poly(styrene-divinylbenzene)

Problem

Use the data of Fig. 4-18 to determine a rate equation for dimerization of isobutylene catalyzed by macroporous cross-linked poly(styrene-sulfonic acid).

Solution

Here is a reaction for which the gel form polymer would be ineffective because it would not be swelled by the reactants. An analysis of the kinetics follows from recognition of the following limiting cases from the plots. At low concentrations of isobutylene (IB),

$$r = k_{39} C_{IB}^2 \tag{4-39}$$

and at high concentrations,

$$r = k_{40} C_{IB} \tag{4-40}$$

The following equation is the simplest representation of all the data:

$$r = \frac{k K_{IB} C_{IB}^2}{1 + K_{IB} C_{IB}} \tag{4-41}$$

This equation takes the form

$$r = k \theta_{IB} C_{IB} \tag{4-42}$$

where

$$\theta_{IB} = \frac{K_{IB} C_{IB}}{1 + K_{IB} C_{IB}} \tag{4-43}$$

The rate of the dimerization reaction is proportional to the concentration of adsorbed isobutylene (which forms the *t*-butyl cation) and to the concentration of isobutylene in the liquid phase.

The result suggests a rate-determining step that is the combination of one adsorbed reactant (a "solvated" carbenium ion) with a reactant molecule from the liquid phase. Such a step is familiar from the discussion of carbenium ion chemistry in Chapter 2. When the rate-determining step

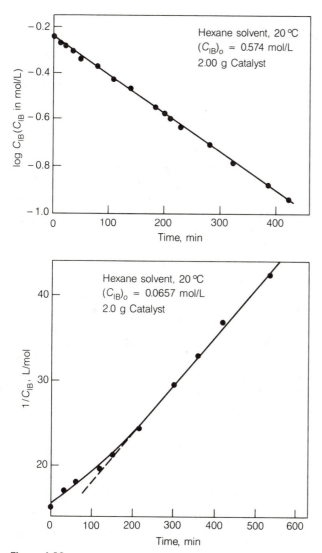

Figure 4-18
Kinetics of oligomerization of isobutylene catalyzed by macro-porous cross-linked poly(styrenesulfonic acid) at 20°C [20]. Reactants were present in the liquid phase.

involves a reactant from a fluid phase and one that is adsorbed, the term **Eley–Rideal (or Rideal) mechanism** is applied; the corresponding kinetics is called Eley–Rideal kinetics. In contrast, with a Langmuir–Hinshelwood mechanism (Example 4-4), all the reactants involved in the rate-determining step are adsorbed. Eley–Rideal kinetics is observed frequently in catalysis by solid acids. ∎

4.8
INTRAPARTICLE TRANSPORT INFLUENCE [21, 22]

Thus far in this chapter it has been implicitly assumed that reactants either met no resistance in finding their way to catalytic groups or, on the other hand, that they were excluded from them almost entirely. Consider now the intermediate case and the influence of mass transport on the rate of catalytic reaction. The following development is for catalyst particles that are polymer gels; it is later extended to porous catalyst particles.

Consider a single spherical particle of a polymer having uniformly distributed catalytic groups (Fig. 4-19). A reactant molecule from the surrounding fluid phase can penetrate into the polymer, and its transport takes place entirely by diffusion, that is, as a result of the random motion of molecules between the polymer chains. The molecules diffuse down a concentration gradient, with the concentration of reactant at the periphery of the particle being higher than that at the center, since reactant is being consumed throughout the particle as a result of the catalytic reaction. The diffusing molecules at the particle periphery or at any point in the polymer have a choice: They can either undergo reaction, giving products, or they can continue to diffuse toward the particle center. If the resistance to diffusion offered by the polymer is great and/or if the activity of the catalyst is great, the reactants will be converted near the periphery, and the concentration gradient will be steep (Fig. 4-20a). On the other hand, if the resistance to diffusion is low and/or if the activity of the catalyst is low, then the concentration profile may be nearly flat (Fig. 4-20b). These ideas are only qualitative; they are developed into a quantitative framework according to the models proposed independently by Thiele [23], Damköhler [24], and Zeldovitch [25].

4.8.1 The Thiele Model: the Influence of Intraparticle Mass Transport on Catalysis [23]

Consider steady-state reaction in a spherical isothermal catalyst particle, assuming for simplicity that the intrinsic kinetics of the reaction is first order and

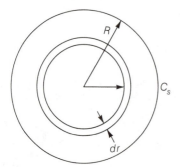

Figure 4-19
Spherical catalyst particle with diffusing reactant. C_s is the concentration of reactant at the particle periphery.

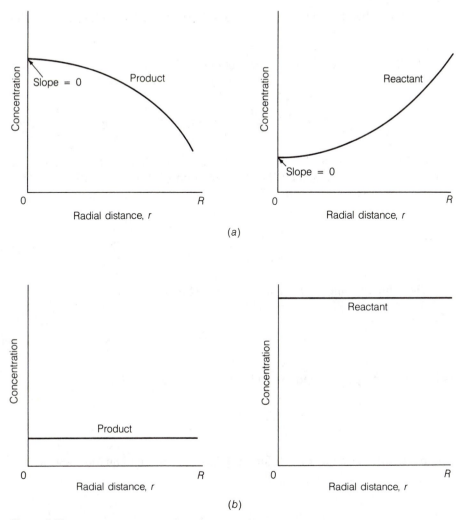

Figure 4-20
Concentration profiles indicating the concentration gradients of diffusing reactants and products in a uniform catalyst particle. (*a*) Significant diffusion resistance: (*b*) Negligible diffusion resistance.

the reaction is irreversible:

$$r = kC \tag{4-44}$$

where k is the rate constant per unit volume of the catalyst particle and C is the concentration of the reactant. Equation 4-44 expresses the intrinsic reaction rate, namely, that in the absence of a mass transport influence.

A mass balance on reactant in a spherical shell of thickness dr (Fig. 4-19) is

rate of diffusion inward at $r + dr$ − rate of diffusion inward at r (4-45)
= rate of consumption of reactant in shell

$$4\pi(r + dr)^2 D \left[\frac{dC}{dr} + \frac{d^2C}{dr^2} \, dr \right] - 4\pi r^2 D \frac{dC}{dr} = 4\pi r^2 \, dr \, kC \qquad (4\text{-}46)$$

where $4\pi r^2$ is the area of the shell at radius r and dC/dr is the concentration gradient. Here, it is assumed that the diffusion of reactant in the particle is described by Fick's law, with D being the diffusion coefficient.

Doing the calculus gives

$$\frac{d^2C}{dr^2} + \frac{2}{r}\frac{dC}{dr} = \frac{kC}{D} \qquad (4\text{-}47)$$

The dimensionless *Thiele modulus* is defined as

$$\phi = R \sqrt{\frac{k}{D}} \qquad (4\text{-}48)$$

(where R is the particle radius) and substituted into Eq. 4-47:

$$\frac{d^2C}{dr^2} + \frac{2}{r}\frac{dC}{dr} = \frac{\phi^2}{R^2}C \qquad (4\text{-}49)$$

The boundary conditions for solution of this differential equation are the following:

$$\text{At } r = 0, \frac{dC}{dr} = 0; \qquad \text{at } r = R, C = C_s,$$

where C_s is the concentration of reactant at the surface (periphery) of the sphere. The solution of Eq. 4-49 for the reactant concentration as a function of particle radius is

$$C = \frac{C_s R \, \sinh(\phi r/R)}{r \sinh \phi} \qquad (4\text{-}50)$$

This equation gives the profiles shown in Fig. 4-21.

The overall rate in the catalyst particle could be determined by integrating over the radius r, but a simpler derivation follows from the fact (from conservation of mass) that at steady state the rate of conversion of reactant in the catalyst particle is equal to the rate of transport of reactant into the particle

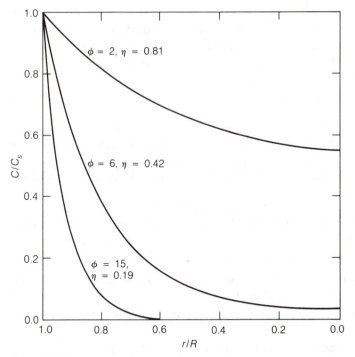

Figure 4-21
Diffusion with reaction in a spherical gelular (or porous) catalyst particle: concentration profiles of the reactant for a first-order isothermal reaction.

(here, rate is written out so as not to be confused with radius r):

$$\text{rate} = 4\pi R^2 D\left(-\frac{dC}{dr}\right)_{r=R} \tag{4-51}$$

This is the rate per particle, not per unit volume. Using Eq. 4-50 to determine the derivative dC/dr and substituting into this equation gives

$$\text{rate} = 4\pi\phi RDC_s\left(\frac{1}{\tanh \phi} - \frac{1}{\phi}\right) \tag{4-52}$$

This solution is now represented in a more convenient way. First, consider the special case for which the catalytic reaction rate everywhere in the particle is equal to the intrinsic rate, that is, the rate that would prevail if the reactant concentration at the particle periphery were the reactant concentration everywhere in the particle. This maximum rate for the total particle is equal to the particle volume times the rate per unit volume:

$$\text{rate}_{\text{max}} = \frac{4}{3}\pi R^3 kC_s \tag{4-53}$$

Now, define a dimensionless **effectiveness factor** η as was done in Chapter 2: η = rate/rate$_{max}$. This is the overall rate divided by that which would prevail if there were no diffusion influence. It follows from this definition and the preceding equations that

$$\eta = \frac{3}{\phi}\left[\frac{1}{\tanh \phi} - \frac{1}{\phi}\right] \tag{4-54}$$

The dependence of the effectiveness factor on the Thiele modulus is shown in the logarithmic plot of Fig. 4-22. This result is for isothermal, irreversible first-order reaction in a sphere. The figure also includes results derived similarly for reactions of other orders.

The shape of the curve is especially simple and important. There are two limiting cases; first, if $\phi \ll 1$, then $\eta \cong 1$ and rate = kC_s. This case corresponds to the absence of any diffusion influence on the reaction rate; it follows from the definition of ϕ that it pertains when

1. The particle radius R approaches zero, that is, the maximum diffusion path becomes very small, and/or
2. When D, the diffusion coefficient of the reactant in the catalyst particle, becomes very large, and/or
3. When the catalytic activity measured by k becomes very small.

The combined effects of k, D, and R are represented in the dimensionless group called the Thiele modulus.

The second limiting case is for large values of ϕ, for which

$$\eta = \frac{3}{\phi} \tag{4-55}$$

The rate is inversely proportional to the particle radius; the rate is proportional to the peripheral surface area of the catalyst particle. In this case,

$$\text{rate} = \eta k C_s = 3k\frac{1}{R}\left(\frac{D}{k}\right)^{1/2}C_s \tag{4-56}$$

or

$$\text{rate} = \frac{3k^{1/2}D^{1/2}}{R}C_s \tag{4-57}$$

This result implies that in the limiting case of large ϕ, the activation energy that would be measured for the reaction is one half that for the intrinsic reaction plus one half that for diffusion; the latter value is usually negligible, roughly a few kcal/mol. This is an extremely important result. It shows that a strong diffusion influence disguises the kinetics; the activation energy measured in a kinetics experiment is not that of the chemical reaction alone. It is crucial in

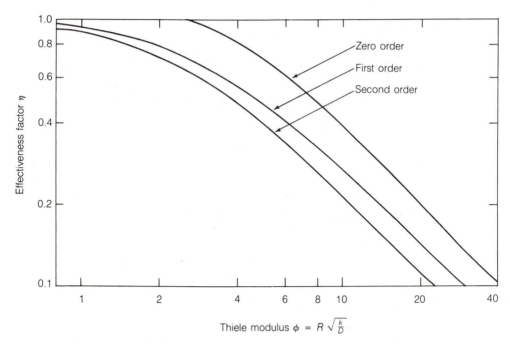

Figure 4-22
Results of the Thiele model: dependence of effectiveness factor on the dimensionless Thiele modulus.

catalytic kinetics to resolve the chemical and physical processes. This is best done by a systematic variation of the Thiele modulus, which is most easily done by the systematic variation of the catalyst particle size R. When there is no effect of the catalyst particle size on the rate of reaction in isothermal particles, the effectiveness factor is unity and the measured kinetics is the intrinsic kinetics. The diffusion also disguises the reaction order when the intrinsic order is different from 1. It also influences selectivity in parallel and sequential reactions. Key results are summarized in Table 4-5; details are given in the chemical reaction engineering literature [22, 26].

Table 4-5
INFLUENCE OF INTRAPARTICLE DIFFUSION IN THE KINETICS OF CATALYTIC
REACTIONS IN THE LIMIT OF LARGE ϕ

Intrinsic Reaction Order	Observed Reaction Order	Observed Activation Energy[a]
0	1/2	$E_{act}/2$
1	1	$E_{act}/2$
2	3/2	$E_{act}/2$

[a] Calculated on the basis of the assumption that the diffusion coefficient is constant (Knudsen diffusion) and the activation energy for diffusion is zero; E_{act} is the intrinsic activation energy of the catalytic reaction.

EXAMPLE 4-11 **Experimental Determination of Mass Transport Influence in Sucrose Inversion Catalyzed by Cross-linked Poly(styrenesulfonic acid)**

Problem

Analyze the data of Table 4-6 on the basis of the Thiele model for simultaneous diffusion and reaction in gel form particles of the sulfonic acid resin.

Solution

Qualitative examination of the data at each temperature shows the expected trend: The reaction rate (measured by the observed first-order rate constant) decreases with increasing particle size, indicating an intraparticle transport effect. A plot of the logarithm of the observed rate constant against the logarithm of the particle radius at each temperature (as suggested by the result of the Thiele analysis, Fig. 4-22) gives a curve very similar to that shown for the first-order reaction in a sphere.

A quantitative estimate of η and ϕ is obtained for a given pair of catalyst particle sizes by "triangulating" on Fig. 4-22. A given ratio of η values (the ratio of particle diameters) determines unique values of η and ϕ. A better fit of all the data is obtained by overlaying Fig. 4-22 on the plots of rate constant vs. particle radius. The result is Fig. 4-23, which demonstrates that the whole set of data is in good accord with the model. ∎

It is helpful to represent the results of the Thiele model in a slightly different form that allows easier estimation of the influence of the diffusion re-

Table 4-6
KINETICS OF SUCROSE INVERSION CATALYZED BY PARTICLES
OF CROSS-LINKED POLY(STYRENESULFONIC ACID) [27]

Catalyst Particle Diameter, mm	Temperature, °C	Observed First-Order Rate Constant, cm³/(equiv. of —SO₃H groups·s)
0.04	50	0.807
0.27	50	0.458
0.55	50	0.277
0.77	50	0.203
0.04	60	2.52
0.27	60	1.21
0.55	60	0.672
0.77	60	0.490
1.09	60	0.688
0.04	70	7.90
0.27	70	2.90
0.55	70	1.46
0.77	70	1.06

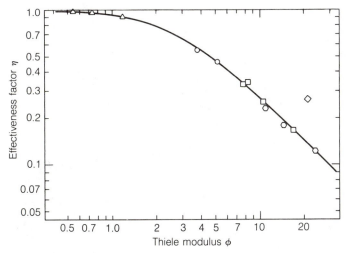

Figure 4-23
Effectiveness factors for sucrose inversion catalyzed by gel-form sulfo-nated poly(styrene-divinylbenzene) at 50–70°C and 1 atm [27]. The reactant was present in aqueous solution, and the polymer was swelled with water and the —SO₃H groups dissociated.

sistance. Define a new dimensionless modulus:

$$\Phi = \phi^2 \eta \qquad (4\text{-}58)$$

With this definition, it is a simple matter to calculate the dependence of η on Φ, as shown in Fig. 4-24. In the limiting cases described above, η as a function of Φ behaves the same way as η as a function of ϕ.

From the definitions of Φ and η it follows that

$$\Phi = \frac{R^2 \times \text{rate}}{DC_s} \qquad (4\text{-}59)$$

where rate is the reaction rate per unit volume of catalyst; this is the rate that can be measured experimentally (not the intrinsic rate except in the special case for which $\eta = 1$). The values of rate, R, and C_s can all be measured straightforwardly; D can also be measured or estimated; therefore, Φ (and hence η) can be estimated. In contrast, the modulus ϕ is not so straightforwardly estimated, because it depends on k, the intrinsic rate constant. Data would have to be obtained for small enough catalyst particles so that $\eta = 1$ (or for a series of particle sizes, so that η and ϕ could be estimated, as in Example 4-11).

The analysis has been done for spherical catalyst beads; it can be done just as easily for cylinders and semi-infinite flat slabs. The results, which are

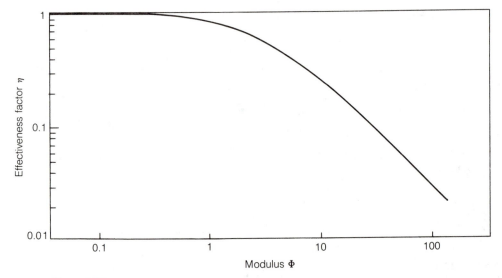

Figure 4-24
Thiele plot, with the effectiveness factor related to the modified Thiele modulus Φ for first-order isothermal reaction.

very similar to those presented for spheres, are given elsewhere [22]. To a good approximation, the results shown for the sphere can be used for particles of arbitrary shape; the length parameter is taken to be that of the equivalent sphere, namely, three times the particle volume divided by the peripheral surface area; for the sphere, this is just the radius R.

The development thus far has been restricted to first-order kinetics, with results cited for zero-order and second-order kinetics. But the kinetics encountered in the examples of this chapter are more complex, representing competitive adsorption and saturation. The Thiele derivation has been extended to account for these more typical forms of rate equations [28, 29]. Results are presented below for one representative case; more results are available in textbooks [22].

Consider kinetics of the form (where r is now again used to designate rate)

$$r = \frac{k K_A K_B C_A C_B}{(1 + K_A C_A + K_B C_B + K_P C_P)^2} \tag{4-60}$$

and where P is a product; the term $K_P C_P$ might be a sum over several products $\Sigma_i K_i C_i$.

The derivation of the η–Φ dependence for the complex rate equation involves computer calculations for the case of flat slab geometry of the catalyst [28]; only partial results are presented here. Several parameters are needed: Φ, E (defined as follows),

$$E = \frac{-D_B C_{Bs}}{v_B D_A C_{As}} - 1 \qquad (4\text{-}61)$$

and KC_{As}, where C_{As} is the concentration of reactant A at the particle periphery and K is defined below; A designates that reactant which gives nonnegative E. The stoichiometric coefficient for A is taken to be -1; v_B is the stoichiometric coefficient of the other reactant (by convention a negative number; the stoichiometric coefficient of a product is positive). The term K is defined as follows:

$$K = \frac{K_A - D_A \Sigma_i K_i v_i / D_i}{\omega} \qquad (4\text{-}62)$$

where

$$\omega = 1 + \Sigma_i K_i \left(C_{is} + \frac{C_{As} v_i D_A}{D_i} \right) \qquad (4\text{-}63)$$

The results of the computations for the case $E = 10$ are shown in Fig. 4-25. The definition of the modulus here is

$$\Phi_L = \frac{L^2 \times \text{rate}}{D C_{As}} \qquad (4\text{-}64)$$

where L is the half-thickness of the slab; for other particle shapes L can be approximated as the ratio of volume to peripheral surface area.

The curves are similar to those for the simple kinetics; however, the complexity of the kinetics leads to some intriguing results, which are different from what was observed with the power law kinetics. Effectiveness factors exceeding unity are predicted for a range of values of Φ when KC_A is 10 or larger. This result follows from the form of the rate equation, Eq. 4-60, which shows that the rate passes through a maximum with increasing C_A. This is a consequence of the competitive adsorption on a fixed number of catalytic groups (i.e., groups that can be saturated). As C_A increases, the fraction of the groups occupied by A (θ_A) increases, but θ_B decreases (B is losing the competition) and the rate is proportional to $\theta_A \theta_B$.

Another point evident from Fig. 4-25 is that there are ranges of variables in which η is multiple valued (dashed lines at top of figure). In these regions, the operation of the catalyst is unstable. There is no one steady-state reaction rate determined by conditions outside the catalyst particle; the value of η observed upon achievement of a steady state depends on the direction from which the steady state is reached, that is, on how the reactor is started up.

In the treatment thus far, the transport of reactant in the catalyst particle has been described as simple diffusion. This is usually a good approximation when the catalyst is a polymer gel, although prediction of the diffusion coefficient may be difficult because it is dependent on the swelling of the polymer

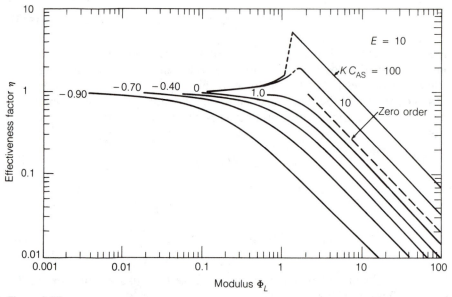

Figure 4-25
Dependence of effectiveness factor on the modulus for saturation kinetics with isothermal reaction in a flat slab. The results are for the case $E = 10$. See text for definitions of the parameters [29].

and the cross-linking. The diffusion coefficient may not be strictly constant (as was assumed in the Thiele model) as a result of these complications.

The Thiele model is now extended to porous catalysts, and the results pertain just as well to most of the porous inorganic solids considered in the following chapters. The key to the analysis is the simple assumption that the transport of reactants in the pores can be described by Fick's law, but instead of a true diffusion coefficient, an effective diffusion coefficient D_{eff} is used. The diffusion is described with equations which are the same as those used for the gel, but it actually occurs only in the pores. There are well-established methods for measuring D_{eff} for transport through porous particles [30]. D_{eff} can also be predicted, often quite accurately, provided that the pore structure is well characterized. The prediction involves, first, estimation of the molecular diffusion coefficient of the reactant in the pores and, second, estimation of the Knudsen diffusion coefficient in the pores. Molecular diffusion predominates when the collisions of molecules with other molecules are much more frequent than the collisions with pore walls. Knudsen diffusion predominates in the opposite case. There are well-established methods for predicting the molecular diffusion coefficient on the basis of the kinetic theory of gases or empirical correlations for liquids [22]. Similarly, Knudsen diffusion coefficients can be predicted from the kinetic theory of gases on the basis of molecular properties and pore geometry [22]. For gases, the molecular diffusion coefficient is inversely proportional to pressure and does not depend on the pore size; the Knudsen diffusion coefficient is independent of pressure and proportional to pore diameter. Recall that a constant diffusion coefficient has been assumed in the analysis.

In straight uniform pores, the resistances to diffusion are additive, and the value of D_{eff} is estimated from the equation

$$\frac{1}{D_{eff}} = \frac{1}{D_{mol}} + \frac{1}{D_K} \tag{4-65}$$

where D_{mol} is the molecular diffusion coefficient (now written with a subscript for bookkeeping purposes) and D_K is the Knudsen diffusion coefficient.

The pores in a catalyst are not straight and cylindrical; rather, they are a labyrinth offering a tortuous path for a diffusing molecule. The effective diffusion coefficients in a real porous catalyst are represented empirically as follows:

$$D_{K\ eff} = \frac{D_K \epsilon}{\tau} \tag{4-66}$$

$$D_{mol\ eff} = \frac{D_{mol} \epsilon}{\tau} \tag{4-67}$$

The void fraction ϵ of the particle has been introduced so that the effective diffusion coefficients describe transport through the total cross section of the porous solid. The parameter τ, the *tortuosity*, accounts for the nonuniformity of the pores; values (determined empirically) are typically between 2 and 8. Calculation of the overall effective diffusivity D_{eff} from $D_{K\ eff}$ and $D_{mol\ eff}$ for the porous catalyst is based on the additivity of the reciprocals of $D_{K\ eff}$ and $D_{mol\ eff}$.

In the foregoing discussion of the estimation of the effective diffusion coefficient, it was implicitly assumed that the transport resulted from the random motion of molecules in the gas or liquid phase filling the catalyst pores. Sometimes matters are not so simple. One complication involves diffusion of molecules on the surface; this is most likely to be important when there are high concentrations of weakly adsorbed molecules. Another complication characteristic of the polymer catalysts specifically follows from the fact that the microspheres (Fig. 4-17) are gel-form polymer, through which reactants and products can diffuse. If the cross-linking of the microspheres is high (≥ 20 percent), then they may be almost impermeable to most reactants, and the catalytic reaction will take place on the microsphere surface, that is, on the internal surface of the porous particle. But if the cross-linking is low, then the reactants may be transported into the microparticles. Analysis of this problem requires accounting for the diffusion in the pores with the *macro effectiveness* factor and for the diffusion in the gelular microspheres with the *micro effectiveness* factor [31].

Significant mass transport limitations within catalyst particles influence the selectivity in catalytic conversions when reactions occur in parallel or in series [22]. For example, in the series reaction $A \rightarrow B \rightarrow C$, an analysis of the reaction/transport processes as in the development of the Thiele model, for the case of two isothermal first-order irreversible reactions occurring in the absence of significant volume change, leads to the results shown in Fig. 4-26. When the effectiveness factor is < 0.3, the conversion to the intermediate B is less than that obtainable in the absence of a significant transport restriction. The result is easy to understand qualitatively: As the resistance to transport of B increases, B is more likely to be converted to C, rather than to be transported from the catalyst particle to the external phase. The consequences are important for a number of reactions, such as the sequential oxidations discussed in Chapter 6, whereby the desired product is an intermediate (a partially oxidized product such as ethylene oxide) and not the total oxidation product CO_2.

4.8.2 The Influence of Intraparticle Heat Transport

In the preceding section effects of mass transport resistance within isothermal catalyst particles were considered. Now consider nonisothermal particles and

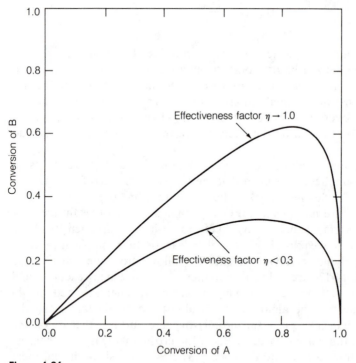

Figure 4-26
Effect of intraparticle transport limitations on the selectivity of a catalyst for the reaction sequence

$$A \xrightarrow{k_1} B \xrightarrow{k_2} C$$

with $k_1/k_2 = 4$. For details, see ref. 22b.

the effects of heat transfer resistance in the particles. The Thiele model was derived by writing an equation for the conservation of mass and describing the mass transport with Fick's law. The model can be extended to incorporate nonisothermal catalyst performance by use of a simple expression for heat conduction (Fourier's law, which is mathematically analogous to Fick's law); an equation for conservation of energy is needed to complement the equation for conservation of mass. The two differential equations are coupled and do not yield an easy analytical solution, but they have been solved numerically for simple kinetics [32].

Results are shown in Fig. 4-27 for a first-order irreversible reaction in a spherical catalyst particle. As before, the diffusion coefficient D is assumed to be constant. The effectiveness factor is shown as a function of the modulus Φ; it depends also on two other dimensionless groups:

$$\gamma = \frac{E_{act}}{RT_s} \tag{4-68}$$

where E_{act} is the intrinsic activation energy for the catalytic reaction, R is the

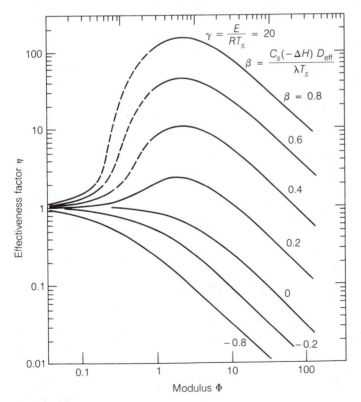

Figure 4-27
Effectiveness factors for first-order catalytic reaction in a sphere: the non-isothermal case. The curves pertain to a value of $\gamma = 20$ [32].

gas constant, and T_s is the absolute temperature at the peripheral surface of the particle, and on

$$\beta = \frac{C_s(-\Delta H_r)D}{\lambda T_s} \qquad (4\text{-}69)$$

In this equation, ΔH_r is the enthalpy of reaction (negative for an exothermic reaction), and λ is the thermal conductivity of the catalyst particle; λ is based on the total cross section of a porous particle, the value being less than that of the nonporous solid itself. For a porous catalyst particle, D_{eff} is used instead of D. The curves shown in Fig. 4-27 are for $\gamma = 20$.

Physically, it is evident how these two parameters arise in the analysis; γ is a measure of the temperature sensitivity of the reaction rate, and β represents the maximum temperature difference that can exist in the catalyst particle relative to the temperature at the peripheral surface, $(T - T_s)_{max}/T_s$.

The effectiveness factor is taken to be unity when both the concentration and temperature gradients are negligible, that is, when $C = C_s$ and $T = T_s$ throughout the particle. The curves in Fig. 4-27 show that values of η greatly exceeding unity can be achieved. This situation is obviously encountered only for exothermic reactions; when the temperature in the interior of the particle exceeds that at the periphery, the rate of reaction at any point in the interior (which is exponentially dependent on temperature) may be much higher than that at the periphery, even though the reactant concentration may be lower than at the periphery. Figure 4-27 shows that η can be multivalued at large values of β and that instability phenomena occur at these large values.

4.9
APPLICATIONS OF POLYMER CATALYSTS

Strong-acid polymers are applied as catalysts in a number of industrial processes, including esterification, phenol alkylation, propylene hydration, bisphenol A synthesis, and methyl t-butyl ether synthesis. The primary advantages of the solid catalysts are the ease of separation from products, lack of corrosion, and lack of soluble acid disposal problems. Virtually all the polymers applied as catalysts are porous cross-linked poly(styrenesulfonic acid). The greater stability of highly cross-linked porous polymers accounts for their dominance over the gel-form polymers.

These catalysts are efficient and have long lifetimes, but there are several disadvantages. They are not resistant to attrition and abrasion; they are therefore typically used in fixed-bed and not stirred reactors. The polymers are subject to deactivation by soluble cations (such as tramp iron) in feed streams, as a result of ion exchange. They constantly slough off low-molecular-weight fragments that contaminate and discolor products of catalytic reactions. In the presence of water, they undergo desulfonation. The lack of stability of the polymer at temperatures greater than about 130°C restricts the applications. Much more robust solid acids, such as zeolites (Chapter 5), are used as catalysts under conditions much more severe than the polymers can withstand; from the technological viewpoint, these inorganic catalysts are far more useful than the polymers.

Polymer-supported transition metal complex catalysts also lose activity by loss of catalytic groups [33, 34]. The deactivation of a hydroformylation catalyst consisting of rhodium complexes attached to phosphine groups in cross-linked polystyrene resulted from this metal loss [33]. Redistribution of the metal from the upstream toward the downstream end of the flow reactor, and out of the reactor, took place as a result of ligand exchange. The leaching of the metal was perhaps critical in preventing industrial application of this catalyst.

EXAMPLE 4-12 Conception of a Process for Bisphenol A Synthesis

Problem

Suggest a flow sheet for bisphenol A synthesis with the sulfonic acid resin catalyst. Identify some of the issues of process engineering.

Solution

The porous poly(styrenesulfonic acid) catalyst can be used in a fixed-bed flow reactor with a liquid feed of phenol and acetone. The product stream

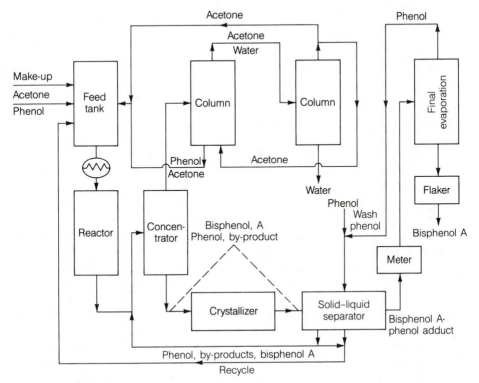

Figure 4-28
Process flow diagram for bisphenol synthesis from phenol and acetone catalyzed by particles of cross-linked poly(styrenesulfonic acid).

can be purified by distillation, with the unconverted reactant recycled to the reactor. It might be best to operate the reactor at relatively low conversions, because one of the products (water) is a strong reaction inhibitor (Eq. 4-32). The reaction temperature would be chosen to represent a trade-off between the rate of the catalytic reaction and that of catalyst deactivation; it might be slightly greater than 100°C. A process flow diagram adapted from the literature [35] is shown in Fig. 4-28.   ■

4.10
EXTRAPARTICLE TRANSPORT INFLUENCE

There may be a significant resistance to transport of reactants in the fluid phase surrounding a catalyst particle. This possibility has been ignored thus far, and usually the resistance to transport in a gas or liquid is much less than that within a gel or a small-pored solid. But for highly active catalysts, the reaction rate may be determined by how fast the reactants get through the fluid phase to the surface.

An analysis of the extraparticle mass transport with reactants in the gas or liquid phase is based on the idea that the rate is characterized by a single parameter, called a *mass transfer coefficient*, defined for a liquid as k_L and for a gas as k_G, and a concentration driving force:

$$r = k_L a(C_B - C_s) \tag{4-70}$$

or

$$r = k_G a(C_B - C_s) \tag{4-71}$$

where k_L and k_G have units, for example, of cm/s, and a has units of cm^2/cm^3. In each equation, the rate of transport of a particular species from the fluid phase to a particle surface (periphery) is assumed to be proportional to the mass transfer coefficient, to the peripheral surface area, and to the driving force. This driving force is the average bulk fluid concentration of the species being transported minus the concentration at the surface of the particle to which it is transported. If the particle were an impermeable and nonporous catalyst with infinite catalytic activity, then the reactant would be converted instantaneously upon reaching the surface, and C_s would be zero.

The transport of molecules in a fluid phase flowing over a fixed bed of particles or around suspended particles in a stirred-tank reactor is complicated, involving convection and diffusion. The fluid develops a thin boundary layer near the surface; this is a region in which there is a sharp velocity gradient over a short distance (typically 1 mm) perpendicular to the flow. The faster the flow of the fluid, the thinner is this boundary layer. The fluid velocity is zero at the surface. Mixing of fluid elements in the bulk is usually relatively rapid [unless the flow is slow (laminar)]; but in the laminar boundary layer there is little mixing, and the transport through the layer is by molecular diffusion.

In the bulk fluid, the transport is independent of diffusion. But very near the surface, the transport rate is proportional to the concentration gradient and the molecular diffusion coefficient. Experiments have shown that for many situations the overall rate of transport from the bulk to the surface is proportional to some fractional power of this molecular diffusion coefficient.

This result forms part of the basis for prediction of the mass transfer coefficient. Physical reasoning and dimensional analysis, beyond the scope of this book, lead to the expectation of generalized correlations involving dimensionless groups:

$$\frac{k_L d_P}{D} = f(N_{Re}, N_{Sc}) \tag{4-72}$$

where d_P is a dimension representative of the flow passage (taken to be the spherical particle diameter or the diameter of the equivalent sphere) and D is the molecular diffusion coefficient for the diffusing species in the fluid.

The dimensionless Reynolds number N_{Re} is defined as

$$N_{Re} = \frac{d_P G}{\mu} \tag{4-73}$$

and the Schmidt number N_{Sc} as

$$N_{Sc} = \frac{\mu}{\rho D} \tag{4-74}$$

where μ is the fluid viscosity, ρ is the fluid density, and G is the mass flow rate of the fluid [e.g., in units of g/(s·cm² of total bed cross section)].

There are several empirical correlations, all more or less the same and based on the relationship originally suggested by Chilton and Colburn, two of the pioneers of chemical engineering:

$$j_D = \frac{k_L \rho}{G} N_{Sc}^{2/3} = f(N_{Re}) \tag{4-75}$$

and (for an ideal gas)

$$j_D = \frac{k_G C_t}{G_M} N_{Sc}^{2/3} = f(N_{Re}) \tag{4-76}$$

where C_t is the total concentration (e.g., mol/cm³) and G_M is the flow rate in moles of mixture per unit time per unit of total bed cross-sectional area. For convenience, the term on the left is usually referred to as j_D, the "j factor." Many data fall near the same line when j_D is plotted against N_{Re}; a better correlation is found when another dimensionless term, the void fraction of the fixed bed (or "porosity") is included (this is the fraction of the bed that is not filled with solid particles, not to be confused with the porosity of the individual particles).

A recommended correlation for gases with $3 < N_{Re} < 2000$ is as follows [22, 36]:

$$\epsilon j_D = \frac{0.357}{N_{Re}^{0.359}} \qquad (4\text{-}77)$$

Recommended correlations for liquids are [22, 37], for $55 < N_{Re} < 1500$:

$$\epsilon j_D = \frac{0.250}{N_{Re}^{0.31}} \qquad (4\text{-}78)$$

and for $0.0016 < N_{Re} < 55$:

$$\epsilon j_D = \frac{1.09}{N_{Re}^{2/3}} \qquad (4\text{-}79)$$

A development similar to that given here allows estimates of temperature differences between the particle surface and the bulk fluid. A heat transfer coefficient h is defined similarly to the mass transfer coefficient:

$$r = ha(T_B - T_s) \qquad (4\text{-}80)$$

where T_B and T_s are the fluid temperatures of the bulk (an average) and at the solid surface, respectively. Correlations similar to those stated above allow estimates of h.

EXAMPLE 4-13   Extraparticle Mass Transport Influence for a Solid-Catalyzed Reaction

Problem
  Determine whether mass transport influenced the data of Fig. 4-8 for ethanol dehydration catalyzed by spherical beads of gel form sulfonic acid resin in a nearly isothermal fixed-bed reactor.

Solution
  Data are presented in Fig. 4-8 for sieved catalyst particles in two size ranges; the reaction rate is clearly not a function of catalyst particle size in this range. Therefore, the effectiveness factor is unity; intraparticle transport resistance was negligible.
  This result suggests that the gas-phase mass transfer resistance was negligible, since the resistance to transport in the solid is greater than that external to the solid. To check this, the correlation of Eq. 4-77 is used; the analysis is done for the simple limiting case for which the value of C_s is assumed to be zero. In this instance, the gas-phase transport would be

rate determining—this situation is comparable to that in which an elementary step in a catalytic cycle is rate determining. The resistance to the catalytic process is all in the gas phase (restricting the transport) in the one case, whereas it is all in the chemistry of the single slow step in the other.

A statement of the conservation of mass is used; a mass balance equation for the reactant over a differential length of the reactor is

$$r \, dz = -G_M \, dY \tag{4-81}$$

where r is the catalytic reaction rate per unit volume, z is the bed length, and Y is the mole fraction of reactant ethanol in the gas stream.

Using the equation that defines the mass transfer coefficient (Eq. 4-71) and the assumption that C_s is zero gives

$$r = k_G a C_A = k_G a Y C_t \tag{4-82}$$

where C_A is the alcohol concentration and a is the catalyst particle surface area (peripheral surface area) in units of cm^2/cm^3 of reactor volume.

Combining the preceding two equations and integrating gives

$$\ln \frac{Y_1}{Y_2} = \frac{k_G a C_t z}{G_M} \tag{4-83}$$

where Y_1 and Y_2 are the mole fractions of reactant in the feed and product, respectively, for a bed length z.

Substituting in the correlation of Eq. 4-77 gives

$$\ln \frac{Y_1}{Y_2} = \frac{0.357 a z}{\epsilon N_{Re}^{0.359} N_{Sc}^{2/3}} \tag{4-84}$$

To carry out the integration, it was assumed that the physical properties of the fluid are constant over the length of the reactor; since the conversion was only about 10 percent, the assumption is good.

An evaluation of the results for the data of Fig. 4-8 requires the bed length (about 8 cm), ϵ (about 0.4, typical for a bed of spheres), a (roughly 30 cm^2/cm^3 of bed volume; the value is different for the different catalyst particle sizes), and the values of N_{Re} and N_{Sc}.

From estimates of the physical properties,

$$N_{Sc} = \frac{\mu}{\rho D} = \frac{1.14 \times 10^{-4} \text{ g/(s·cm)}}{(1.43 \times 10^{-3} \text{ g/cm}^3)(1.06 \times 10^{-1} \text{ cm}^2/\text{s})} \tag{4-85}$$

$$= 0.75 \tag{4-86}$$

N_{Re} is estimated from the viscosity, an approximate average value of d_P (1 mm), and an approximate average value of G [0.9×10^{-3} g ethanol/(cm^2·s), for an observed conversion of 10 percent]:

$$N_{Re} \cong 0.8 \tag{4-87}$$

This is less than the lower limit for which Eq. 4-77 is recommended, so that the result will not be exact.

Using these values in Eq. 4-84 gives a value of Y_1/Y_2 that indicates that more than 99.9 percent of the reactant would be converted if external mass transfer determined the rate. (This estimate is based on the assumption that the reaction is irreversible.) Since the observed conversion is only 10 percent, it follows that the rate was not influenced by external mass transfer. ■

If external-phase mass transport is rate determining, then the temperature dependence of the rate is similar to the temperature dependence of a molecular diffusion coefficient, with an activation energy of perhaps a few kcal/mol. The activation energy of a catalytic reaction is then a diagnostic for the role of external-phase mass transport. If the activation energy is significantly more than a few kilocalories per mole, then the external mass transport resistance is negligible. Another diagnostic is a measurement of whether a change in mass velocity (G) (or linear velocity) at a given space velocity changes the rate; the mass transfer coefficient increases with the mass velocity through the bed because the mixing is greater. If the observed rate at a given space velocity is independent of G, the rate is not determined by external-phase mass transport.

There is an important distinction between inter- and intraparticle transport resistances: the former cannot be rate determining but the latter can. This point is clarified by an electrical circuit analogy—the external phase mass transport of the reactant, the adsorption of reactant, the steps in the catalytic cycle, the desorption of the product, and the external transport of the product take place in sequence; their resistances are in series. If one resistance in a series circuit is much larger than all the others, then, to a good approximation, the current (analogous to the reaction rate) is determined by the one large resistance for a given applied voltage (analogous to the reactant concentration). The intraparticle transport, on the other hand, is characterized by a resistance in parallel with those characterized by the adsorption, desorption, and chemical reaction steps. Therefore, the rate cannot be described by this resistance alone (as is shown by the results of the Thiele model). The rate of reaction in the interior of a catalyst particle is influenced by the intraparticle transport if $\eta < 1$, but the rate of reaction at the exterior surface cannot be influenced by the resistance to transport within the particle.

REFERENCES

1. Frilette, V. J., Mower, E. B., and Rubin, M. K., *J. Catal.*, **3**, 25 (1964).
2. Gates, B. C., and Rodriguez, W., *J. Catal.*, **31**, 27 (1973).

3. Schwab, G.-M., *Catalysis from the Standpoint of Chemical Kinetics* (translated by H. S. Taylor and R. Spence), Van Nostrand, New York, 1937.

4. Scott, J. P., Ph.D. thesis, University of Delaware, 1985.

5. Kabel, R. L., and Johanson, L. N., *AIChE J.,* **8,** 621 (1962).

6. Gates, B. C., and Johanson, L. N., *J. Catal.,* **14,** 69 (1969).

7. Mullarkey, T. B., M. S. thesis, Pennsylvania State University, 1970.

8. Gates, B. C., Ph. D. thesis, University of Washington, 1966.

9. Thornton, R., and Gates, B. C., *Proc. 5th Int. Congr. Catal.,* (Palm Beach) Vol. 1, p. 357, 1973.

10. Gates, B. C., Wisnouskas, J. S., and Heath, H. W., Jr., *J. Catal.,* **24,** 320 (1972).

11. Wesley, R. B., and Gates, B. C., *J. Catal.,* **34,** 288 (1974).

12. Bonds, W. D., Jr., Brubaker, C. H., Jr., Chandrasekaran, E. S., Gibbons, C., Grubbs, R. H., and Kroll, L. C., *J. Am. Chem. Soc.,* **97,** 2128 (1975).

13. Affrossman, S., and Murray, J. P., *J. Chem. Soc. B,* **1966,** 1015.

14. Regen, S. L., Besse, J. L., and McLick, J., *J. Am. Chem. Soc.,* **101,** 116 (1979).

15. Batchelder, R. F., Gates, B. C., and Kuijpers, F. P. J., *Proc. 6th Int. Congr. Catal.* (London), Vol. 1, p. 499, 1977.

16. Reinicker, R. A., and Gates, B. C., *AIChE J.,* **20,** 933 (1974).

17. Gammill, B. B., Ladewig, G. R., and Ham, G. E., U. S. Patent 3,634,341 (1972).

18. Arai, H., and Yashiro, M., *J. Mol. Catal.,* **3,** 427 (1978).

19. Webber, K. M., Gates, B. C., and Drenth, W., *J. Mol. Catal.,* **3,** 1 (1977/78).

20. Haag, W. O., *Chem. Eng. Prog. Symp. Ser.,* **63,** No. 73, 140 (1967).

21. Weisz, P. B., *Science,* **179,** 433 (1973).

22. (a) Satterfield, C. N., *Mass Transfer in Heterogeneous Catalysis,* M.I.T. Press, Cambridge, MA, 1970.
    (b) Wheeler, A., *Adv. Catal.,* **3,** 249 (1951).

23. Thiele, E. W., *Ind. Eng. Chem.,* **31,** 916 (1939).

24. Damköhler, G., *Chem. Ing.,* **3,** 430 (1937).

25. Zeldovitch, Ya. B., *Acta Physicochim. U.S.S.R.,* **10,** 583 (1939).

26. Froment, G. F., and Bischoff, K. B., *Chemical Reactor Analysis and Design,* 2nd Edition, Wiley, New York, 1990.

27. Gilliland, E. R., Bixler, H. J., and O'Connell, J. E., *Ind. Eng. Chem. Fundam.,* **10,** 185 (1971).

28. Roberts, G. W., and Satterfield, C. N., *Ind. Eng. Chem. Fundam.,* **4,** 288 (1965).

29. Roberts, G. W., and Satterfield, C. N., *Ind. Eng. Chem. Fundam.,* **5,** 317 (1966).

30. Baiker, A., New, M., and Richarz, W., *Chem. Eng. Sci.,* **37,** 643 (1982).

31. Dooley, K. M., Williams, J. A., Gates, B. C., and Albright, R. L., *J. Catal.,* **74,** 361 (1982).

32. Weisz, P. B., and Hicks, J. S., *Chem. Eng. Sci.,* **17,** 265 (1962).
33. Lang, W. H., Jurewicz, A. T., Haag, W. O., Whitehurst, D. D., and Rollmann, L. D., *J. Organomet. Chem.,* **134,** 85 (1977).
34. Garrou, P. E., and Gates, B. C., in *Synthesis and Separations Using Functional Polymers*, P. Hodge and D. C. Sherrington, eds., Wiley, Chichester, 1988, p. 123.
35. Konrad, F., in *Benzene and its Industrial Derivatives*, Ernest Benn, London, 1973.
36. Petrovic, L. J., and Thodos, G., *Ind. Eng. Chem. Fundam.,* **7,** 274 (1968).
37. Wilson, E. J., and Giankopolis, C. J., *Ind. Eng. Chem. Fundam.,* **5,** 9 (1966).

FURTHER READING

A detailed treatment of polymer catalysis is given in the monograph edited by P. Hodge and D. C. Sherrington, *Synthesis and Separations Using Functional Polymers*, Wiley, Chichester, 1988. A guide to the synthesis of polymer-supported catalysts was published by Lieto et al., *CHEMTECH,* **13,** 46 (1983). The concepts of competitive diffusion/reaction phenomena, which extend far beyond the examples here, are set out in a review by Weisz [21]. Satterfield's text [22] is a how-to guide to evaluating the influence of transport phenomena in catalysis by solids; Thiele published an account of the early history [*Am. Scientist,* **55**(2), 176 (1967)]. The ideas of chemical reaction engineering introduced here are elaborated in K. G. Denbigh and J. R. C. Turner's *Chemical Reactor Theory*, Cambridge University Press, 1971, and Froment and Bischoff's advanced text [26].

PROBLEMS

4-1 A solid polymer incorporating the following groups

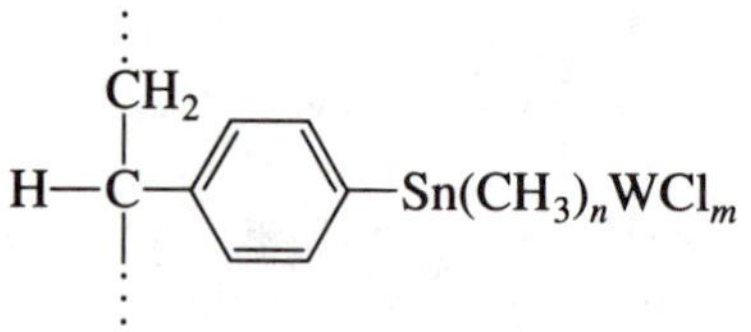

$$H\text{---}\underset{|}{\overset{\overset{\displaystyle CH_2}{|}}{C}}\text{---}\langle\!\!\!\rangle\text{---}Sn(CH_3)_n WCl_m$$

(where n and m are unknown) catalyzes olefin metathesis. WCl_6 in solution is catalytically inactive for the reaction, but WCl_6 in solution with R_4Sn (where R is alkyl) is active. Explain the role of Sn in the polymer catalyst.

4-2 Examine the initial rate data in Table 4-7 for the hydrogenation of ethylene,

$$CH_2\!\!=\!\!CH_2 + H_2 \rightarrow CH_3\text{---}CH_3 \tag{4-48}$$

catalyzed by a polymer-supported metal cluster, $\circledP\text{---}PPh_2Ir_4(CO)_{11}$,

Table 4-7
KINETICS OF ETHYLENE HYDROGENATION IN THE PRESENCE
OF SUPPORTED Ir_4 CARBONYL

Flow Rate, cm^3/s							
H_2	C_2H_4	He	Conversion, %	$10 \times r^a$	P_{H_2}, atm	$P_{C_2H_4}$, atm	T, °C
0.20	0.20	0.65	0.27	0.78	0.199	1.99	40
0.42	0.20	0.45	0.039	0.12	0.411	0.196	40
0.60	0.20	0.24	0.057	1.65	0.605	0.201	40
0.71	0.20	0.13	0.066	1.90	0.715	0.201	40
0.84	0.20	0.00	0.074	2.13	0.846	0.201	40
0.27	1.23	0.00	0.0079	1.41	0.191	0.872	40
0.27	1.00	0.23	0.0086	1.25	0.192	0.711	40
0.27	0.80	0.46	0.0091	1.06	0.203	0.604	40
0.27	0.44	0.81	0.010	0.641	0.191	0.311	40
0.20	0.020	0.65	0.027	0.78	0.199	1.99	53
0.42	0.20	0.45	0.053	1.53	0.414	0.195	53
0.60	0.20	0.24	0.075	2.18	0.60	0.20	53
0.71	0.20	0.13	0.082	2.36	0.71	0.20	53
0.84	0.20	0.00	0.010	2.48	0.84	0.20	53
0.27	1.23	0.00	0.010	1.788	0.19	0.87	53
0.27	1.00	0.23	0.010	1.454	0.19	0.71	53
0.27	0.80	0.46	0.010	1.16	0.20	0.60	53
0.27	0.44	0.81	0.011	0.898	0.19	0.31	53
0.84	0.20	0.00	0.106	3.05	0.20	0.20	64
0.71	0.20	0.13	0.095	2.14	0.20	0.20	64
0.60	0.20	0.24	0.082	2.37	0.20	0.20	64
0.50	0.20	0.35	0.072	2.07	0.20	0.20	64
0.42	0.20	0.45	0.056	1.63	0.19	0.19	64
0.34	0.20	0.51	0.048	1.40	0.20	0.20	64
0.20	0.20	0.65	0.029	0.84	0.20	0.20	64
0.27	1.23	0.00	0.011	1.96	0.87	0.87	64
0.27	1.00	0.23	0.011	1.60	0.71	0.71	64
0.27	0.80	0.46	0.04	1.28	0.60	0.60	64
0.27	0.44	0.81	0.014	0.70	0.31	0.31	64

^{a} Units: molecules/(Ir atom·s).

used in an isothermal flow reactor (Lieto, J., Ph.D. thesis, University of Delaware, 1979). Represent the data with rate equations of the power law form ($r = kP_{CH_2=CH_2}^x P_{H_2}^y$) and also the Langmuir–Hinshelwood form. Before the quantitative work, provide a statement of the qualitative conclusions derived from plots of the data. Were the conversions differential?

4-3 A macroporous sulfonic acid resin has been used to catalyze the reaction of benzene and propylene to give isopropylbenzene at 55°C and 1 atm [11]. Use the data in Table 4-8 to determine a rate equation; interpret the equation.

4-4 Superacids in solution are so corrosive and difficult to handle that there is little prospect of their practical application as catalysts, but the patent

Table 4-8
KINETICS OF BENZENE PROPYLATION

$10^8 \times$ Rate, mol/(equiv. of —SO$_3$H groups·s)	$C_{benzene}$, mol/L	$C_{propylene}$, mol/L	$C_{n\text{-heptane}}$, mol/L
1.48	14.49	0.224	—
0.944	12.56	0.267	1.98
0.833	11.54	0.277	3.00
0.420	8.16	0.309	6.45
0.175	4.31	0.474	10.22
0.140	1.97	0.591	12.58
1.66	14.55	0.206	—
1.75	14.56	0.067	—
1.54	14.55	0.024	—
0.820	14.52	0.006	—
0.480	14.49	0.005	—

literature reports numerous possibilities using supported superacids as alternatives. How would you prepare such solid superacids in the form of solid polymers?

4-5 Concentration profiles are shown in Fig. 4-20 for reactants and products in catalyst particles in the case of a first-order, isothermal reaction. (a) Sketch the profiles qualitatively for the case of a zero-order reaction. (b) Derive the Thiele model for the zero-order reaction and compute the profiles.

4-6 The polymer referred to by the trade name Nafion® is a fluorocarbon ether containing —CF$_2$CF$_2$OSO$_3$H groups. This polymer is reported to be active for the isobutane–propylene alkylation, whereas the sulfonated poly(styrene-divinylbenzene) is not [McClure, J. D., U.S. Patent 4,022,847 (1977)]. Explain the difference in activities of the two polymer catalysts. Qualitatively, what difference would you expect in the activities of the two polymers for the benzene–propylene alkylation?

4-7 Consider the ethanol dehydration kinetics data of Example 4-4. Estimate the maximum temperature at the center of a catalyst particle. Is this significantly different from the nominal reaction temperature of 120°C? Assume an apparent activation energy of 20 kcal/mol for the catalytic reaction.

4-8 In an investigation of the dehydration of isobutyl alcohol catalyzed by sulfonated poly(styrene-divinylbenzene) at low partial pressures of alcohol, Thornton and Gates [*J. Catal.*, **34**, 275 (1974)] observed both isobutylene and isobutane as products. Formation of isobutane was accompanied by the formation of polymeric product (tar) deactivating the catalyst. At high alcohol partial pressures, the paraffin formation was suppressed, but the olefin formation was not. What do these results imply about the acidity of the catalyst? Suggest mechanisms for formation of the olefin and paraffin products.

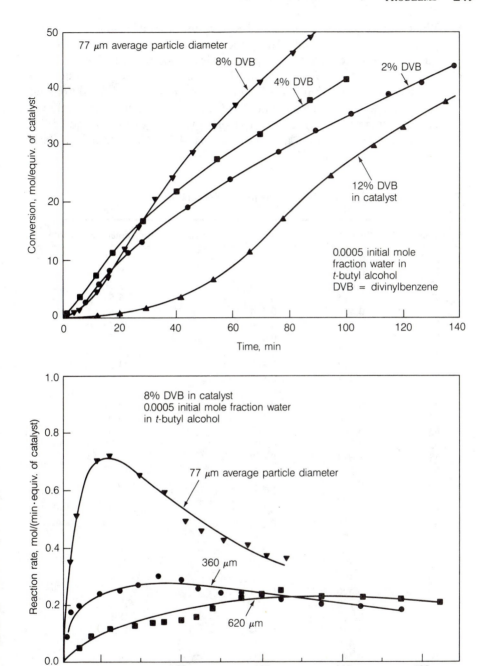

Figure 4-29
Conversions of *t*-butyl alcohol at 75.6°C in the presence of 200–400 mesh particles of gel form sulfonic acid resin catalysts. DVB is the cross-linking agent divinylbenzene.

4-9   What side products are expected in the synthesis of bisphenol A from phenol and acetone catalyzed by the sulfonated polymer? What consequences do these have for the process design sketched in Fig. 4-28?

4-10   Heath and Gates [*AIChE J.*, **18**, 321 (1972)] measured conversions in the dehydration of *t*-butyl alcohol in the presence of beads of sulfonated poly(styrene-divinylbenzene) at 75.6°C and 1 atm. Dehydrated catalyst particles were transferred directly from a vacuum oven into the liquid-phase reactant. The data are shown in Fig. 4-29. The increasing rate is

Table 4-9
KINETICS OF ETHYLENE HYDROGENATION IN THE PRESENCE
OF SUPPORTED RuPt₂ CARBONYL CLUSTERS

Temperature, °C	P_{H_2}, atm	$P_{CH_2=CH_2}$, atm	$10^2 \times$ Rate, molecules/(cluster·s)
73	0.859	0.192	8.51
73	0.732	0.196	8.15
73	0.618	0.196	7.38
73	0.514	0.195	6.55
73	0.429	0.194	5.88
73	0.351	0.196	5.23
73	0.207	0.196	3.73
73	0.189	0.885	4.83
73	0.212	0.862	4.99
73	0.207	0.691	5.69
73	0.184	0.682	5.16
73	0.189	0.568	5.04
73	0.188	0.326	5.02
83	0.864	0.187	15.05
83	0.729	0.195	14.28
83	0.614	0.198	13.01
83	0.518	0.197	11.97
83	0.427	0.193	10.72
83	0.353	0.197	9.15
83	0.219	0.201	6.95
83	0.175	0.890	8.00
83	0.194	0.717	9.87
83	0.187	0.561	9.66
83	0.195	0.440	8.29
83	0.191	0.318	8.18
98	0.866	0.181	20.9
98	0.722	0.193	19.8
98	0.617	0.195	18.6
98	0.510	0.194	16.6
98	0.426	0.193	15.2
98	0.345	0.193	14.4
98	0.207	0.197	10.6
98	0.195	0.883	9.39
98	0.192	0.709	10.47
98	0.187	0.562	11.31
98	0.193	0.436	11.53
98	0.191	0.319	10.47

Table 4-10
KINETICS OF ETHYLENE HYDROGENATION IN THE PRESENCE
OF SUPPORTED AuOs$_3$ CARBONYL CLUSTERS

Temp., °C	P_{H_2}, atm	$P_{C_2H_4}$, atm	P_{He}, atm	Conversion, %	$10^3 \times$ Rate, molecules/ (AuOs$_3$ cluster·s)
73	0.931	0.104	—	0.0095	2.32
73	0.818	0.104	0.113	0.0085	2.08
73	0.708	0.103	0.199	0.0070	1.71
73	0.623	0.104	0.312	0.0057	1.39
73	0.526	0.103	0.412	0.0053	1.29
73	0.411	0.105	0.527	0.0040	0.97
73	0.310	0.104	0.632	0.0032	0.77
73	0.230	0.105	0.711	0.0024	0.59
73	0.106	0.106	0.835	0.0008	0.20
73	0.935	0.104	—	0.0095	2.32
73	0.835	0.204	—	0.0080	3.71
73	0.646	0.387	—	0.0066	4.81
73	0.526	0.515	—	0.0040	4.77
73	0.406	0.635	—	0.0031	4.61
73	0.328	0.712	—	0.0029	4.59
73	0.166	0.873	—	0.0011	2.03
83	0.940	0.105	—	0.0189	4.61
83	0.828	0.105	0.116	0.0167	4.06
83	0.717	0.104	0.229	0.0139	3.38
83	0.631	0.105	0.317	0.0121	2.94
83	0.533	0.105	0.418	0.0098	2.39
83	0.416	0.107	0.534	0.0081	1.97
83	0.233	0.106	0.722	0.0045	1.09
83	0.107	0.107	0.848	0.0019	0.47
83	0.941	0.105	—	0.0189	4.61
83	0.831	0.202	—	0.0132	6.12
83	0.643	0.386	—	0.0111	8.10
83	0.523	0.512	—	0.0065	7.76
83	0.404	0.631	—	0.0049	7.29
83	0.311	0.726	—	0.0038	4.77
83	0.200	0.834	—	0.0018	3.33
83	0.104	0.934	—	0.0008	1.74
92	0.941	0.104	—	0.0310	7.55
92	0.831	0.105	0.116	0.0259	6.30
92	0.684	0.105	0.264	0.0225	5.48
92	0.533	0.105	0.419	0.0178	4.33
92	0.417	0.107	0.535	0.0149	3.62
92	0.316	0.105	0.642	0.0115	2.80
92	0.233	0.106	0.724	0.0078	1.89
92	0.108	0.108	0.849	0.0032	0.77
92	0.940	0.104	—	0.0310	7.55
92	0.841	0.205	—	0.0235	10.89
92	0.650	0.391	—	0.0208	15.20
92	0.530	0.519	—	0.0133	15.87
92	0.409	0.640	—	0.0090	13.39
92	0.315	0.734	—	0.0064	10.91
92	0.203	0.844	—	0.0047	8.70
92	0.105	0.944	—	0.0015	3.29

suggestive of autocatalysis, but the dependence on particle size and catalyst cross-linking is indicative of a physical effect. Explain the data.

4-11 The kinetics data summarized in Table 4-9 were obtained for the hydrogenation of ethylene in the presence of a RuPt$_2$ carbonyl cluster bound to a phosphine-functionalized poly(styrene-divinylbenzene) (McQuade, K. J., M.Ch.E. thesis, University of Delaware, 1980). Determine a rate equation. What inferences can you draw from the form of the equation?

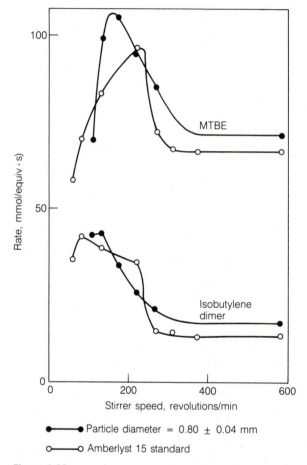

Particle diameter = 0.80 ± 0.04 mm

Amberlyst 15 standard

Figure 4-30
Effects of stirrer speed and catalyst particle size on the rate of synthesis of methyl t-butyl ether (MTBE) from methanol and isobutylene in a continuous stirred-tank reactor. The catalyst was macroporous sulfonated poly(styrene-divinylbenzene), and separate experiments were done with catalyst having various cross-linkings to demonstrate that the resistance to mass transport in the microparticles of the polymer was negligible. $C_{methanol}$ = 0.89 mol/L; $C_{isobutylene}$ = 9.2 mol/L; C_{MTBE} < 0.053 mol/L [Rehfinger, A., Hoffman, U., *Chem. Eng. Sci.*, 45, 1605 (1990)].

4-12 The kinetics data summarized in Table 4-10 were measured for the hydrogenation of ethylene catalyzed by a $AuOs_3$ cluster bonded to phosphine-functionalized poly(styrene-divinylbenzene) (McQuade, K. J., M.Ch.E. thesis, University of Delaware, 1980). Determine a rate equation and interpret it.

4-13 Demonstrate the correctness of the assertion on p. 238 that β is equal to $(T - T_s)_{max}/T_s$.

4-14 Use data presented in this chapter to estimate the enthalpy of adsorption of ethanol on the cross-linked poly(styrenesulfonic acid).

4-15 Derive an equation for the dependence of the effectiveness factor on the Thiele modulus for a first-order reaction at large values of the modulus.

4-16 In the presence of a catalyst prepared by combining $AlCl_3$ with a macroporous, sulfonated poly(styrene-divinylbenzene), *n*-butane is isomerized to give isobutane, and at the same time *n*-butane is disproportionated to give equimolar amounts of C_3 and C_5 hydrocarbons. Postulate a mechanism for the disproportionation reaction.

4-17 In processes for synthesis of methyl *t*-butyl ether from methanol and isobutylene catalyzed by macroporous sulfonic acid resin, it is important to minimize side products and catalyst deactivation. What are the expected side products? How does the catalyst lose activity?

4-18 How does the effectiveness factor depend on the Thiele modulus for an isothermal reaction that is of order -1 in the reactant? The results of Fig. 4-25 may be helpful.

4-19 The reaction rate data of Fig. 4-30 were measured for methyl *t*-butyl ether (MTBE) synthesis from methanol and isobutylene at 60°C. The results show the effects of stirrer speed in a continuous stirred-tank reactor containing liquid reactants and particles of macroporous sulfonic acid resin catalyst. The catalyst was used in two particle size ranges, a standard size range and one with particles 10 percent larger than the standard. Interpret the results; what do they imply about mass transport resistance external to the catalyst particles? What do they imply about intraparticle transport resistance? What do they imply about the form of the intrinsic kinetics? How is isobutylene dimer formed?

5

Catalysis in Molecular-Scale Cavities

5.1
STRUCTURES OF CRYSTALLINE SOLIDS

A distinctive characteristic of polymeric solids as catalysts is their solutionlike character and their ability to conform to the geometry of reactant molecules. As polymers become more highly cross-linked and rigid, they lose this solutionlike character, and their physical properties approach those of inorganic solids, the focus of this chapter. Most solid catalysts used on a large scale are inorganic, but only those of a small group, zeolites, consist of crystalline solids. Zeolites are the primary subject of this chapter.

Crystals consist of atoms, ions, or molecules arranged in a periodic array. The regularity of a crystalline array causes incident radiation having a wavelength nearly the same as the interatomic distances (i.e., X-rays) to be diffracted; interpretation of the diffraction pattern with Bragg's law determines the crystal structure. Crystal structures of solids are determined routinely by X-ray diffraction, with computers facilitating the calculations.

A summary of crystal structures observed in nature is given in Fig. 5-1 and Table 5-1. The most common ideal structures are close packings of atoms or ions—cubic close packing and hexagonal close packing. These arrangements are illustrated in Fig. 5-2 by the close packings of spheres. Consider a single close-packed layer: this has a hexagonal symmetry, with each sphere surrounded by six nearest neighbors. It is geometrically impossible to pack the layer more closely than this, since the maximum number of circles that can be tangent to a given circle of the same diameter is six.

Now consider how to place a second close-packed layer on top of the first. The layers will be most compact if the spheres of the second layer fit into

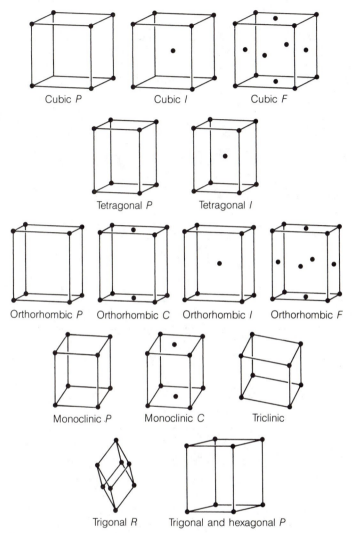

Cubic *P* Cubic *I* Cubic *F*

Tetragonal *P* Tetragonal *I*

Orthorhombic *P* Orthorhombic *C* Orthorhombic *I* Orthorhombic *F*

Monoclinic *P* Monoclinic *C* Triclinic

Trigonal *R* Trigonal and hexagonal *P*

Figure 5-1
Crystal structures observed in nature: the 14 Bravais or space lattices.
The cells shown are the conventional cells, which are not always the
primitive cells.

the cusps between spheres of the first layer. In other words, each sphere of
the second layer is positioned with its center directly above the center of the
triangle defined by the centers of three neighboring spheres in the first layer.
This arrangement is depicted in Fig. 5-2.

In positioning the third layer most compactly on top of the second, there
is a choice. One possibility is to place the spheres of the third layer in cusps
of the second so that the center of a sphere of the third layer is directly above
the center of a sphere of the first layer. If the first layer is designated A and

Table 5-1
CRYSTAL LATTICES

Crystal Structure	Number of Lattices	Lattice Symbols	Conventional Cell Axes and Angles[a]
Triclinic	1	P	$a \neq b \neq c$ $\alpha \neq \beta \neq \gamma$
Monoclinic	2	P, C	$a \neq b \neq c$ $\alpha = \gamma = 90° \neq \beta$
Orthorhombic	4	P, C, I, F	$a \neq b \neq c$ $\alpha = \beta = \gamma = 90°$
Tetragonal	2	P, I	$a = b \neq c$ $\alpha = \beta = \gamma = 90°$
Cubic	3	P or sc I or bcc F or fcc	$a = b = c$ $\alpha = \beta = \gamma = 90°$
Trigonal	1	R	$a = b = c$ $\alpha = \beta = \gamma < 120°; \gamma \neq 90°$
Hexagonal	1	P	$a = b \neq c$ $\alpha = \beta = 90°; \gamma = 120°$

[a] For the crystal axes a, b, and c, the angle between b and c is α, that between a and c is β, and that between a and b is γ.

the second B, then the sequence is ABAB . . . (Fig. 5-3a). The second possibility is to place the spheres of the third layer in the cusps of the second layer offset from those of the ABAB . . . structure; the result is the ABCABC . . . structure (Fig. 5-3b). In each of these structures (Figs. 5-3, 5-4, and 5-5), there is an ideal close packing, and each sphere (except those at a surface) has 12 nearest neighbors touching it—six in the same plane, three above, and three below. The difference between the two structures is that in the ABAB . . . structure a triangle defined by the centers of three spheres in the third layer is directly above such a triangle in the first layer, whereas in the ABCABC . . . structure, the one triangle is rotated 180° with respect to the other.

The ABAB . . . structure is a *hexagonal close-packed* structure, and the ABCABC . . . structure is a *cubic close-packed* or *face-centered cubic* structure. The hexagonal close-packed structure has already appeared in Fig. 2-49,

First layer

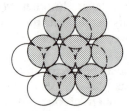

First two layers

Figure 5-2
Close packings of spheres.

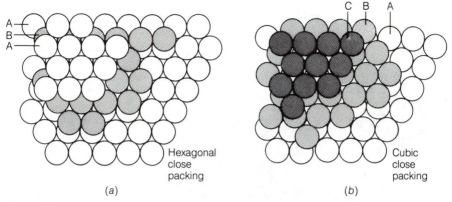

Figure 5-3
Structures formed by close packing of uniform spheres. (*a*) Hexagonal close
packing (ABAB . . .). (*b*) Cubic close packing (ABCABC . . .). The different
layers are illustrated with different shadings.

representing the rhodium atoms of the metal cluster $[Rh_{13}(CO)_{25}H_3]^{3-}$. This
cluster anion might be looked upon as a fragment of rhodium metal having
chemisorbed carbonyl and hydride ligands, but bulk rhodium metal has a face-
centered cubic structure. There are, however, many metals (e.g., osmium) that
have hexagonal close-packed structures. There are many arrangements of
atoms and ions in crystals besides the two common close packings (Fig. 5-1).
For example, the ions of rock salt, NaCl, are arranged in a simple cubic packing;
this structure is better described as two interpenetrating face-centered cubic
lattices, one of Na$^+$ ions and one of Cl$^-$ ions. Some other packings are mentioned in the chapters that follow.

Figure 5-4
Illustration of the hexagonal close-packed structure of spheres.

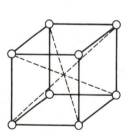

Figure 5-5
Illustration of the cubic close-packed structure of spheres.

Figure 5-6 is helpful in illustrating two common symmetries (or surroundings) encountered in the crystalline state. Both the simple cubic lattice and the face-centered cubic lattice have positions at the centers of the cubes. The latter, which might be filled by atoms or ions, are referred to as *octahedral holes*, designated with the symbol O_h, which means the symmetry is octahedral (or cubic). This symmetry is familiar from the discussion of transition metal complexes (Fig. 2-27); coordinatively saturated compounds like [RhH$_2$Cl(PPh$_3$)$_3$] exemplify octahedral surroundings of a metal.

Another surrounding familiar from Fig. 2-27 is the tetrahedral surrounding, designated as T_d. There are many transition metal complexes having the

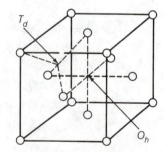

Simple cubic lattice
showing cubic hole
in the center

Face-centered cubic lattice
showing octahedral hole, O_h,
and one of the tetrahedral
holes, T_d

Figure 5-6
Simple cubic and face-centered cubic packings of spheres.

metal in tetrahedral surroundings like the carbon in methane. Tetrahedral positions are shown in Fig. 5-6.

5.2
STRUCTURES OF ZEOLITES
(CRYSTALLINE ALUMINOSILICATES)

The crystalline solids of primary catalytic interest, called aluminosilicates, incorporate Al, Si, and O. Naturally occurring minerals and many solids prepared in the laboratory exemplify this class. Several of the more than 100 synthetic aluminosilicates find wide application as industrial catalysts. The zeolites are structurally unique in having cavities or pores with molecular dimensions as a part of their crystalline structures. These bear catalytic sites. Having microscopic cavities, the zeolites are comparable to enzymes, which have catalytic groups within molecular-scale clefts, and they are also suggestive of the donut-shaped cyclodextrin molecule represented in Fig. 2-53. Because the zeolites have well-defined crystalline structures, the catalytic groups in them are relatively well understood.

5.2.1 Silicalite and ZSM-5

Before consideration of the structures of zeolites, it is helpful to consider a group of solids having the simpler composition SiO_2. Later, the group will be expanded by replacement of some of the Si atoms with Al. The compound SiO_2 is stable in the gas phase, having a structure with two $Si{=}O$ bonds resembling that of CO_2. There is a large free-energy driving force for self-association of the SiO_2 molecules, with conversion of the $Si{=}O$ bonds into $Si{-}O$ bonds. Many geometries of the resulting condensed structures exist, two of the best known being quartz and cristobalite, and another being silicalite, which is closely related to a zeolite. All of these are crystalline, with each Si atom in tetrahedral surroundings, sharing bridging O atoms with neighboring Si atoms. The noncrystalline material silica also has the composition SiO_2. It is discussed in Chapter 6.

The SiO_4 tetrahedron in a crystalline network is often represented in the following ways:

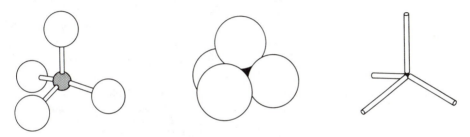

The second structure (a space-filling model based on close-packed spheres) is the most accurate representation. But the third, in which the Si atom is represented as a point and the O atom as a line, is the most convenient for showing the complex structures of zeolites. The SiO_4 unit is a building block, and the ways these are assembled define the structures of the crystalline solids with composition SiO_2.

The SiO_4 tetrahedra can be combined in many arrays with sharing of O atoms [1]. When they are arranged as shown in Fig. 5-7a, the result is the secondary building block of silicalite. Here, to emphasize the geometry, the third form of representing the structure is used, showing the Si atom as an intersection of lines; the O atom is located at the center of each line. When some of the Si atoms in this structure are replaced by Al atoms, the resulting aluminosilicate is referred to as ZSM-5, a zeolite and an industrially important catalyst [2, 3, 4].

In silicalite and ZSM-5, the tetrahedra are linked to form the chain-type building block shown in Fig. 5-7b. The chains can be connected to form a layer, as shown in Fig. 5-8. Rings consisting of five O atoms are evident in this structure; the name *pentasil* is therefore used to describe it. Also evident in Fig. 5-8 are rings consisting of 10 oxygen atoms; these are important because they provide openings in the structure large enough for passage of even rather large molecules. The layers can be linked in two ways, the neighboring layers being related either by the operation of a mirror or an inversion. The former pertains to the zeolite ZSM-11, the latter to silicalite or ZSM-5; intermediate structures constitute the pentasil series.

The three-dimensional structure of silicalite (and ZSM-5) is represented in Fig. 5-9a. The 10-membered rings provide access to a network of intersecting pores within the crystal. The pore structure is depicted schematically in Fig.

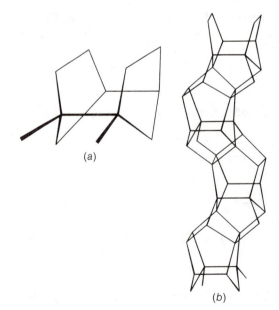

(a)

(b)

Figure 5-7
(a) The secondary building block of silicalite, formed from SiO_4 tetrahedra. (b) The chain-type building block formed from the secondary building blocks by sharing of oxygens by linked SiO_4 tetrahedra.

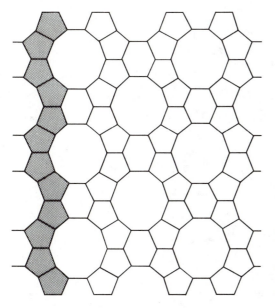

Figure 5-8
Schematic diagram of silicalite layers, formed by linking of the chains shown in Fig. 5-7b through sharing of oxygen in linked SiO_4 tetrahedra. The shaded portion denotes one of the chains.

5-9b; there is a set of straight, parallel pores intersected by a set of perpendicular zigzag pores. Many molecules are small enough to penetrate into this intracrystalline pore structure, where they may be catalytically converted.

The crystalline nature of ZSM-5 is evident in the electron micrograph of Fig. 5-10. With a high-resolution electron microscope, it is even possible to see evidence of the regularity in the pore structure of this material [6].

ZSM-5 can be synthesized from an aqueous gel prepared from sodium aluminate, silica sol, NaOH, H_2SO_4, and tetrapropylammonium bromide. After

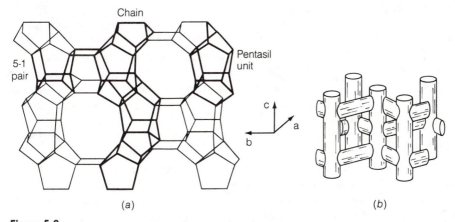

(a)   (b)

Figure 5-9
Representation of three-dimensional structure of silicalite (ZSM-5). (a) Structure formed by stacking of sequences of layers shown in Fig. 5.8. (b) Schematic representation of the intracrystalline pore structure.

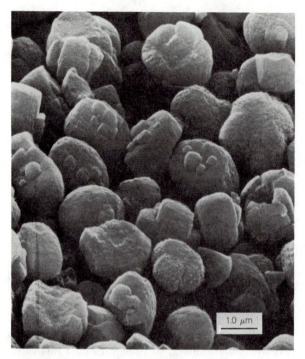

Figure 5-10
Crystals of the zeolite ZSM-5 viewed with a scanning electron
microscope. (Courtesy of Professor C. Lyman.)

standing at 95°C for 10 to 14 days, the mixture gives a yield of 80 to 90
percent ZSM-5; the composition of the solid is approximately
$1.8(TPA)_2O \cdot 1.2Na_2O \cdot 1.3Al_2O_3 \cdot 100SiO_2 \cdot 7H_2O$ [7]. The tetrapropylammonium
bromide may play the role of a template whereby the zeolite crystallizes around
it. The size, shape, and polarity of the presumed template regulate the course
of the crystallization, which is determined by kinetics and is not well understood. The crystals of zeolite are typically nonuniform in composition, as is
illustrated by a gradient in the Si/Al ratio that was determined by X-ray emission
spectroscopy with a scanning electron microscope (Fig. 5-11). Other kinds of
nonuniformities and imperfections in the structure are shown by transmission
electron microscopy [6].

The aluminosilicate structure is ionic, incorporating Si^{4+}, Al^{3+}, and O^{2-}.
When some of the Si^{4+} ions in the SiO_4 tetrahedra in this framework are replaced by Al^{3+} ions, as in ZSM-5 and other zeolites, an excess negative charge
is generated. A compensating source of positive charge must be added, namely,
cations in addition to the framework Si^{4+} and Al^{3+}. These nonframework cations play a central role in determining the catalytic nature of zeolites. The
zeolites are ion exchangers, comparable to the sulfonated resins of Chapter 4.
Bringing an aqueous salt solution in contact with the zeolite leads to incorporation of cations from the salt into the zeolite, replacing some of the non-

framework cations initially present. Ion exchange is the simplest and most important method for modifying the properties of a zeolite. The names of zeolites specify the cations: for instance, in NaZSM-5, the cations are Na^+, in HZSM-5, they are H^+.

Some cations (e.g., Ti, Ga, and Fe) have been incorporated in zeolite frameworks.

5.2.2 Zeolite A

To understand the structure of zeolite A (which is synthesized easily by mixing a solution of sodium aluminate and NaOH with one of sodium meta-silicate [7]), consider the arrangement of 24 primary building blocks, namely, SiO_4 + AlO_4 tetrahedra, to form the truncated octahedron shown in Fig. 5-12. This secondary building block is called a sodalite cage; its geometry is easy to visualize when the oxygen ions are represented as lines and the Si and Al ions as points of intersection (Fig. 5-12).

When the sodalite cages are arranged in a regular array so that each square face of a truncated octahedron (i.e., each ring consisting of four oxygen ions) is shared by two sodalite cages, the result is the structure of the mineral sodalite, shown in Fig. 5-13. (What is the crystallographic arrangement of the sodalite cages? See Fig. 5-1). The figure shows that the largest aperture into any enclosed volume is a six-membered oxygen ring that opens into a sodalite cage.

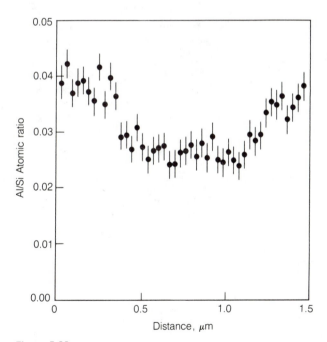

Figure 5-11
Compositional variation across a crystal of the zeolite ZSM-5 [5].

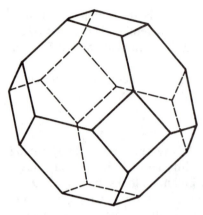

Figure 5-12
A truncated octahedron formed from 24 SiO$_4$ + AlO$_4$ tetrahedra.

The ring is described primarily by the oxygen ions; a space-filling model shows that oxygen ions are larger than the cations, which almost seem to be buried between them. These rings have a diameter of about 0.26 nm, and a cage is large enough to hold a sphere with a diameter of 0.66 nm. Therefore, the cages can accommodate small molecules such as H$_2$ and H$_2$O, but access of larger molecules is geometrically excluded. Sodalite is therefore of virtually no interest as a catalyst.

It is possible, however, to obtain a catalytically interesting structure, called zeolite A, by stacking the sodalite cages in a more widely spaced manner, as shown in Fig. 5-14. Now, the sodalite cages are connected by bridging oxygen ions between the four-membered oxygen rings. This structure has larger apertures than sodalite, namely, eight-membered oxygen rings, each having a diameter of 0.42 nm and opening into a cavity, called a *supercage*, surrounded by eight sodalite cages; this cavity is large enough to contain a sphere with a

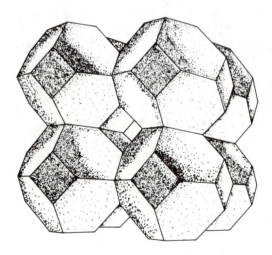

Figure 5-13
The sodalite structure, composed of truncated octahedra with shared square faces.

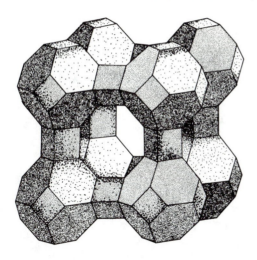

Figure 5-14
Structure of zeolite A. The sodalite cages are connected by bridging oxygen ions between the four-membered oxygen rings. The cavity (supercage) is large enough to contain a sphere with a 1.14-nm diameter.

diameter of 1.14 nm. Supercages are characteristic of zeolite A, among others; silicalite does not have supercages.

5.2.3 Zeolites X and Y (Faujasites)

The zeolites finding the largest-scale application in catalysis belong to the family of faujasites, including zeolite X and zeolite Y. Having 0.74-nm apertures (12-membered oxygen rings) and a three-dimensional pore structure, they admit even hydrocarbon molecules larger than naphthalene. Their chief application is in catalytic cracking of petroleum molecules (primarily in the gas oil fraction), giving smaller, gasoline-range molecules.

The framework structure of zeolites X and Y (Fig. 5-15) is closely related to that of zeolite A. The sodalite cages in faujasites are arranged in an array with greater spacing than in zeolite A. Each sodalite cage is connected to four other sodalite cages; each connecting unit is six bridging oxygen ions linking the hexagonal faces of two sodalite units, as shown in Fig. 5-15. The bridging oxygens form what is called a hexagonal prism. The high-resolution electron micrograph of Fig. 5-16 shows the regularity of the pores in a crystal of zeolite Y.

The supercage in this structure, surrounded by 10 sodalite units, is large enough to contain a sphere with a diameter of about 1.2 nm. The three-dimensional pore structure is large enough to admit reactant molecules like the hydrocarbons in gas oil, but the 0.74-nm pore apertures are still small enough that some transport restrictions are expected.

There is a range of faujasite compositions, with a typical unit cell formula being $Na_j[(AlO_2)_j(SiO_2)_{192-j}] \cdot zH_2O$, where z is about 260. The value of j is between 48 and 76 for zeolite Y and between 77 and 96 for zeolite X. X-Ray diffraction and NMR data have determined the structures of faujasites, showing, for example, that the unit cell dimension decreases slightly as the Si/Al

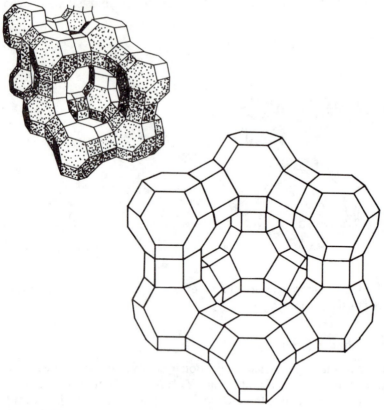

Figure 5-15
Structure of faujasite.

10 nm

Figure 5-16
High-resolution transmission electron micrograph of crystalline
NaY zeolite. (Courtesy of B. Tesche.)

ratio increases. The X-ray data have also determined the exact positions of cations present to balance the excess negative charge of the AlO_4 tetrahedra. Four cation sites are illustrated in Fig. 5-17. Type I sites are located at the centers of the hexagonal prisms, type I' sites are located in the sodalite cages across the hexagonal faces from type I sites, type II sites are located in the supercages near the unjoined hexagonal faces, and type II' sites are located in the supercages, farther from the hexagonal faces than the type II sites.

Since cations such as H^+ are largely responsible for the catalytic activity of a zeolite, their locations are important. Cations in type II and II' sites are readily implicated in catalysis, since they are accessible to reactants. The type I and I' sites, being less accessible, are less important for catalysis.

Most zeolites are synthesized in the sodium form, the common starting materials for synthesis of zeolite Y being sodium aluminate, NaOH, and silica sol (a colloidal silica suspension), and the product typically has the approximate composition $Na_2O \cdot Al_2O_3 \cdot 5.3SiO_2 \cdot 5H_2O$ [6]. Preparing the hydrogen form of zeolite Y is not so simple, because the faujasite framework collapses when in contact with strongly acidic solutions. This difficulty can be circumvented by exchanging Na^+ with NH_4^+ and raising the temperature, causing the ammonium ions (now in the zeolite) to decompose into NH_3 gas, which leaves the zeolite, and H^+ ions, which remain in place. (This is the standard method of preparing any hydrogen-form zeolite.) The positions of the H^+ ions in faujasite are suggested in Fig. 5-17; the difficulty of locating these small ions by X-ray diffraction leaves the assignments somewhat uncertain.

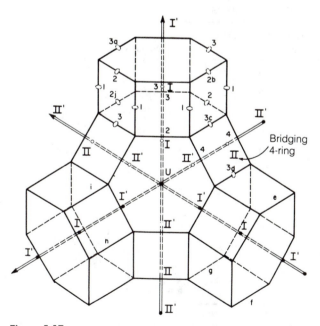

Figure 5-17
Locations of cation sites in X and Y zeolites.

5.3
FAMILIES OF ZEOLITES

5.3.1 Pore Sizes and Molecular Sieving

The zeolites all contain intracrystalline pores and apertures having dimensions approximately equal to those of many of the molecules converted in catalytic processes. The zeolites are classified according to the sizes of these apertures (Table 5-2). The table includes the number of oxygen atoms in the aperture of each zeolite (8, 10, or 12) and the aperture dimension (the smallest being about 0.4 nm for zeolite A and the largest being 0.74 nm for faujasite). The size of the aperture is dependent on the sizes of the nearby cations, which may partially block it.

The average channel sizes of some of these zeolites are summarized in Fig. 5-18, along with the sizes of the cavities (supercages) and the critical molecular dimensions of a number of hydrocarbons that are potential reactants in zeolite-catalyzed reactions. This figure illustrates the property of zeolites that accounts for the name *molecular sieve*. Isoparaffins, for example, are too large to enter the pores of zeolite A, whereas straight-chain paraffins fit. This crucial size difference is exploited in industrial processes for gas purification by molecular sieve adsorbents. A column is packed with particles of molecular sieve and a mixture of gases flows through; the small molecules are held up

Table 5-2
ZEOLITES AND THEIR PORE (APERTURE) DIMENSIONS[a]

Zeolite	Number of Oxygens in the Ring	10 × Aperture Dimensions, nm
Chabazite	8	3.6 × 3.7
Erionite	8	3.6 × 5.2
Zeolite A	8	4.1
ZSM-5 (or silicalite)	10	5.1 × 5.5; 5.4 × 5.6
ZSM-11	10	5.1 × 5.5
Heulandite	10	4.4 × 7.2
Ferrierite[b]	10	4.3 × 5.5
Faujasite	12	7.4
Zeolite L	12	7.1
	12	7.0
Mordenite	12	6.7 × 7.0
Offretite	12	6.4

[a] The framework oxygen is assumed to have a diameter of 0.275 nm.
[b] There are also apertures with eight-membered oxygen rings in this zeolite.

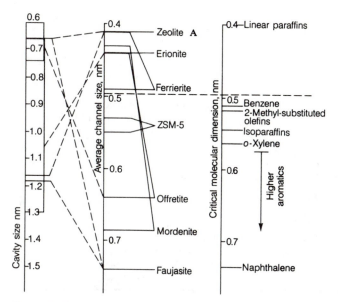

Figure 5-18
Pore dimensions of zeolites and critical dimensions of some hydrocarbons [8].

in the column because they enter the zeolite pores, but the large molecules
flow directly through the column without this holdup.

5.3.2 Aluminum Content and Acidity

Zeolites are further grouped into families on the basis of composition, namely,
the Si/Al ratio (Table 5-3). Since the ion exchange capacity is equal to the
concentration of Al^{3+} ions in the zeolite, the structures with low Si/Al ratios
can have higher concentrations of catalytic sites than the others. There are
other transitions in properties exemplified by the acid form zeolites (those
incorporating H^+ as the exchangeable ion [9]). The zeolites with high concentrations of H^+ are hydrophilic, having strong affinities for polar molecules small
enough to enter the pores. The zeolites with low H^+ concentrations (in the
limit, silicalite, for example) are hydrophobic, taking up organic compounds
(e.g., ethanol) from water–organic mixtures; the transition occurs at a Si/Al
ratio near 10 [9]. The stability of the crystal framework also increases with
increasing Si/Al ratios; decomposition temperatures of the different zeolites
range from roughly 700°C to 1300°C [9]. Zeolites with high Si/Al ratios are
stable in the presence of concentrated acids, but those with low Si/Al ratios
are not; the trend is reversed for basic solutions.

There are also trends in the strength of acidity depending on the Si/Al
ratio. To understand the acidity of zeolites, it is helpful to consider a simplified
representation of the interior regions of the pores, where adsorbates or reac-

Table 5-3
ACID FORM ZEOLITES CLASSIFIED BY THEIR Si/Al RATIOS[a]

Si/Al Atomic Ratio	Zeolites	Properties
Low (1–1.5)	A, X	Relatively low stability of framework; low stability in acid; high stability in base; high concentration of acid groups with moderate acid strength
Intermediate (2–5)	Erionite Chabazite Clinoptilolite Mordenite Y	
High (~10 to ∞)	ZSM-5 Erionite[b] Mordenite[b] Y[b]	Relatively high stability of framework; high stability in acid; low stability in base; low concentration of acid groups with high acid strength

[a] Adapted from ref. 7.
[b] Formed by chemical framework modification (dealumination); the Al is partially removed by treatment with $SiCl_4$, for example.

tants interact with the catalyst. To simplify the representation, the SiO_4 and AlO_4 tetrahedra are strung out, with a loss of the depiction of the three-dimensional geometry. In this simplified representation, the exchangeable cations are placed near AlO_4 tetrahedra, because the negative charges are predominantly located there. A segment of a zeolite in the sodium form is represented as follows:

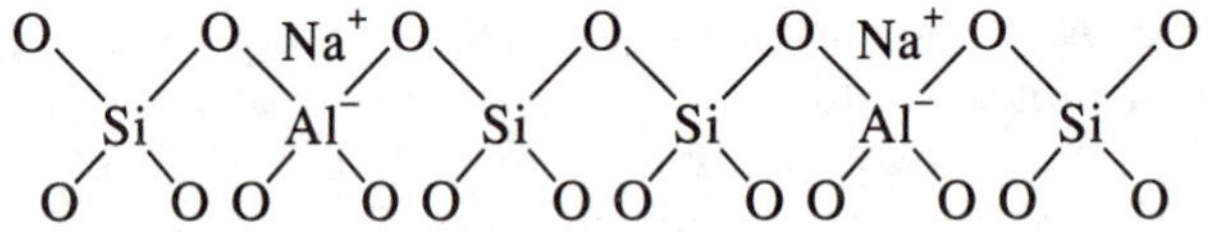

When the Na^+ is exchanged for Ca^{2+}, the structure is written as follows:

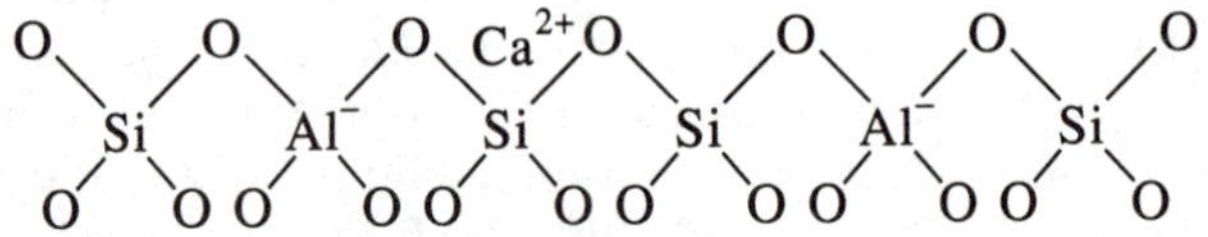

The single Ca^{2+} ion balances the charge of two AlO_4^- tetrahedra. In the actual zeolite, the negative charge is not localized on one or two tetrahedra but is distributed over the framework of oxygen ions. The distribution of negative charge in the framework may be important in catalysis in stabilizing cationic intermediates such as carbenium ions; the framework surrounds the cation, approximately as was illustrated for enzymes in Chapter 3 and for sulfonated polymer catalysts in Chapter 4.

The hydrogen form zeolite is represented more explicitly as follows, although it should be recognized that the proton is mobile within the structure:

Infrared spectra of zeolites such as A, X, and Y provide evidence of —OH groups in various positions. These are associated with neighboring SiO_4 tetrahedra. The interactions of the —OH groups with bases like pyridine, measured by infrared spectroscopy, demonstrate that the —OH groups located near AlO_4 tetrahedra are strong Brønsted acids.

There are wide distributions of proton donor strengths among the Brønsted acid groups in zeolites. When zeolites have low densities of proton donor groups, the proton donor strengths are high. For example, HY or HZSM-5 zeolite with a low density of acid groups is like an ideal solution of dispersed noninteracting protons in a solid matrix. But, depending on the synthesis conditions, this analogy may be strained; for example, there may be amorphous material (aluminum oxide) present in the zeolite and nonuniformities of structure ranging from cracks and faults in the crystals (visible by electron microscopy) to local variations in the surroundings of individual Si and Al ions even in perfect crystals; the latter are detected by ^{29}Si and ^{27}Al NMR spectroscopy [10]. NMR results show that Al occupies certain sites preferentially in ZSM-5; it is present in five-membered rings but not in four-membered rings [11].

NMR data also show that there are no near-neighbor AlO_4^- tetrahedra in zeolites; they are always interspersed with SiO_4 tetrahedra. (This is Löwenstein's rule; what does it imply about the structure of zeolite A with a Si/Al ratio of 1?)

Zeolites with low Al contents are prepared by the process of dealumination, whereby a reactant such as $SiCl_4$ extracts Al from the framework, making $AlCl_3$. Zeolite Y prepared this way (or by treatment in steam) is called ultrastable, being applied in cracking catalysts (described below). The greatest proton donor strengths are exhibited by the zeolites containing the lowest concentrations of AlO_4^- tetrahedra (such as HZSM-5 and ultrastable HY); these are superacids, which at high temperatures (about 500°C) are able to protonate even paraffins. Evidently the proton donor strength is greatest for groups that are associated with AlO_4^- tetrahedra having the smallest number of Al neighbors [12]. As the Al distribution is not uniform (except for zeolites with Si/Al ratios of 1), a range of acid strengths is expected.

The nonuniformities in strengths of proton donor sites in zeolites have been measured by temperature-programmed desorption of adsorbed bases; the bases on the most strongly acidic sites require the highest temperatures for desorption. Infrared spectra of adsorbed bases, such as NH_3 and pyridine, provide indications of the nature of the adsorption site (adsorbed pyridinium ion, for example, being an indication of proton donor sites). Bases such as the Hammett indicators are less appropriate because they are too large to find access to many of the ion exchange sites; even pyridine may be too large. NMR and electron spin resonance (ESR) spectra of adsorbed bases also help elucidate the nature of the acidic sites [12].

There are various proton donor groups in zeolites. Addition of water to a H^+ form zeolite, for example, leads to the formation of H_3O^+. Cations such as La^{3+} react with water to give acidic La—OH groups. Al—OH groups have also been identified in zeolites, some possibly associated with noncrystalline impurities.

When a hydrogen form zeolite is heated to high temperature, water is driven off and coordinatively unsaturated Al^{3+} ions are formed. These are strong Lewis acids; a base like pyridine is typically more strongly bonded to these sites than to strong proton-donor sites, as shown by infrared spectra and temperature-programmed desorption. Formation of the Lewis acid sites is depicted schematically as follows:

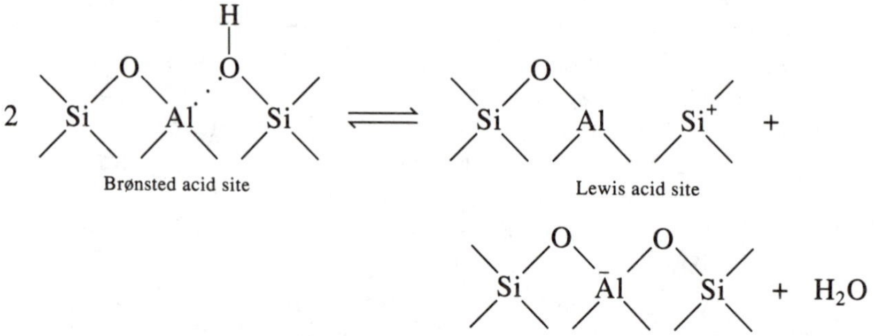

According to this stoichiometry, one Lewis acid site is formed from two Brønsted acid sites; the result has been confirmed by measurements of the numbers of adsorbed pyridinium ions (indicating Brønsted acid sites) and Al^{3+}-coordinated pyridine molecules (indicating Lewis acid sites) [13]. The structures shown above are oversimplifications. There are bond-breaking and bond-forming processes that lead to a rapid interchange of structures such as these; there are no rigid fixed sites in the sense that the structures might seem to imply. There is recent evidence from NMR spectroscopy that casts doubt on the structural picture presented above [14].

5.4
ADSORPTION AND DIFFUSION IN ZEOLITES

The void spaces in the crystalline structures of zeolites provide a high capacity for adsorbates, referred to as guest molecules. The sorption capacity is a conveniently measured property that is used to identify zeolites. Isotherms characterizing the equilibrium sorption of CF_4 in sodium form faujasite are shown in Fig. 5-19. The shapes of these isotherms are typical for nonpolar guest molecules in zeolites. The more condensable the guest molecule (or the lower the temperature), the more rectangular is the isotherm.

Sorption of polar molecules is more complicated, and chemisorption is influenced strongly by the nature of the cations and the interactions between the cations and guest molecules. The specificity of the interactions can be used to advantage in separation processes. Some guest molecules react with zeolites to change the zeolite structures; for example, adsorption of H_2 on nickel form zeolite A at elevated temperatures leads to reduction, migration, and aggregation of the nickel, and even to destruction of the zeolite framework. Guest molecules also change the configuration of the aluminosilicate framework slightly. This framework is to some degree flexible and solventlike, although not so much as the cross-linked polymers considered in Chapter 4.

Adsorption in the pores cannot take place unless the guest molecules are small enough to fit through the apertures. Transport of molecules in the narrow zeolite pores is often slow, since the molecules may barely fit through the apertures. A three-dimensional pore structure (as in faujasites) facilitates the transport. The countertransport of molecules leaving the zeolite pore structure can severely hinder the entry of the guest molecules replacing them. Cations that interact strongly with the guest molecules can also hinder the transport severely. The transport processes are poorly understood. Transport may be roughly described as the hopping of molecules from site to site over significant

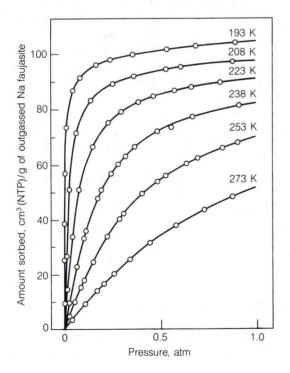

Figure 5-19
Equilibrium adsorption of CF_4 in sodium form faujasite [15].

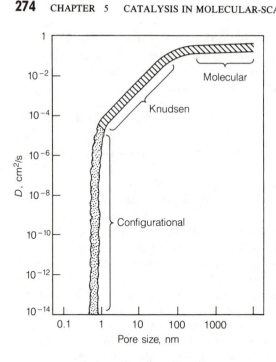

Figure 5-20
Approximate effective diffusion coefficients of molecules in porous solids [16]. When the pore size is about the same as the critical dimension of the molecule being transported, a decrease of only 0.1 nm in pore size can reduce the transport rate by orders of magnitude.

energy barriers; it is much different from molecular diffusion or diffusion in pores that have diameters much larger than the molecular dimensions.

A qualitative representation of transport in the pores of solids is shown in Fig. 5-20. Here, the effective diffusion coefficient is shown as a function of the pore size. For larger pores, the interactions between the molecules predominate, and molecular diffusion (described by Fick's law) describes the transport. When the pores are smaller, the interactions with the pore walls (Knudsen diffusion) predominate, as discussed in Chapter 4. When the pores are so small that the molecules can barely pass through them, the effective diffusion coefficient is strongly dependent on the aperture size. This is often the situation with zeolites, and the term *configurational diffusion* is applied. Fick's law is often not a good representation, although it is typically used for simplicity and lack of a better model.

The sharp dependence of transport rate on pore size provides a basis for understanding the separation of molecules having different sizes. The separation of two molecules may be efficient even if both fit into the pores of the zeolite, provided that the ratio of apparent diffusion coefficients is large. In the limit, this ratio will approach infinity; this is true of molecular sieving, whereby one kind of molecule enters the pores and the other does not.

5.5
THE SOLVENTLIKE NATURE OF ZEOLITE PORES

Guest molecules in zeolites are enveloped by the small pores. The zeolite pores are like solvents; the microenvironments of zeolites are in this sense comparable to those of the clefts of enzymes and the spaces between the chains of swellable synthetic polymers. The structure and composition of the zeolite pore therefore strongly influence the reactivity of a guest molecule.

The ionic character of zeolites such as A and Y implies the presence of strong electric fields imposed on the guest molecules. Zeolites show properties varying from those of weak to strong electrolytes; the latter zeolites have strong affinities for ionic materials such as salts, and neutral guest molecules may be ionized upon occlusion into the zeolite [17].

Zeolites act as catalysts in part by virtue of this strong solventlike character. Potassium form Y zeolite catalyzes cracking of hexane at 500°C [17]. This reaction can also be catalyzed by acids (Chapter 2), but the zeolite sample was free of acid. The product distribution is characteristic of a free-radical reaction rather than one proceeding through carbenium ion intermediates. The products are different from those obtained without a catalyst: Hydrogen abstraction from hydrocarbons by radical intermediates is favored in the presence of the zeolite, whereas β-scission of the radicals is unaffected. In other words, the zeolite causes an acceleration of bimolecular steps but not of unimolecular steps in the free-radical reaction. The implication is that in the microenvironment of the solventlike cages of the zeolite, the reactants are more highly concentrated than in the gas phase, and the bimolecular processes are therefore favored. The hydrocarbon reactants are concentrated in the zeolite as a consequence of the high affinity of the strong electrolyte for the hydrocarbon, which becomes polarized upon adsorption [17]. This interpretation implies that the much less polar environment of silicalite pores exerts a much weaker influence on the hydrocarbon. Consequently, silicalite, in contrast to the potassium form Y zeolite, acts like inert quartz chips, having no measurable effect on the hydrogen abstraction step [17].

Many experimental results indicate that the solventlike properties of zeolite pores and their resistance to transport of reactants and products influence the catalytic properties. In the following sections, the catalytic properties of zeolites are therefore explained on the basis of chemistry like the molecular chemistry of Chapter 2 combined with the influence of the pore microenvironment and the transport limitations.

5.6
CATALYSIS BY ZEOLITES

5.6.1 Acid–Base Catalysis

Zeolites are catalysts that offer the advantages of high densities of catalytic sites combined with stability at high temperatures; the latter is an advantage characterizing many inorganic solids and accounts for their wide applications as industrial catalysts. The industrial applications of zeolites involve acid-catalyzed reactions. Virtually all the reactions catalyzed by acids in solutions and in polymer matrices are catalyzed by zeolites incorporating acidic groups [18]. This generalization is based on experimental evidence for hundreds of catalytic reactions, including almost all those mentioned in Chapter 2. The exceptions involve reactants too large to enter the zeolite pores and products too large to form in, or leave, the pores. Some of the points emphasized in this chapter are the contrasts between the zeolites and the acid catalysts considered in the preceding chapters. The contrasts are explained by the solventlike nature of the zeolite pores and the transport restrictions. The constraints of zeolite pores allow reactivity patterns that are unattainable in solution catalysis.

CRACKING

Some of the simplest lessons in zeolite catalysis are demonstrated by the cracking of paraffins catalyzed by zeolites with low densities of acidic groups, associated with AlO_4^- tetrahedra. Recall the statement that HZSM-5 can be regarded as an "ideal solution" of acidic groups; if these are far enough apart from each other, they may act almost independently in catalysis. Results suggesting such a simple interpretation are shown in Fig. 5-21. There is a direct proportionality between the activity of the zeolite for cracking of n-hexane and the concentration of Al^{3+} ions in the zeolite, which can be measured by Al NMR spectroscopy. A similar pattern has been observed for other reactions and other zeolites (e.g., Fig. 5-22). Because each acidic group is associated with an AlO_4^- tetrahedron, the activity is proportional to the concentration of acidic groups in each zeolite. This result implies that the acidic groups are the catalytic sites.

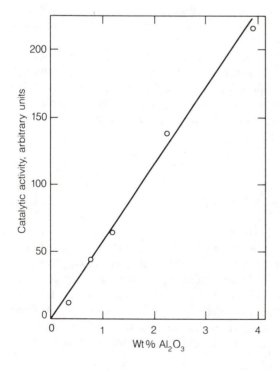

Figure 5-21
Dependence of activity for n-hexane cracking at 538°C on the Al content of HZSM-5 catalyst [19].

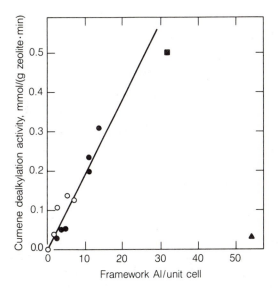

Figure 5-22
Dependence of activity for cracking of cumene at 290°C on the framework Al content of HY zeolite [20].

These results are strikingly simple. They suggest that the distribution of acid strengths may be almost independent of the concentration of acidic sites, provided that this concentration is low. Since the strengths of the acidic sites decrease on average as the concentration increases, and since some reactions (e.g., cracking) require strongly acidic sites, curves such as those shown in Figs. 5-21 and 5-22 pass through maxima if the Al contents are increased to high values.

The acidic sites in HZSM-5 are strong; at high temperatures, this zeolite is a superacid, capable of protonating paraffins and thereby initiating catalytic cracking. This point is illustrated by product distribution data obtained for cracking of n-butane at 475–525°C [21] and for cracking of larger paraffins at lower temperatures [22] (Table 5-4). The product distribution data, extrapolated to zero conversion, correspond well to the products expected from the decomposition of the carbonium ion formed by protonation of the reactant paraffin at the most highly substituted carbon atom (Fig. 5-23); the products include H_2, paraffins, and olefins. [Can you explain the products observed with n-hexane extrapolated to zero conversion (Table 5-4)?] These results demonstrate the relevance of superacid solution chemistry (p. 49 ff.) to catalysis by a solid acid at a much higher temperature.

As shown in Chapter 2, however, many complicating side reactions intrude when carbonium ions are formed. These ions decompose to give hydrogen and carbenium ions, which may be deprotonated, and may undergo isomerization, disproportionation, and alkylation. The data referred to in the preceding paragraph were obtained at low conversions to minimize these side reactions. They are not typical of the product distributions usually observed in cracking of paraffins.

A simplified version of the more typical pattern of cracking involves pro-

Table 5-4
PRODUCTS OBSERVED IN THE CONVERSION OF PARAFFINS
CATALYZED BY HZSM-5

	n-Butane [21]		3-Methylpentane [22]			*n*-Hexane [22]	
Temperature, °C	496	496	350	450	450	450	450
Hydrocarbon partial pressure, atm	1	0^a	1	1	0^a	1	0^a
Conversion, %	17	0^a	9	25	0^a	25	0^a
Product distribution, mol/100 mol converted:							
H_2	6	15	7	27	31^a	$—^b$	6
CH_4	9	20	3	16	34^a	3	12
C_2H_6	13	17	6	23	35^a	8	33
C_2H_4	14	15	—	—	—	—	—
C_3H_8	40	0	55	56	0^a	70	40
C_3H_6	10	16	—	—	—	—	—
$n\text{-}C_4H_{10}$	—	—	⎧ 44 ⎫	13	0^a	23	9
$i\text{-}C_4H_{10}$	—	—	⎩ ⎭	17	0^a	12	—
C_4H_8	8	17	—	—	—	—	—
	$\overline{100}$	$\overline{100}$	$\overline{115}$	$\overline{153}$	$\overline{100}$	$\overline{\geq116}$	$\overline{100}$

^{a} Extrapolated value.
^{b} Not analyzed.

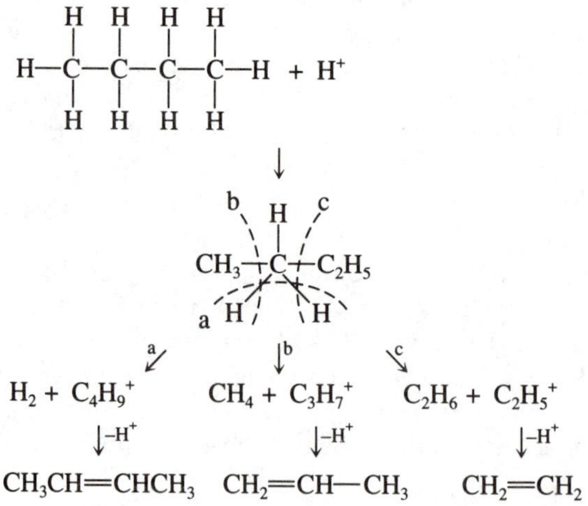

Figure 5-23
Pattern of decomposition of carbonium ion formed by protonation of *n*-butane postulated to form during cracking catalyzed by HZSM-5 [21]. Reaction a results from bond breaking indicated by the dashed line a, etc. The carbenium ions react further, being deprotonated to give olefins, which may undergo secondary reactions.

tonation of an olefin (a much better base than a paraffin) to form a carbenium ion, which undergoes β-scission to give a smaller carbenium ion and another olefin; this is a chain reaction, appearing to be autocatalytic:

Initiation:

$$\text{olefin} + H^+ \rightarrow R^+ \tag{5-2}$$

Chain propagation $\begin{cases} R^+ + R'H \rightarrow RH + R'^+ & (5\text{-}3) \\[1em] R'^+ \rightarrow \text{olefin} + R^+ & (5\text{-}4) \end{cases}$

where R'H is the reactant paraffin and RH and olefin are the products. The chain carrier R^+ is a relatively stable carbenium ion such as the *t*-butyl cation.

These reactions are extremely fast in comparison with protonation of a paraffin, unless the olefin concentration is very low. When cracking conversions are high, olefin concentrations are high because olefins are products, and this latter mechanism of cracking predominates (Table 5-4). Olefin impurities in a feed stream are often sufficient to ensure the predominance of this mechanism, and the mechanism involving protonation of paraffins to give carbonium ions is a rarity. Cracking involving the carbonium ion gives methane and ethane as major products; cracking by the more typical pattern gives very little of these and much more C_3–C_6 paraffins and olefins.

Cracking in industrial practice is carried out with catalysts containing Y zeolite (some stabilized by incorporation of rare earth ions and some by extraction of part of the Al from the framework) embedded in a matrix of an amorphous solid acid such as silica–alumina (SiO_2–Al_2O_3, described in Chapter 6) [23]. Reactant paraffins and other components of gas oil and heavier petroleum fractions are catalytically converted into smaller paraffins and olefins, many boiling in the gasoline range. The reactions mentioned above occur, with many side reactions.

Important side reactions are isomerization, disproportionation, and formation of high-molecular-weight aromatic products called coke. Solid deposits of coke accumulate in the catalyst; the deposits clog the pores and cover the catalytically active sites. Coke is a catalyst poison. Coke formation is extremely fast under conditions of industrial cracking of gas oil (~500°C and 1 atm), as shown by the data of Fig. 5-24. The catalyst is therefore deactivated rapidly in an industrial reactor (called a catalytic cracker). It is separated from the oil after only about 3 s of contact and is sent to a separate reactor (a regenerator), where the coke is burned off with air. A key to the success of this scheme is the efficient handling of the solid catalyst. In the process (Fig. 5-25), small particles of the catalyst are in a "fluidized" state; they are suspended as a mass of chaotically moving particles that are kept in motion by the force of the flowing vapors. In the cracking reactor itself, the particles are entrained, that is, carried rapidly through a tubular reactor to a disengaging unit, where

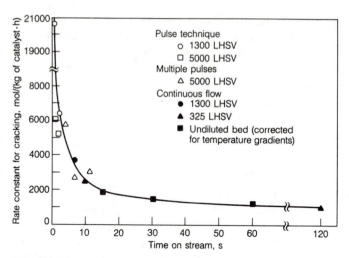

Figure 5-24
Deactivation of zeolite X catalyst containing rare earth and hydrogen ions as a result of coke formation during *n*-hexadecane cracking at 482°C [24]. LHSV is liquid hourly space velocity.

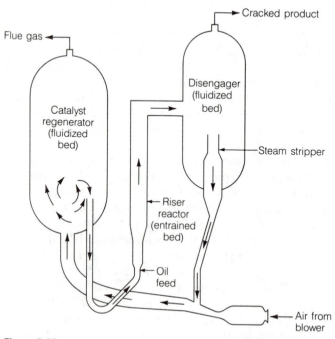

Figure 5-25
Schematic diagram of a process for catalytic cracking of petroleum fractions. The flow pattern of the catalyst particles is indicated by arrows.

a gas–solid separation is carried out. The solid particles then flow continuously to the fluidized-bed regenerator, where the coke is burned off, and then they flow back to the reactor to be mixed with gas oil (and some steam to disperse and rehydroxylate the zeolite; cf. Eq. 5-1). Since cracking is an endothermic reaction, energy must be continuously provided; it comes from the exothermic combustion of the coke and is carried by the large mass of solid catalyst cycling around the unit. As the catalyst structure is slowly degraded in the process, catalyst is continuously removed and replaced with fresh particles.

Olefins are products of paraffin cracking. The origin of coke can be understood on the basis of the reactivity of olefins in acidic zeolites. The pattern is illustrated schematically by the conversion of ethylene in the presence of a rare-earth exchanged zeolite X at 213°C and 1 atm [25]. This simple olefin undergoes a complex set of reactions explained by the carbenium ion chemistry of Chapter 2, as illustrated schematically in Fig. 5-26. These reactions include C–C bond forming (giving higher-molecular-weight products) and hydride transfer steps, which, when combined with proton transfer steps, account for the production of a mixture of saturated (paraffinic) and unsaturated (aromatic) products. As aromatics with more and more rings form via alkylation, ring closure, and dehydrogenation, the product takes on the character of coke. The catalyst used for ethylene conversion undergoes rapid deactivation. The same types of reactions account for the deactivation of cracking catalysts and many others used in hydrocarbon conversions.

The hydrogen transfer reactions are crucial to the success of the cracking process with zeolite catalysts. Zeolites have high activities for bimolecular reactions transferring hydrogen, for example, from a cycloparaffin to a carbenium ion, at least in part because of the solventlike character of the pores, which accounts for high local concentrations of the reactants. For this reason, the zeolites are more active for hydrogen transfer than amorphous aluminosilicates, which lack the small pores. A consequence of this difference is that zeolites have replaced the amorphous materials as industrial cracking catalysts. The improvement is illustrated in Fig. 5-27, which shows the product distributions obtained from a petroleum fraction with each kind of catalyst.

Part of the uniqueness of zeolite catalysts is their selectivity for the following class of reaction:

$$\text{olefins } + \text{ cycloparaffins} \rightarrow \text{paraffins } + \text{ aromatics} \qquad (5\text{-}5)$$

which is accounted for by hydrogen (hydride and proton) transfers [26].

The details of the hydrogen transfer reactions are not fully understood, but the essential idea is summarized schematically in Fig. 5-28. The crucial role of the small zeolite pores is to favor the bimolecular reactions that convert olefins (with other structures including cycloparaffins and even condensed ring structures that are coke precursors) into relatively unreactive gasoline-range paraffins + aromatics. The reactants are concentrated in the solventlike pores.

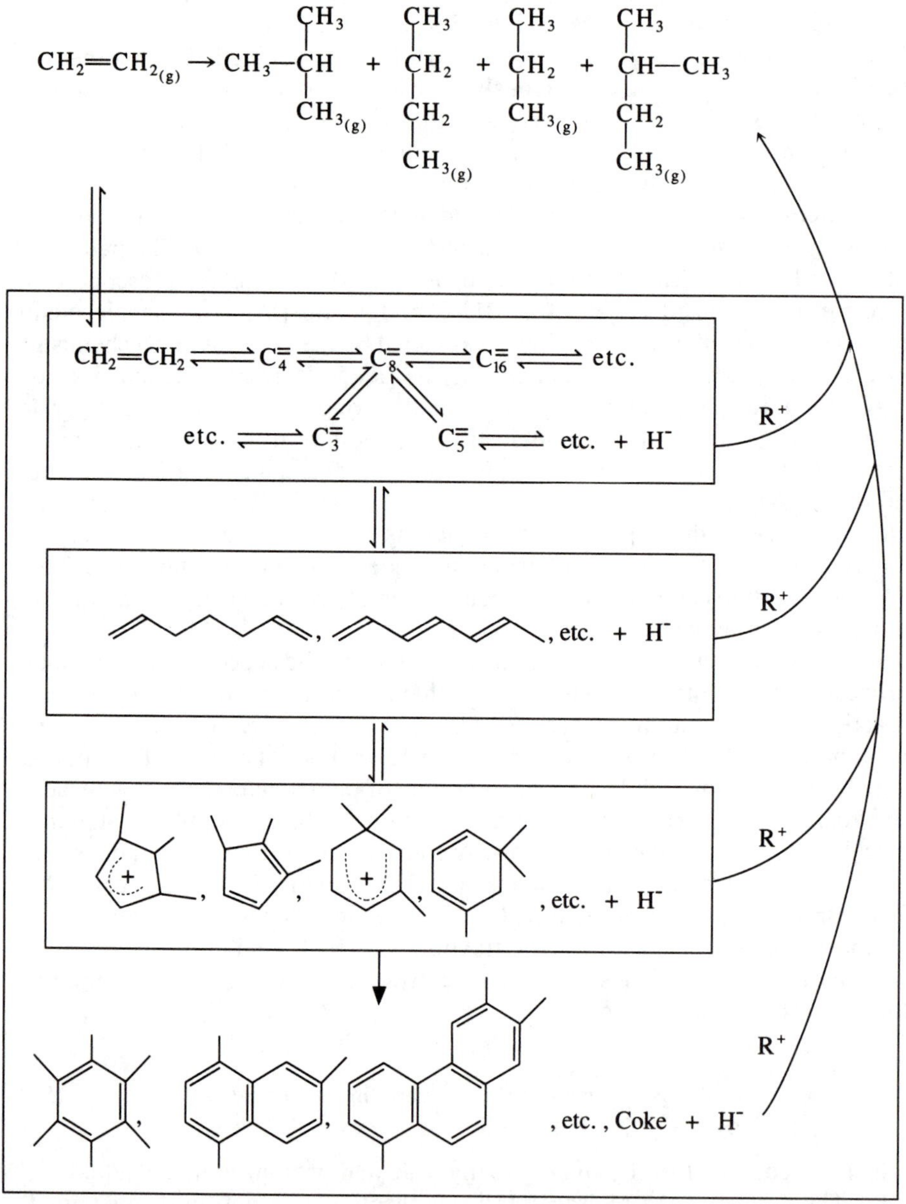

Figure 5-26
The redistribution of hydrogen in hydrocarbons formed from low-molecular-weight olefins [25]. The ring closures result in formation of polycyclic aromatics, ultimately including coke. The reactions shown in the large rectangle take place in the zeolite pores. Many hydrocarbons are precursors of coke. C_n olefins are denoted $C_n^=$.

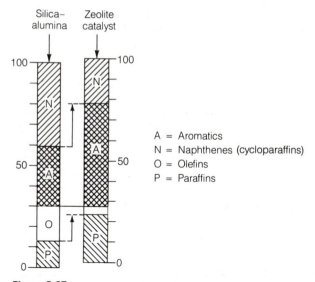

Figure 5-27
Distributions of products obtained in cracking of a petroleum fraction with an amorphous silica–alumina and with a zeolite catalyst [16].

Consequently, the olefin reacts to give desired products and has less opportunity to undergo cracking to the undesired C_1–C_4 products. More of the valuable gasoline-range products are obtained with the zeolite; less olefin is obtained, which is an advantage because olefin rapidly undergoes side reactions, including formation of cracked products having molecular weights too low for gasoline. Olefins, however, increase the octane number of gasoline; there is a trade-off between yield and octane number.

OTHER REACTIONS OF OLEFINS

Many other reactions of olefins are catalyzed by acidic zeolites [27]. These proceed through carbenium ion intermediates, and often the patterns of reactivity are just those predicted on the basis of the solution catalysis, combined with transport and steric effects. These are discussed in the following section.

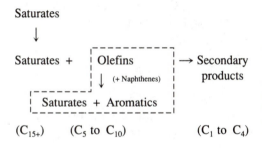

Figure 5-28
Schematic representation of some of the reactions occurring during catalytic cracking [16]. The zeolite catalyst favors the bimolecular reactions that convert olefins into less reactive and desirable gasoline components, thereby minimizing the further cracking into undesirable C_1–C_4 products.

There are, however, some contrasts between the reactions in solutions and in zeolites that are not related to the restrictions of the zeolite pore structure; rather, they illustrate the effects of adsorption of reactants and pertain to reactions catalyzed by surfaces generally, including those of the porous polymers or amorphous oxides.

These points are illustrated by a comparison of two aromatic reactants in ethylation [28]. In acidic solutions, ethylene reacts with phenol more rapidly than with benzene, since phenol has the more electron-rich aromatic ring undergoing electrophilic attack by an ethyl cation (formed as a result of protonation of ethylene by the catalyst). In the presence of the zeolite, however, phenol reacts less rapidly than benzene. These results are explained by competitive adsorption phenomena [28]. The ethylene must be protonated for the catalytic cycle to proceed; the protonation is represented as an adsorption step:

$$H^+ + CH_2{=}CH_2 \rightarrow (CH_3CH_2^+)_{ads} \tag{5-6}$$

where the proton donor group of the zeolite is abbreviated as H^+. It would be more exact to represent the carbenium ion as "solvated" by the polar (and anionic) environment of the zeolite pore. There is an analogy to the "solvation" of the carbenium ion by the sulfonated polymer catalyst, shown on p. 208. Carbenium ions do not form in measurable concentrations in the zeolite; they are highly reactive intermediates.

The adsorbed carbenium ion reacts with an impinging aromatic molecule from the environment of a zeolite pore, leading to alkylation. But when the polar phenol molecule replaces benzene in the zeolite pore, it competes successfully with the less-polar olefin for the acidic sites, forming adsorbed phenol:

$$ArOH + H^+ \rightarrow (ArOH_2^+)_{ads} \tag{5-7}$$

Since phenol is more strongly adsorbed than ethylene, it usurps the catalytic sites, hinders protonation of the olefin, and therefore blocks the catalytic cycle. At higher temperatures ($>200°C$), the coverage of the sites by phenol is less, and some ethylene can adsorb and be alkylated. The idea of inhibition of reaction by a reactant (or by any other molecule having a strong affinity for the catalytic groups) is the same as that introduced in Chapter 2.

The interpretation of phenol's role as an inhibitor has been confirmed in separate experiments involving phenol and benzene reacting with deuterated zeolite (i.e., zeolite with —OD groups instead of —OH groups [28]). The results show that phenol is strongly adsorbed under reaction conditions and undergoes H–D exchange, both at the —CH and —OH positions.

EXAMPLE 5-1   Kinetics and Mechanism of Zeolite-Catalyzed Alkylation of Aromatics

Problem

The kinetics of isopropylation and of ethylation of toluene catalyzed by rare-earth-exchanged zeolite Y have been reported to be of the following form [29]:

$$r = \frac{kC_O C_T}{1 + K_O C_O} \qquad (5\text{-}8)$$

where O = olefin and T = toluene.

What does this result imply about the mechanism of the zeolite-catalyzed alkylation?

Solution

The form of the rate equation is indicative of Rideal kinetics, as was observed for isobutylene oligomerization (Example 4-10) and for benzene propylation (Section 4.7) catalyzed by macroporous sulfonic acid polymers. The equation can be written as

$$r = k\theta_O C_T \qquad (5\text{-}9)$$

where the Langmuir model has been assumed to represent adsorption of olefin and θ_O is the fractional coverage of the sites with olefin. This equation is consistent with a rate-determining step that is the reaction of the adsorbed carbenium ion with the toluene impinging from the surrounding pore volume of the zeolite. In the presence of a strong acid, the formation of the carbenium ion is expected to be rapid, and therefore this is just the result expected on the basis of the analogous solution catalysis. It also agrees with the patterns observed with the sulfonic acid polymer; it is expected to pertain to strong acids with some generality.

This interpretation is evidently consistent with the description of phenol alkylation given in the preceding paragraphs. What form of kinetics would you expect for phenol ethylation catalyzed by the zeolite?

STERIC AND TRANSPORT EFFECTS: MOLECULAR SIEVING AND SHAPE-SELECTIVE CATALYSIS

The discussion of zeolite catalysis is based on acid–base chemistry analogous to that described in Chapter 2, but thus far there has been no consideration of the steric and transport restrictions imposed by the narrow pore structure. The effects of these physical restrictions are often dramatic, overturning the patterns of reactivity expected on the basis of solution catalysis alone. The effects are exploited in many industrial processes; some of them are used as examples to illustrate the principles stated in the following pages.

Two types of constraints are imposed by the zeolite pores:

1. A transport restriction such as that described by the Thiele model (Section 4.8.1): Transport of reactants or products may be affected strongly by the pores and, in the limiting case, one kind of molecule may be too large to fit through the pores, although another may fit. This is the idea of molecular sieving whereby, for example, small molecules in a

mixture may enter the pores and be catalytically converted whereas larger molecules pass through the reactor unconverted.

2. A different kind of restriction related to the steric requirements of the catalysis rather than to the transport of reactants and products: If there is not sufficient space in the zeolite pores to allow formation of the transition state for a step in a catalytic cycle, then that cycle is suppressed. If it is suppressed at the expense of another cycle, the term *restricted transition state selectivity* is applied.

These ideas of shape-selective catalysis are developed in the pages that follow.

Consider a mixture of two reactants, one small enough to enter the pores of a zeolite catalyst and the other too large to enter. A classic example is *n*-butanol in a mixture with isobutanol in the presence of zeolite A incorporating Ca^{2+} ions [30]. The straight-chain alcohol fits into the zeolite pores and undergoes catalytic dehydration much faster than the branched-chain homologue, which does not fit.

Another example involves cracking of straight- and branched-chain paraffins [30]. The data shown in Table 5-5 demonstrate that the straight-chain reactant is selectively converted in the presence of the small-pore zeolite A, whereas the intrinsic reactivity of the branched-chain reactant is greater, as is confirmed by the data obtained with the amorphous catalyst $SiO_2-Al_2O_3$.

These two examples are extreme cases demonstrating molecular sieving in catalysis. The reactivity pattern expected on the basis of solution chemistry alone is reversed by the zeolite. The limiting case of no reaction of the larger reactant is not quite reached because some reaction occurs on the small external surface of the zeolite crystallites.

Transport in zeolite pores also influences selectivity in the less extreme case in which both reactants can enter the pores, but one is transported more rapidly than the other. The framework for treating the transport-influenced kinetics, the Thiele model, was developed in Chapter 4, Section 4.8.1. Some data that allow an evaluation of this model for zeolites form the basis for the following examples.

Table 5-5
DEMONSTRATION OF SHAPE-SELECTIVE CRACKING OF PARAFFINS
BY ZEOLITE A [30]

Reactant	No Catalyst[b]	Conversion, %[a]		Silica–alumina
		4A Zeolite	5A Zeolite	
n-Hexane	1.1	1.4	9.2	12.2
3-Methylpentane	<1.0	<1.0	<1.0	28.0

[a] Reaction conditions: 500°C, contact time = 7 s.
[b] Conversion determined with reactor packed with quartz chips.

EXAMPLE 5-2 Diffusion with Reaction in Zeolite Pores

Problem

The zeolite mordenite has acidic catalytic sites in pores that are nearly straight, nonintersecting tubes with elliptical apertures having dimensions of 0.67×0.70 nm (Table 5-2). Rates of the dehydration reaction of methanol to give dimethyl ether catalyzed by mordenite crystals in various size ranges have been measured at 205°C and 1 atm, with the following results [31]:

Catalyst Pore Length, μm	$10^4 \times$ Rate of Methanol Dehydration, mol/(g of catalyst·s)
5.9 ± 3.2	7.33 ± 0.94
11.3 ± 4.3	6.17 ± 0.48
16.6 ± 5.7	4.94

What do these data imply about the accessibility of the interior pore structure to the reactant?

Solution

The qualitative interpretation of these results is straightforward: As the crystallites of zeolite become longer (i.e., as the pores, which are accessible only from the ends of the crystallites, become longer), the catalytic activity per unit mass (or per pore) decreases. In other words, the effectiveness factor becomes smaller as the pore becomes longer, and a comparison of the data with the Thiele model is suggested.

The first two data points (which have greater precision than the third) allow a comparison with the Thiele model for first-order reaction in an infinite flat slab (why is this the proper geometry?),

$$\eta = \frac{\tanh \phi_L}{\phi_L} \qquad (5\text{-}10)$$

which can be derived in the same way as Eq. 4-54 of Chapter 4, provided that slab geometry is assumed. The ratio of the two rate values is equal to the ratio of the two values of η, and the ratio of the two values of pore length is equal to the ratio of the two values of ϕ_L. Therefore, Eq. 5-10 can be solved for η and ϕ_L, which are plotted in Fig. 5-29. Also shown on the figure are the lines corresponding to Eq. 5-10 and the third experimental value, calculated from the pore length and the values of k and D determined by the first two points ($k = 190$ s^{-1}; $D = 7 \times 10^{-5}$ cm^2/s).

Since all three data points agree well with the Thiele model, it is concluded that it may be an acceptable first approximation. The value of D for methanol in the pore is several orders of magnitude less than that for self-diffusion of methanol in the pure liquid. The diffusion of methanol

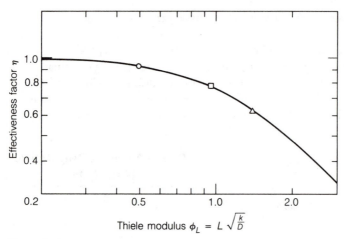

Figure 5-29
Effectiveness factors determined for methanol dehydration catalyzed by
H-mordenite [31].

is evidently hindered by the pores and the counterdiffusing product ether
and water molecules. ■

The opportunities for influencing selectivity by the choice of zeolite structure are evident from the strong dependence of effective diffusivity in the configurational diffusion range on pore dimensions, as is shown qualitatively in
Fig. 5-20. Small differences in the dimensions of molecules can dramatically
affect the transport rates. The data of Fig. 5-30 illustrate a three-order-of-mag-

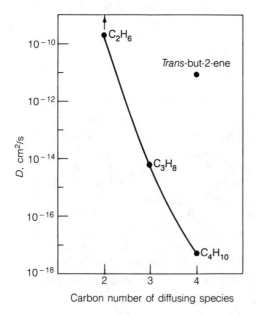

Carbon number of diffusing species

Figure 5-30
Transport of small hydrocarbons in hydrogen form zeolite T [32]. Zeolite T is mostly
offretite with a minor intergrowth of erionite.

nitude difference in the diffusivities of *trans*-but-2-ene and *n*-butane. (Can you explain why the difference is so great?)

When the transport of a species in a zeolite pore is restricted, the effectiveness factor for conversion of that species may be small. This is demonstrated by a dependence of the rate of the catalytic reaction of this species on the Thiele modulus, which is most easily varied by variation of the catalyst particle size (pore length).

In contrast, restricted transition state selectivity is a steric effect, which is influenced by the local pore geometry and not by the transport resistance. It is independent of the catalyst particle size (pore length) but dependent on the pore diameter. The distinction is illustrated in the following example.

EXAMPLE 5-3 Shape-Selective Cracking Catalyzed by HZSM-5 [33]
Problem

Analyze the data that characterize cracking of hydrocarbons by HZSM-5 (Table 5-6); determine the nature of the selectivity.

Solution

Some qualitative results are obvious. The olefins are more reactive than the paraffins, as is expected on the basis of the solution acid–base chemistry, since the olefins are much more easily protonated. Some observed rate constants are independent of catalyst particle size (e.g., for *n*-hexane cracking); the effectiveness factors are therefore unity. Some reactions, however, are characterized by significant intracrystalline transport restrictions (e.g., 2,2-dimethylbutane cracking).

Effectiveness factors are calculated for the reactions influenced by the transport restrictions. Since the dominant diffusion path involves the straight pores, which are larger in cross section than the zigzag pores (Fig. 5-9b), the crystallites are approximated as (infinite) flat slabs, as in Example 5-2. Solving for η and ϕ as in Example 5-2 gives the results summarized in Table 5-6.

Some patterns in the data are indications of intracrystalline transport limitations, as follows:

1. Among the hexanes, the linear and monobranched compounds are subject to negligible transport resistance, but the bulkier dibranched compound is subject to significant resistance. The same pattern holds true for the nonanes.

2. Among the hexenes, all of the compounds are subject to a significant transport resistance. They are intrinsically more reactive than the hexanes; hence, for a given skeletal structure (e.g., *n*-hexane and 1-hexene), the Thiele modulus is greater for the olefin than for the paraffin.

Calculation of the values of η and ϕ allows calculation of the values of the intrinsic rate constants, which are included in Table 5-6. A remarkable result emerges from the comparison of these intrinsic reactiv-

Table 5-6
REACTIVITIES OF HYDROCARBONS IN CRACKING CATALYZED BY HZSM-5 WITH A Si/Al RATIO OF 72 AT 538°C AND 1 ATM [33]

Reactant	Observed First-Order Rate Constant, s^{-1}		Effectiveness Factor		Intrinsic First-Order Rate Constant, s^{-1}
	0.025 μm[a]	1.35 μm	0.025 μm	1.35 μm	
n-Hexane C—C—C—C—C—C	29	28	1	1	29
3-Methylpentane C—C—C—C—C with C branch	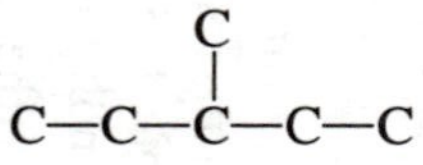 19	20	1	1	19
2,2-Dimethylbutane C—C—C—C with C branches	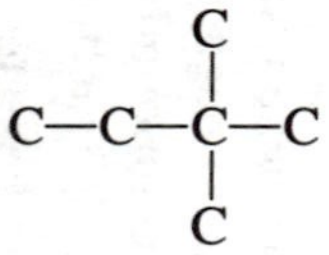 12	3.6	1	0.30	12
n-Octane C—C—C—C—C—C—C—C	54	—	—	—	—
2-Methylheptane C—C—C—C—C—C—C with C branch 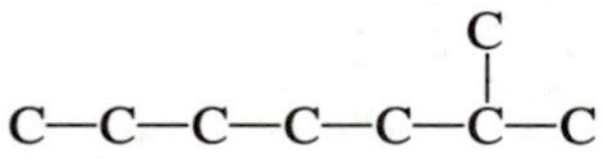	37	—	—	—	—

		Structure				
n-Nonane	93	C—C—C—C—C—C—C—C—C	93	1	1	93
2,2-Dimethylheptane	63	C—C—C—C—C—C (with C, C branches)	8.4	1	0.13	63
n-Dodecane	663	C—C—C—C—C—C—C—C—C—C—C—C	662	1	1	663
1-Hexene	7530	C—C—C—C—C=C	6480	1	0.86	7530
3-Methylpent-2-ene	7420	C—C—C=C—C (with C branch)	3610	1	0.50	7420
3,3-Dimethylbut-1-ene	4350	C—C—C=C (with C branches)	141	0.86	0.028	4950

IS-5P

[a] Zeolite crystallite size.

291

ities. Among the hexanes, the order of reactivity is *n*-hexane > 3-methylpentane > 2,2-dimethylbutane; among the hexenes, the order of reactivity is 1-hexene ≥ 3-methylpent-2-ene > 3,3-dimethylbut-1-ene. This order in each case is different from what would be expected simply on the basis of the solution acid–base chemistry; the reactants giving the carbenium ion most readily in solution are not the most reactive in the zeolite.

The transport effects have been resolved in the calculation of the intrinsic reactivities; it is evident that the zeolite imposes a selectivity separate from the transport effect; this is therefore an example of a restricted transition state selectivity. Both effects suppress the cracking of more highly branched hydrocarbons relative to the cracking of the less highly branched hydrocarbons. Both effects tend to reverse the order of reactivity determined by simple acid–base chemistry. ■

The above example shows that the effect of the transport and steric limitations is to suppress the reactions of the branched hydrocarbons. Branched hydrocarbons are preferred motor fuel components because they have relatively high octane numbers; straight-chain paraffins have lower octane numbers than the branched isomers. HZSM-5 is used in processes to convert preferentially the paraffins in a mixture having low octane numbers [34, 35]. The data of Fig. 5-31 illustrate how the less desirable compounds are selectively removed. These facts explain why HZSM-5 is used as a component of cracking catalysts designed to give products with high octane numbers.

Another practical example of shape-selective catalysis by HZSM-5 in-

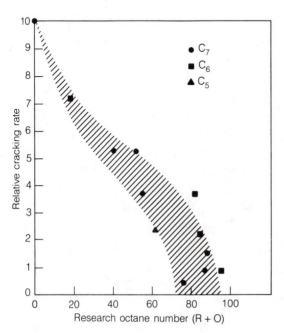

Figure 5-31

Selective cracking of low-octane-number paraffins by the zeolite HZSM-5 [34, 35]. Alkylation of aromatics accompanies cracking in the process.

volves the disproportionation of toluene, shown schematically in Fig. 5-32 [34]. The desired product is *para*-xylene; the other isomers are less valuable. The para isomer is formed in excess of the equilibrium concentration, and the selectivity results from the restriction of transport of the ortho and meta isomers. This is a third type of shape selectivity, resulting from the transport of the products rather than the reactants; the undesired products are trapped within the zeolite and react away to give the desired product, which escapes rapidly.

Several modifications of the zeolite have been successful in increasing the selectivity for *para*-xylene to >90 percent [36]. These include increasing the zeolite crystallite size, introducing cations or other inorganic matter into the pore structure, and closure of a fraction of the pore mouths. (Can you explain why each of these changes improved selectivity?)

Restrictions on the transition states that might be formed are important in the reactions that accompany the benzene–ethylene alkylation to give ethylbenzene. A high selectivity for ethylbenzene is observed because of the suppression of the side reactions [37, 38]. HZSM-5 is consequently used as the catalyst in industrial processes for ethylbenzene manufacture.

The alkylbenzene formed in all these processes can be further converted in both isomerization and disproportionation reactions. The zeolites selectively suppress the bimolecular disproportionation reaction over the isomerization, which is believed to be unimolecular [34]. The data of Fig. 5-33 show the selectivity for disproportionation vs. isomerization of xylene catalyzed by various zeolites. These data make clear why HZSM-5 is the preferred catalyst for xylene isomerization.

HZSM-5 is also used in a commercial process for conversion of methanol into gasoline-range hydrocarbons (including aromatics) plus water [40]. The reaction sequence involves dehydration of the methanol to give dimethyl ether followed by a series of reactions that proceed through carbenium ion intermediates. The largest products (e.g., durene, 1,2,4,5-tetramethylbenzene) correspond to the highest-boiling compounds in gasoline. Restricted transition state selectivity accounts for the favorable product distribution.

These examples of catalysis by HZSM-5 represent some of the most successful industrial applications of catalysis; they are remarkably well understood. More important, the development of processes with these catalysts has emerged in large measure from thought processes like those stated in this chapter. It is proper to talk about design of the zeolite catalysts. A vivid contrast will be evident in the next chapter, which is primarily concerned with amorphous solid catalysts (the typical ones used in industrial practice), which are still developed largely by Edisonian methods.

INFLUENCE OF STERIC AND TRANSPORT EFFECTS ON CATALYST DEACTIVATION

Zeolite catalysts are deactivated by the coke that accumulates in the pores. The pore structure strongly influences the rate of coke accumulation, as explained by steric and transport effects.

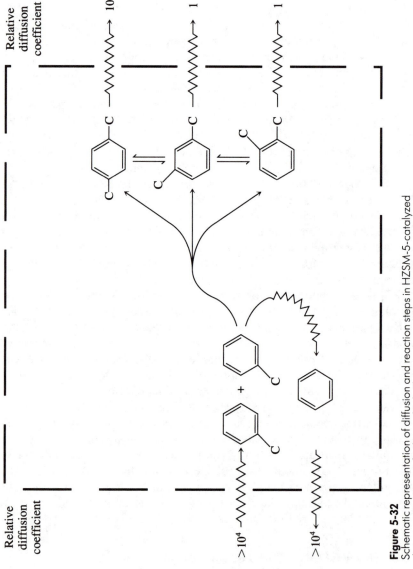

Figure 5-32
Schematic representation of diffusion and reaction steps in HZSM-5-catalyzed disproportionation of toluene [34]. The numbers are approximate values of relative diffusion coefficients into and out of the pores of the zeolite.

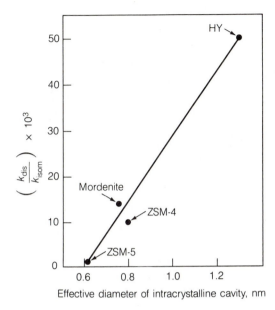

Figure 5-33
Selectivities of acidic zeolites for disproportionation and isomerization of xylene. The HZSM-5 catalyst is preferred,
minimizing the bimolecular disproportionation reaction by virtue of restricted
transition state selectivity [39].

The zeolite ZSM-5, for example, is resistant to coke formation because of steric constraints on the transition states, and this is one of the reasons why it finds wide industrial application [41]. The restricted transition state selectivity has been suggested to prevent secondary reactions of alkylaromatics [41].

The shape selectivity is even more pronounced for zeolites having smaller pores and less pronounced for those having larger pores; ZSM-5 is often the optimum. Coke yields for a variety of zeolites have been correlated with a measure of the shape selectivity in paraffin cracking (the ratio of the rate constant for *n*-hexane cracking to that for 3-methylpentane cracking), as shown in Fig. 5-34. The pore geometry is of most importance in determining the rate of coke formation in zeolites; the Si/Al ratio and crystallite size are less important. Feed composition is also important, but this is not evident from the data.

The larger-pore zeolites such as X and Y (represented by some of the points in the upper left of Fig. 5-34) are especially susceptible to deactivation by coke [27]. The coke formed in the supercages may be trapped, being too large to diffuse out through the smaller apertures. This is an example of a reverse-shape selectivity, which is strongly disadvantageous.

5.6.2 Catalysis by Zeolites Containing Metal Complexes and Clusters

TRANSITION METAL COMPLEXES

The solution chemistry of acid–base catalysis provides the platform for understanding catalysis by hydrogen form zeolites, and the complications caused by the zeolite pore geometry can drastically alter the patterns of reactivity. A

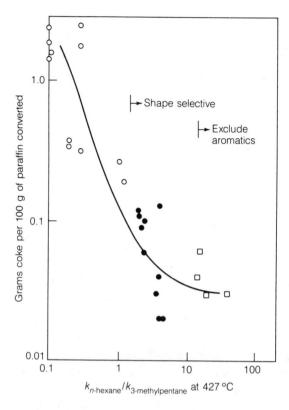

Figure 5-34
Shape-selective coke formation in ze-olites having various pore dimensions [41d]. The reactant mixture contained equal weights of n-hexane, 3-methyl-pentane, 2,3-dimethylbutane, benzene, and toluene. The H_2/hydrocarbon ratio (presumably molar) was 3/1, and the pressure 15 atm.

similar statement applies to transition metal complex catalysis; there are, how-ever, far fewer examples of catalysis by metal complexes in zeolites and none of industrial significance.

A relatively well-understood example is methanol carbonylation cata-lyzed by Rh(I) complexes in faujasites [42]. The catalysts were prepared by partial exchange of Na form X or Y zeolite with Rh cations. The optimum Rh loading is about one ion per unit cell. This loading corresponds to an average spacing of 2.5 nm between Rh ions, which is consistent with the inference that they are isolated and act independently. The kinetics of methanol carbonylation by the zeolite was found to be of the form [42]

$$r = kP_{CH_3I} \tag{5-11}$$

when the reaction was carried out with CH_3OH and CH_3I (the promoter) in the vapor phase. Since the rate is expressed per unit mass of zeolite, the form of the kinetics is equivalent to that observed for methanol carbonylation cat-alyzed by Rh complexes in solution [Eq. 2-146, p. 99]. The activation energy is nearly the same for the zeolite-catalyzed reaction and that occurring in so-lution, suggesting that nearly the same catalytic cycle takes place in the zeolite supercage and in solution. The rhodium in the zeolite is, however, an order of magnitude less active than that in solution, which suggests that the Rh sites in the zeolite are not uniformly accessible (a transport restriction) and/or that the activity of the Rh in the zeolite cage is less than that of the Rh in solution (comparable to a solvent or cage effect). (What experiments would you suggest to test these possibilities?)

There are other examples of reactions catalyzed by transition metal ions in zeolites, including oligomerization and hydroformylation of olefins and oxidation of CO and of hydrocarbons [42]. A reaction of hexane to give benzene (dehydrocyclization, considered again in Chapter 6) has been found to be catalyzed by zeolite X containing Na and Te ions [43, 44], although the activity is much less than that of a platinum catalyst. A thorough determination of the zeolite structure by X-ray diffraction and other techniques led to the inference of the catalytic site shown in Fig. 5-35. The site is a Te^{2+} ion located in a supercage and coordinated to two Na^+ ions, one in site II and one in site III. The suggestion of the catalytic site is remarkable for its preciseness; only for a crystalline material could the surroundings of the Te ion be determined so exactly.

Copper ions (Cu^+ and Cu^{2+}) in ZSM-5 are active in a redox cycle for the catalytic decomposition of NO to give N_2 and O_2 [45]. The chemistry has been characterized by microbalance measurements of oxygen adsorption and desorption combined with measurements of the kinetics of the catalytic reaction [46]. The structure of the catalytic site is not known.

An oxidation catalyst trapped in a faujasite has been made by a ship-in-a-bottle synthesis [47]. Iron complexes with large multidentate phthalocyanine ligands formed in the supercages and were too large to migrate through the apertures.

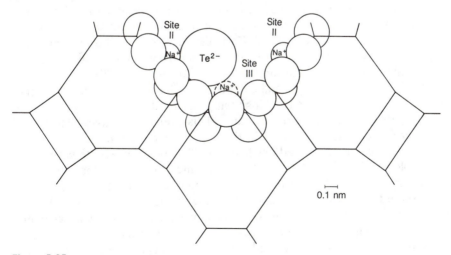

Figure 5-35
Catalytic site for dehydrocyclization of n-hexane: a Te^{2-} ion coordinated to two Na^+ ions in the faujasite supercage [43]. The solid lines represent the Si–Al framework; circles represent oxygen ions.

The design of a bifunctional zeolite catalyst has been inferred from known solution chemistry. An example familiar from Chapters 2 and 4 is the oxidation of ethylene to give acetaldehyde, which requires Pd^{2+} to oxidize the ethylene and Cu^{2+} to reoxidize the reduced Pd. Zeolite Y containing Pd^{2+} and Cu^{2+} is an effective bifunctional catalyst for this reaction [48]. The optimum Cu^{2+}/Pd^{2+} ratio appears to be about 6.

A limitation of the zeolites as supports for metal complex catalysts may be the difficulty of getting the catalyst into the zeolite; since some complexes are too large to pass through the apertures, they must be formed in the cages. (Is this true for the methanol carbonylation catalyst?) On the other hand, the zeolite matrices provide high capacities for the metal ions and the advantages of site isolation considered in Chapter 4.

METAL CLUSTERS AND AGGREGATES

When faujasite containing Rh ions is subjected to 80 atm of $CO + H_2$ at 130°C, a restructuring of the metal takes place, giving species inferred from the infrared spectrum to be $Rh_6(CO)_{16}$ in the supercage [49]. Other relatively stable metal clusters, including $Ir_6(CO)_{16}$, have been prepared similarly [50]. These clusters are formed in "ship-in-a-bottle" syntheses; they become trapped in the zeolite cages. This would seem to be a way to stabilize the clusters for catalysis, but there is little evidence that these neutral clusters have much activity.

Metal clusters, suggested to be Na_4^{3+} in the sodalite cages of zeolite Y, were formed by introduction of sodium azide, NaN_3, into the sodium form of the zeolite, followed by heating [51]. The framework of the resulting zeolite is strongly basic; the zeolite catalyzes hydrocarbon conversions much as they occur in basic solutions [52]. The zeolite has been used for ship-in-a-bottle syntheses of osmium and rhodium carbonyl clusters, believed to be anions. These clusters in the zeolite are more stable than neutral clusters in zeolites, provided that an environment of stabilizing ligands $(CO + H_2)$ is present; they catalyze the conversion of CO and H_2 into hydrocarbons in the case of osmium [53a] and alcohols + hydrocarbons in the case of rhodium [53b].

If a faujasite containing metal complexes such as $[Pt(NH_3)_4]^{2+}$ is treated with a reducing agent such as H_2 at high temperature, the metal may be reduced and aggregated inside the pore structure. This preparation gives highly dispersed zero-valent metals trapped in the cages, which are active for a great range of reactions, as discussed in Chapter 6. The metals in zeolite pores may be unstable at high temperatures, however, migrating out of the pores and forming larger aggregates (a process called sintering, discussed in Chapter 6). Formation in the cages of aggregates that have diameters greater than the aperture diameter does not necessarily prevent sintering because small fragments may form and migrate out of the pores.

Small platinum clusters formed in zeolite L appear to be both stable and nearly uniform in structure. The samples are prepared by introducing platinum amine nitrate followed by reduction in hydrogen at 500°C [54]. The reduction converts the platinum into small clusters, estimated to have five or six atoms each on the basis of EXAFS spectroscopy (a technique described in Chapter 6) [54]. This catalyst is highly selective for conversion of *n*-hexane into benzene, a dehydrocyclization reaction described in Chapter 6, and it has been investigated for possible industrial application.

The structures of the platinum clusters in the zeolite are not known with certainty, but the clusters are very small and have been observed by electron microscopy to be in the zeolite pores and virtually absent from the regions outside the pores. A model of a plausible structure of a platinum cluster in zeolite L is depicted on the front of the dust jacket of this book.

A shape-selective hydrogenation catalyst was devised that converted only ethylene in a mixture with propylene at 65 to 260°C [55]. The catalyst was made from mordenite, as Na$^+$ ions were first removed by exchange with H$^+$, then platinum-containing ions were adsorbed (these are too large to penetrate the sodium form mordenite), and then the sodium ions were reintroduced. The intracrystalline structures had sufficient space for transport of ethylene, ethane, and propylene, but propane was not formed. Therefore, both olefins and hydrogen could reach the platinum sites, but after hydrogenation, only the ethane could escape.

5.6.3 Summary

Many of the lessons of Chapter 2 were based on functional group chemistry, since families of catalysts used in solution are active for broad classes of reactions, for example, for hydrogenation of a double bond. A subtle kind of selectivity was introduced with the examples of L-dopa synthesis (pp. 85–91) and propylene polymerization (p. 105), whereby the molecular architecture of the catalyst allows formation of one product isomer almost to the exclusion of another. The enzymes epitomize selectivity that is sensitive to the molecular structure of the reactant, including the positions of the reactive groups and the size, shape, and symmetry of the molecules. The enzymes are so selective because each one has a complex molecular architecture just meeting the requirements of a particular reactant. Man-made catalysts are less selective because they are crudely constructed. But the zeolites, having molecular-scale pockets resembling those of the enzymes, evoke selectivities based on dimensions of the reactants and can be viewed as a first step in the practical design of selective catalysts. The large-scale applications of zeolite catalysts are primarily acid-catalyzed hydrocarbon conversions.

5.7
NONZEOLITE MOLECULAR SIEVES

Several families of new molecular sieves have been reported with elements other than just Al and Si in the tetrahedral framework positions [56, 57]. These are similar in structure to the zeolites, and some duplicate zeolite frameworks. Aluminophosphate molecular sieves have Al and P in the framework with oxygen; these are referred to with the acronym AlPO. Structures with Si, Al, and

P are referred to as SAPOs. Many other elements have been incorporated in the frameworks.

These new molecular sieves include some acids that catalyze reactions much as some zeolites do, but for the most part these materials lack the acid strengths and stabilities of zeolites, and important applications have not yet been reported. There is a new large-pore molecular sieve with rings of 18 oxygen atoms [58], which gives hope that new cracking catalysts may be found for some of the largest molecules in petroleum. However, large-pore molecular sieves with the acidity and stability required for cracking processes have not been found.

5.8
CLAYS AND OTHER LAYERED MATERIALS

There are many inorganic solids with regular structures that incorporate spaces having molecular dimensions. Sheet silicates (often called clays), like zeolites, are cation exchangers. Smectites have layer lattice structures in which two-dimensional oxyanions are separated by layers of (hydrated) cations [59, 60]. The spaces between the layers are accessible even to large molecules, the incorporation being referred to as intercalation. There is a rich chemistry of the intercalated species, including catalysis, similar to that occurring in zeolites [59, 60].

Like polymers, the smectites are able to swell. The extent of expansion of the layers depends on the nature of the swelling agent, the exchange cation, the layer charge, and the location of the layer charge [59, 60]. Under some conditions, the layers of smectites can be separated by tens of nanometers of water, and the layers can even be completely dispersed. The swelling makes the materials difficult to apply as catalysts, but it also obviously offers some opportunities for regulation of shape-selective catalysis.

Only a few commercial applications of clay catalysts exist, but in the early days of catalytic cracking (the 1940s), acidic clays were used on a large scale. Recent work has revived interest in these materials as catalysts, as large cation pillars have been used to hold the layers apart and allow entry of the large molecules found in the heaviest fractions of petroleum. These molecules are too large to enter zeolite pores, and the clays are an intriguing alternative [61].

Transition metal complexes have been incorporated between the layers and have been observed to catalyze the reactions expected on the basis of solution behavior [59]. The dramatic selectivity effects of zeolites have not been observed, however. The layered silicate hectorite (Fig. 5-36) was used, for example, as a host for rhodium phosphine complexes to catalyze olefin hydroformylation [61]. The positively charged phosphonium ligand $Ph_2PCH_2CH_2P^+Ph_2CH_2Ph$ was used to allow an electrostatic bond with the negatively charged layers of the hectorite, which was needed to stabilize the catalyst.

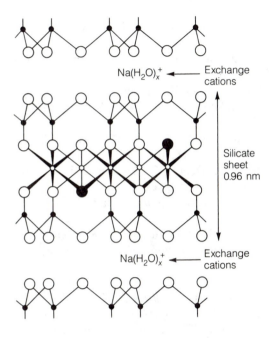

Na(H₂O)ₓ⁺ ← Exchange cations

Silicate sheet 0.96 nm

Na(H₂O)ₓ⁺ ← Exchange cations

Figure 5-36
Schematic representation of the structure of the clay hectorite, $Na_{0.66}$ $(Li_{0.66}Mg_{5.34})$ $(Si_{8.00})O_{20}(OH, F)_4$ [57]: O, oxygen; ●, hydroxyl or fluorine; ●, Si in tetrahedral positions; ○ Li or Mg in octahedral positions.

Many other inorganic materials have molecular-scale spaces and pores. These include graphite. Compounds such as $AlCl_3$ intercalated in graphite have been used as catalysts; it is likely that the graphite simply holds this Lewis acid in a reservoir, with the catalysis occurring outside the layers. Carbons with small pores have molecular sieving properties and are of interest as supports for shape-selective catalysts [62].

There is a class of materials called transition metal chalcogenides (e.g., TaS_2, NbS_2, MoS_2, and WS_2) that also have layer structures. MoS_2 is important as a component of catalysts used in hydroprocessing of fossil fuels, as described in Chapter 6.

REFERENCES

1. Smith, J. J., *Adv. Chem. Ser.*, **101**, 171 (1971).
2. Olson, D. H., Kokotailo, G. T., Lawton, S. L., and Meier, W. M., *J. Phys. Chem.*, **85**, 2238 (1981).
3. Flanigen, E. M., Bennet, J. M., Grose, R. W., Cohen, J. P., Patton, R. L., Kirchner, R. M., and Smith, J. V., *Nature*, (London) **217**, 512 (1978).
4. Topsøe, N.-Y., Pedersen, K., and Derouane, E. G., *J. Catal.*, **70**, 41 (1981).
5. Lyman, C. E., Betteridge, P. M., and Moran, E. F., *ACS Symp. Ser.*, **218**, 199 (1983).
6. Thomas, J. M., *Proc. 8th Int. Congr. Catal.* (Berlin), Vol. 1, p. 31, 1984.
7. Rollmann, L. D., in F. R. Ribeiro, A. E. Rodrigues, L. D. Rollmann,

and C. Naccache, eds., *Zeolites: Science and Technology*, Martinius Nijhoff, The Hague, 1984, p. 109.

8. Derouane, E. G., in M. S. Whittingham and A. J. Jacobson, eds., *Intercalation Chemistry*, Academic Press, New York, 1982, p. 101.

9. Flanigen, E. M., in F. R. Ribeiro, A. E. Rodrigues, L. D. Rollmann, and C. Naccache, eds. *Zeolites: Science and Technology*, Martinius Nijhoff, The Hague, 1984, p. 3.

10. Thomas, J. M., Klinowski, J., Rawdos, S., Anderson, M. W., Fyfe, C. A., and Gobbi, G. C., *ACS Symp. Ser.*, **218**, 159 (1983).

11. Gabelica, Z., Nagy, J. B., Bodart, P., Debras, G., Derouane, E. G., and Jacobs, P. A., in F. R. Ribeiro, A. E. Rodrigues, L. D. Rollmann, and C. Naccache, eds. *Zeolites: Science and Technology*, Martinius Nijhoff, The Hague, 1984, p. 193.

12. Barthomeuf, D., in F. R. Ribeiro, A. E. Rodrigues, L. D. Rollmann, and C. Naccache, eds. *Zeolites: Science and Technology*, Martinius Nijhoff, The Hague, 1984, p. 317.

13. Ward, J. W., *J. Catal*, **11**, 251 (1968).

14. Pfeifer, H., Freude, D., and Hunger, M., *Zeolites*, **5**, 274 (1985).

15. Barrer, R. M., in F. R. Ribeiro, A. E. Rodrigues, L. D. Rollmann, and C. Naccache, eds., *Zeolites: Science and Technology*, Martinius Nijhoff, The Hague, 1984, p. 237.

16. Weisz, P. B., *CHEMTECH*, **3**, 498 (1973).

17. Rabo, J. A., in F. R. Ribeiro, A. E. Rodrigues, L. D. Rollmann, and C. Naccache, eds., *Zeolites: Science and Technology*, Martinius Nijhoff, The Hague, 1984, p. 291.

18. Hölderich, W., Hesse, M., and Naumann, F., *Angew. Chem., Int. Ed. Engl.*, **27**, 226 (1988).

19. Olson, D. H., Haag, W. O., and Lago, R. M., *J. Catal*, **61**, 390 (1980).

20. DeCanio, S. J., Sohn, J. R., Fritz, P. O., and Lunsford, J. H., *J. Catal.*, **101**, 132 (1986).

21. Krannila, H., Gates, B. C., and Haag, W. O., paper presented at Catal. Soc. Meeting, Dearborn, Michigan, 1989.

22. Haag, W. O., and Dessau, R. M., *Proc. 8th Int. Congr. Catal.* (Berlin), Vol. 2, p. 305, 1984.

23. Venuto, P. B., and Habib, E. T., Jr., *Catal. Rev–Sci. Eng.*, **18**, 1, (1978).

24. Nace, D. M., *Ind. Eng. Chem. Prod. Res. Dev.*, **8**, 24 (1969).

25. Venuto, P. B., in G. V. Smith, ed., *Catalysis in Organic Synthesis*, Academic Press, New York, 1977.

26. Pine, L. A., Maher, P. K., and Wachter, W. A., *J. Catal*, **85**, 466 (1984).

27. Venuto, P. B., and Landis, P. S., *Adv. Catal.*, **18**, 259 (1968).

28. Venuto, P. B., and Wu, E. L., *J. Catal.*, **15**, 205 (1969).

29. Haag, W. O., cited in Venuto, P. B., *Adv. Chem. Ser.*, **102**, 260 (1971).

30. Weisz, P. B., Frilette, V. J., Maatman, R. W., and Mower, E. B., *J. Catal.*, **1**, 307 (1962).

31. Swabb, E. A., and Gates, B. C., *Ind. Eng. Chem. Fundam.*, **11**, 540 (1972).

32. Riekert, L., *AIChE J.*, **17**, 446 (1971).
33. Haag, W. O., Lago, R. M., and Weisz, P. B., *Faraday Disc. Chem. Soc.*, **72**, 317 (1981).
34. Weisz, P. B., *Proc. 7th Int. Congr. Catal.* (Tokyo), Vol. 1, p. 1, 1981.
35. Chen, N. Y., Garwood, W. E., Haag, W. O., and Schwartz, A. B., paper presented at Am. Inst. Chem. Engineers Meeting, San Francisco, 1979.
36. Chen, N. Y., Kaeding, W. W., and Dwyer, F. G., *J. Am. Chem. Soc.*, **101,** 6783 (1979).
37. Dwyer, F. G., Lewis, J. P., and Schneider, F. H., *Chem. Eng.*, Jan. 5, 1976, p. 90.
38. Csicsery, S. M., in J. A. Rabo, ed., *Zeolite Chemistry and Catalysis*, American Chemical Society, Washington, D.C., 1976, p. 680.
39. Olson, D. H., and Haag, W. O., *ACS Symp. Ser.*, **248**, 280 (1984).
40. Chang, C. D., *Catal. Rev.–Sci. Eng.*, **25**, 1 (1983).
41. (a) Walsh, D. E., and Rollman, L. D., *J. Catal.*, **56,** 195 (1979); (b) Rollmann, L. D., *ibid*, **47,** 113 (1977); (c) Walsh, D. E., and Rollman, L. D., *ibid.*, **49,** 369 (1977); (d) Rollmann, L. D., and Walsh, D. E., *ibid.*, **56,** 139 (1979).
42. Maxwell, I. E., *Adv. Catal.*, **31,** 1 (1982).
43. Mikovsky, R. J., Silvestri, A. J., Dempsey, E., and Olson, D. H., *J. Catal.*, **22,** 371 (1971).
44. Olson, D. H., Mikovsky, R. J., Shipman, G. F., and Dempsey, E., *J. Catal.*, **24,** 161 (1972).
45. Iwamoto, M., Yokoo, S., Sakai, K., and Kagawa, S., *J. Chem. Soc., Faraday Trans. I*, **77,** 1629 (1981).
46. (a) Li, Y., and Hall, W. K., *J. Phys. Chem.*, **94,** 6145 (1990); (b) Li, Y., and Hall, W. K., *J. Catal.,* **129**, 202 (1991).
47. Herron, N., Stucky, G. D., and Tolman, C. A., *J. Chem. Soc., Chem. Commun.*, **1986,** 1521.
48. Arai, H., Yamashiro, T., Kubo, T., and Tominaga, H., *Bull. Japan. Petrol. Inst.*, **18,** 39 (1976).
49. Mantovani, E., Palladino, N., and Zanobi, A., *J. Mol. Catal.*, **3,** 285 (1977).
50. Bergeret, G., Lefebvre, F., and Gallezot, P., in Y. Murakami, A. Iijima, and J. W. Ward, eds., *New Developments in Zeolite Science and Technology*, Elsevier, Amsterdam, 1986, p. 803.
51. Fejes, P., Kiricsi, I., Hannus, I., Tihanyi, T., and Kiss, A., in B. Imelik, C. Naccache, Y. Ben Taarit, J. C. Vedrine, G. Coudurier, and H. Praliaud, eds. *Catalysis by Zeolites*, Elsevier, Amsterdam, 1980, p. 135.
52. Martens, L. R., Vermeiren, W. J., Huybrechts, D. R., Grobet, P. J., and Jacobs, P. A., *Proc. 9th Int. Congr. Catal.* (Calgary), Vol. 1, p. 420, 1988.
53. (a) Zhou, P.-L., and Gates, B. C., *J. Chem. Soc. Chem. Commun.*, **1989,** 347; (b) Lee, T. J., and Gates, B. C., *Catal. Lett.*, **8,** 15 (1991).
54. Vaarkamp, M., von Grondelle, J., Miller, J. T., Sajkowski, D. J., Mod-

ica, F. S., Lane, G. S., Zajac, G. W., Gates, B. C., and Koningsberger, D. C., *Catal. Lett.*, **6**, 369 (1990).

55. Chen, N. Y., and Weisz, P. B., *Chem. Eng. Progr. Symp. Ser.*, **63** (73), 86 (1967).

56. Flanigen, E. M., Patton, R. L., and Wilson, S. T., *Stud. Surf. Sci. Catal.*, **37**, 13 (1988).

57. Rabo, J. A., in J.-M. Basset, B. C. Gates, J.-P. Candy, A. Choplin, M. Leconte, F. Quignard, and C. Santini, eds., *Surface Organometallic Chemistry: Molecular Approaches to Surface Catalysis*, Kluwer, Dordrecht, 1988, p. 245.

58. Davis, M. E., Montes, C., Hathaway, P. E., Arhancet, J. P., Hasha, D. L., and Garces, J. M., *J. Am. Chem. Soc.*, **111**, 3919 (1989).

59. Pinnavaia, T. J., *Science*, **220**, 365 (1983).

60. Clearfield, A., in J.-M. Basset, B. C. Gates, J.-P. Candy, A. Choplin, M. Leconte, F. Quignard, and C. Santini, eds., *Surface Organometallic Chemistry: Molecular Approaches to Surface Catalysis*, Kluwer, Dordrecht, 1988, p. 271.

61. Quayle, W. H., and Pinnavaia, T. J., *Inorg. Chem.*, **18**, 2840 (1979).

62. Foley, H. C., *ACS Symp. Ser.*, **32**, 1067 (1988).

FURTHER READING

Details of zeolite structure and chemistry (but not catalysis) are available in D. W. Breck's *Zeolite Molecular Sieves*, Wiley, New York, 1974, and R. M. Barrer's *Hydrothermal Chemistry of Zeolites*, Academic Press, New York, 1982, and *Zeolite and Clay Minerals as Sorbents and Molecular Sieves*, Academic Press, New York, 1978. A set of reviews of zeolites, including catalysis, has appeared in *Zeolites: Science and Technology*, edited by F. R. Ribeiro, A. E. Rodrigues, L. D. Rollmann, and C. Naccache, Martinius Nijhoff, The Hague, 1984. Many structural details of zeolites emerge from NMR spectroscopy, as summarized in G. Engelhardt and D. Michel's *High-Resolution Solid-State NMR of Silicates and Zeolites* (Wiley, Chichester, 1987).

PROBLEMS

5-1 Estimate the number of proton donor groups per unit volume in an H form faujasite with a Si/Al atomic ratio of unity. How does this compare with the value for the cross-linked poly(styrenesulfonic acid) discussed in Chapter 4? What does this comparison imply about the economic advantage of the polymer catalyst?

5-2 The zeolite HZSM-5 has been used to catalyze the cracking of *n*-butane at 496°C, with the distribution of products depending on the conversion, as shown in Table 5-7 [21]. The feed was pure *n*-butane at atmospheric pressure. Explain the distribution of products in terms of plausible catalytic cycles. Why is there a change in the mechanism as the conversion increases? Check the internal consistency of the data by determining the H/C ratio in each product and comparing it with the value for the feed.

Table 5-7
CONVERSION OF n-BUTANE CATALYZED BY HZSM-5

Product	Product Distribution, Extrapolated to Zero Conversion, mol%	Product Distribution at 17% Conversion, mol%
H_2	15	6
CH_4	20	9
C_2H_6	17	13
C_3H_8	0	40
C_2H_4	15	14
C_3H_6	16	10
C_4H_8	17	8

5-3 Cracking of n-butane was investigated by Bizreh and Gates [*J. Catal.*, **88**, 240 (1984)] with a pulse microreactor. This is a tubular flow reactor like that described in Example 4-1 (Chapter 4), except that the steadily flowing stream is an inert carrier gas, and the reactant is introduced with a syringe as small pulses; the products of the whole pulse are analyzed with an on-line gas chromatograph. The apparatus is especially useful for rapid determination of catalytic activity, for investigation of catalysts that deactivate rapidly, and for characterization of reactions in the near-absence of products. The data are shown in Table 5-8. What reactions occurred? What mechanism of cracking best describes the data? Why was the pulse microreactor well suited to the investigation?

5-4 In cracking catalyzed by zeolites, methane and ethane are usually minor

Table 5-8
PRODUCT DISTRIBUTION IN n-BUTANE CONVERSION CATALYZED
BY HZSM-5[a]

T, °C	Conversion	Product Distribution, mol%							
		CH_4	C_2H_6	C_2H_4	C_3H_8	C_3H_6	C_4H_8	C_5H_{12}	C_5H_{10}
498	0.024	18	17	21	0.1	19	21	—	2
448	0.0063	16	16	20	0.3	18	26	—	3
410	0.0023	13	13	18	0.1	14	39	—	2
373	0.00075	13	11	18	4	12	42	—	—
342	0.00019	17	11	33	14	16	9	—	—
498	0.023	19	18	22	0.1	20	18	—	3

[a] The catalyst was held on flowing He at 500°C for 2 h, then in He (without flow) for 4 h additionally, prior to injection of the first reactant pulse.

products and C_3–C_6 paraffins and olefins are major products. Explain why.

5-5 The phenol–acetone condensation reaction to give bisphenol A is catalyzed by acid form faujasites [27]. What form of kinetics would you expect for this reaction in the absence of transport effects?

5-6 In Example 5-2, the relation between the effectiveness factor and the Thiele modulus for a first-order reaction in an isothermal infinite flat slab was used:

$$\eta = \frac{\tanh \phi_L}{\phi_L} \tag{5-10}$$

Derive this equation.

5-7 For the following mixture of hydrocarbons, predict the selectivity for cracking relative to n-hexane at 538°C and 1 atm; assume that the catalyst is HZSM-5 with a Si/Al ratio of 72. Plot the selectivity as a function of catalyst particle size from 0.01 to 3.0 μm: n-nonane, n-dodecane, 3-methylpentane, 2-methylheptane, 2,2-dimethylbutane, and 2,2-dimethylheptane.

5-8 The zeolite H-mordenite is characterized by parallel, nonintersecting pores with apertures that are 12-membered oxygen rings; these apertures have openings with diameters of about 0.7 nm. Side pockets open off the main channels (perpendicular to them) and have diameters of 0.4 nm. H-Mordenite has been used to catalyze the dehydration reaction of t-butyl alcohol at 1 atm and 75°C:

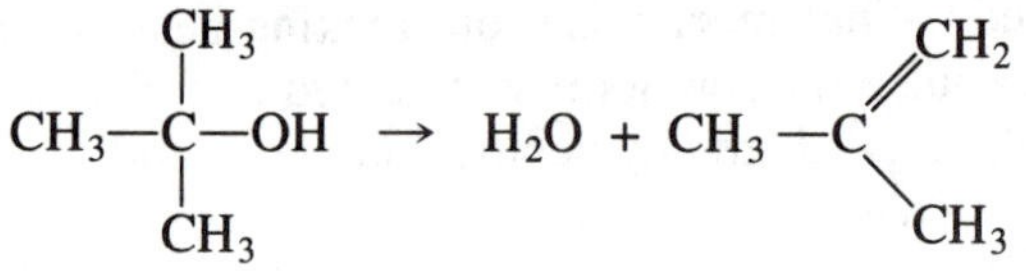

$$CH_3-\underset{\underset{CH_3}{|}}{\overset{\overset{CH_3}{|}}{C}}-OH \;\rightarrow\; H_2O + CH_3-C\overset{\diagup CH_2}{\underset{\diagdown CH_3}{}} \tag{5-11}$$

Table 5-9
MORDENITE-CATALYZED DEHYDRATION OF t-BUTYL ALCOHOL

Composition of Reactant Solution, mol%	Mean Catalyst Pore Length, μm	$10^7 \times$ Initial Reaction Rate, mol/(g of catalyst·s)
100% t-Butyl alcohol	11.3 ± 4.3	6.1
	5.9 ± 3.2	7.4 ± 0.5
90% t-butyl alcohol, 10% n-heptane	5.9 ± 3.2	8.0
90% t-butyl alcohol, 10% n-heptane[a]	5.9 ± 3.2	6.0
90% t-butyl alcohol, 10% methylcyclohexane	5.9 ± 3.2	3.2

[a] Paraffin brought in contact with catalyst at 75°C to fill pores before addition of t-butyl alcohol.

The data shown in Table 5-9 were obtained from a batch reactor initially containing pure liquid t-butyl alcohol (or an alcohol–paraffin solution), to which was added a small number of carefully dried crystallites of H-mordenite at time zero [Ignace, J. W., and Gates, B. C., *J. Catal.*, **29**, 292 (1973)]. Did reaction occur within the mordenite pores or only on the outside surfaces of the crystallites? What is the basis for your answer? What are the roles of the paraffins in these experiments? Set up a calculation to estimate the effective diffusivity of t-butyl alcohol in an H-mordenite pore; state the parameter values you know from what is given.

5-9 It has been suggested [33] that restricted transition state selectivity plays a greater role in the cracking of 3-methylpentane than in the cracking of 3-methylpent-2-ene. What is the evidence, and why is the suggestion plausible?

5-10 The selectivity for cracking of 2,2-dimethylheptane (2,2-DMH) has been measured with respect to the cracking of n-hexane (n-H) in the presence of ZSM-5 catalysts at 538°C [33]. The catalysts had various particle sizes and concentrations of proton donor groups, varied in part by ion exchange with Na^+. The activity for n-hexane conversion was found to be virtually independent of the zeolite crystallite size. The observed rate constants are shown in Table 5-10. How does the selectivity for conversion of the branched paraffin relative to n-hexane,

$$S = \frac{k_{2,2\text{-DMH}}}{k_{n\text{-H}}}$$

depend on the crystallite size?

5-11 What catalyst would you recommend for the toluene–methanol alkylation to give *para*-xylene? Why? What side reactions are expected to be important, and how can they be suppressed?

Table 5-10
ACTIVITY DATA FOR CRACKING OF n-HEXANE
AND 2,2-DIMETHYLHEPTANE

Crystallite Size, μm	$k_{n\text{-H}}$, s^{-1}	$k_{obs\ 2,2\text{-DMH}}$, s^{-1}
0.025	0.85	1.9
0.025	29	63
0.10	25	53
0.10	27	57
1.35	1.8	2.5
1.35	29	8.4
1.35	36	7.6
1.80	43	6.6

5-12 Consult ref. 40 and prepare a summary of the chemistry of methanol conversion into gasoline-range hydrocarbons catalyzed by H-ZSM-5. How are aromatics formed from methanol?

5-13 Use the data presented in this chapter to develop an equation relating the activity for cracking of 2,2-dimethylbutane by HZSM-5 to the Al content and crystallite size of the catalyst.

5-14 Methyl *t*-butyl ether formation from methanol and isobutylene is catalyzed by HZSM-5. The activity of this zeolite is markedly less than that of an equal volume of macroporous sulfonated poly(styrene-divinylbenzene), but the selectivity is better, as lower yields of butene oligomers are formed. Explain these results.

5-15 Consider the data of Problem 5-10 and estimate the fraction of the catalyst activity in each case associated with the external surface of the crystallites rather than the intrapore surface.

5-16 Write a set of reactions starting with isobutylene and cyclohexane to illustrate the pattern of hydrogen transfer that gives paraffins and aromatics in catalytic cracking. Begin with protonation of isobutylene followed by hydride transfer to *t*-butyl cation from cyclohexane.

5-17 Ultrastable zeolite Y is used in relatively large amounts in cracking catalysts that are chosen to give high-octane-number gasoline. Explain why.

5-18 The zeolite HZSM-5 catalyzes isomerization and cracking of *n*-butane at 500°C. The solid superacid made from SbF_5 and macroporous sulfonic acid resin catalyzes the isomerization at 100°C, indicating that it is a stronger acid than the HZSM-5. However, the resin and other superacids do not catalyze measurable cracking at 100°C. Why not?

5-19 What catalyst would you choose in attempts to do shape-selective olefin hydroformylation? For leads, see Takahashi, N., et al., *J. Catal.*, **117**, 348 (1989) and Fukuoka, A., et al. *Appl. Catal.*, **50**, 295 (1989).

5-20 The *n*-butane cracking reaction catalyzed by HZSM-5 at 476–523°C at low conversions and pressures less than atmospheric has been found to be first order in the reactant *n*-butane [21]. The formation of each product is first order in *n*-butane. The magnitude of the enthalpy of adsorption of *n*-butane on the catalyst has been measured to be 15 kcal/mol [21]; this adsorption is exothermic. The products of the *n*-butane cracking at very low conversions are H_2, CH_4, C_2H_6, C_2H_4, C_3H_6, and C_4H_8. The rates of formation of H_2 and butenes are equal; the rates of formation of ethane and ethylene are equal; and the rates of formation of methane and propylene are equal. The dependence on temperature of the logarithm of the first-order rate constant for each reaction determines an activation energy of 34 kcal/mol. (a) If the catalyst is assumed to have a nearly uniform set of catalytic sites (proton donor sites), and the decomposition of the adsorbed carbonium ion intermediate is assumed to be rate determining, what is the expected form of a Langmuir–Hinshelwood rate equation? How does this simplify to be consistent with the observed kinetics? (b) What is the activation energy for the conver-

Table 5-11
PRODUCT DISTRIBUTION IN ISOBUTANE CONVERSION CATALYZED BY HZSM-5 AT 450°C

Inverse Space Velocity, (g of catalyst· min/mL of isobutane at NTP)	Con- version, %	Hydrocarbon Product Distribution, mol%[a]						
		CH₄	C₂H₄	C₃H₈	C₃H₆	But-1-ene	Isobutylene	But-2-ene
0.003	0.21	40.4	0.00	7.3	40.0	0.88	5.3	0.0
0.005	0.31	35.2	0.30	5.3	35.2	2.4	7.6	1.8
0.007	0.37	35.7	0.39	4.8	35.7	1.5	8.5	1.5
0.011	0.55	32.5	0.62	3.7	31.1	2.6	7.4	5.8

[a] No measurements were made to detect hydrogen.

sion of the carbonium ion to give methane? Draw an energy profile for the reaction, including the adsorption and reaction steps.

5-21 Acidic zeolites have been tried as solid alkylation catalysts (e.g., for the isobutane–propylene alkylation), but they have not been successful. Can you infer why not?

5-22 Consider the data in Table 5-11 for conversion of isobutane catalyzed by HZSM-5 at 450°C (Stefanadis, C., Haag, W. O., and Gates, B. C., *J. Mol. Catal.*, **67**, 363 (1991). What reactions occur? What mechanisms can you propose? Why are the products different from those of *n*-butane conversion under nearly the same conditions?

6

Catalysis on Surfaces

6.1
INTRODUCTION

The most widely used industrial catalysts are inorganic solids. Zeolites and clays are the first of these to be considered in this book (Chapter 5) because they have well-defined (crystalline) structures and solventlike environments in their molecular-scale pores. Since catalysis occurs within these solventlike environments, the zeolites and clays provide a transition between solutions and gel-form organic solids (which have much of the character of solutions) and the more typical solids, which catalyze reactions on their surfaces.

The more typical solid catalytic materials include metals, metal oxides, and metal sulfides, sometimes used in combination with each other. Important industrial catalysts in these groups are listed with their applications in Table 6-1. About half the elements in the periodic table are used in one industrial catalyst or another, and an individual catalyst may contain 10 or more elements. Like the zeolites and clays, these solids are porous, but the pores are larger and nonuniform. The pores are the void spaces between aggregated primary particles, which are usually small crystallites of the solid.

The pore structure is a labyrinth like that in the macroporous poly(styrene-divinylbenzene) shown in Fig. 4-17. The pore volume may typically take up about one half the volume of a porous catalyst particle, and the internal surface area is often large. Since the pores of a typical catalyst are large in comparison with molecular dimensions, they do not usually provide an environment to "solvate" a molecule adsorbed on the surface. Rather, the surface and the adsorbed molecule are virtually independent of the neighboring surfaces; the chemistry is two-dimensional. The essential interactions for catalysis involve the surface and the adsorbed reactants. To understand surface catalysis, therefore, it is helpful to begin with the structure and reactivity of surfaces.

Table 6-1

SOME LARGE-SCALE INDUSTRIAL PROCESSES CATALYZED BY SURFACES OF INORGANIC SOLIDS

Catalyst	Reaction
Metals (e.g., Ni, Pd, Pt, as powders or on supports) or metal oxides (e.g., Cr_2O_3)	$C=C$ bond hydrogenation, e.g., olefin $+ H_2 \rightarrow$ paraffin
Metals (e.g., Cu, Ni, Pt)	$C=O$ bond hydrogenation, e.g., acetone $+ H_2 \rightarrow$ isopropanol
Metal (e.g., Pd, Pt)	Complete oxidation of hydrocarbons, oxidation of CO
Fe (supported and promoted with alkali metals)	$3H_2 + N_2 \rightarrow 2NH_3$
Ni	$CO + 3H_2 \rightarrow CH_4 + H_2O$ (methanation) $CH_4 + H_2O \rightarrow 3H_2 + CO$ (steam reforming)
Fe or Co (supported and promoted with alkali metals)	$CO + H_2 \rightarrow$ paraffins $+$ olefins $+ H_2O + CO_2$ ($+$ oxygen-containing organic compounds) (Fischer–Tropsch reaction)
Cu (supported on ZnO, with other components, e.g., Al_2O_3)	$CO + 2H_2 \rightarrow CH_3OH$
Re $+$ Pt (supported on η-Al_2O_3 or γ-Al_2O_3 promoted with chloride)	Paraffin dehydrogenation, isomerization and dehydrocyclization (e.g., n-heptane $\rightarrow$ toluene $+ 4H_2$) (naphtha reforming)
Solid acids (e.g., SiO_2–Al_2O_2, zeolites)	Paraffin cracking and isomerization
γ-Al_2O_3	Alcohol $\rightarrow$ olefin $+ H_2O$
Pd supported on acidic zeolite	Paraffin hydrocracking
Metal-oxide-supported complexes of Cr, Ti, or Zr	Olefin polymerization, e.g., ethylene $\rightarrow$ polyethylene
Metal-oxide-supported oxides of W or Re	Olefin metathesis, e.g., 2 propylene $\rightarrow$ ethylene $+$ butene
V_2O_5 or Pt	$2SO_2 + O_2 \rightarrow 2SO_3$
Ag (on inert support, promoted by alkali metals)	Ethylene $+ \frac{1}{2}O_2 \rightarrow$ ethylene oxide (with $CO_2 + H_2O$)
V_2O_5 (on metal oxide support)	Naphthalene $+ \frac{9}{2}O_2 \rightarrow$ phthalic anhydride $+ 2CO_2 + 2H_2O$ o-Xylene $+ 3O_2 \rightarrow$ phthalic anhydride $+ 3H_2O$
Bismuth molybdate, uranium antimonate, other mixed metal oxides	Propylene $+ \frac{1}{2}O_2 \rightarrow$ acrolein Propylene $+ \frac{3}{2}O_2 + NH_3 \rightarrow$ acrylonitrile $+ 3H_2O$
Mixed oxides of Fe and Mo	$CH_3OH + O_2 \rightarrow$ formaldehyde (with $CO_2 + H_2O$)

Table 6-1 (*continued*)

Catalyst	Reaction
Fe_3O_4 or metal sulfides	$H_2O + CO \rightarrow H_2 + CO_2$ (water gas shift reaction)
$\begin{cases} Co-Mo/\gamma\text{-}Al_2O_3 \text{ (sulfided)} \\ Ni-Mo/\gamma\text{-}Al_2O_3 \text{ (sulfided)} \\ Ni-W/\gamma\text{-}Al_2O_3 \text{ (sulfided)} \end{cases}$	Olefin hydrogenation Aromatic hydrogenation Hydrodesulfurization Hydrodenitrogenation

6.2
SURFACE STRUCTURES

Structures of solid surfaces are notoriously complex and difficult to elucidate. Surfaces of most catalytic solids have microscopic and even macroscopic regions with different compositions, phases, and structures, each with a variety of imperfections. The structures are difficult to determine under the best of conditions (e.g., in a vacuum) and are often almost impossible to determine under the conditions of a catalytic reaction. Because surface catalysis rests on such a weak structural foundation, it is only poorly understood in comparison with solution catalysis, and the best available models of most catalytic sites and reaction intermediates are vague and tentative. This chapter introduces the part of catalysis that, with enzymes, offers the most challenging scientific problems; it also accounts for the greatest technological applications.

6.2.1 Single-Crystal Surfaces of Metals

The simplest solid surfaces are those of single crystals. These are illustrated here with metals, because pure metal surfaces are among the simplest of all, having only one kind of atom. The surface structure depends both on the bulk structure and on the manner in which the bulk is terminated; it can be cleaved to expose various planes having different configurations of atoms.

Several of these structures are illustrated in Fig. 6-1. For convenience, the planes are designated with the Miller indices (*hkl*). The Miller indices for a crystal face are the reciprocals of the intersections of the plane with the *x*, *y*, and *z* axes, respectively, with the finite values set to give the smallest integers. One of the easiest surfaces to visualize is the (111) plane of the cubic close-packed solid (Fig. 6-1); the plane intersects the *x*, *y*, and *z* axes at points equidistant from the origin. This plane was encountered in Chapter 5 in the demonstration of how to build up the three-dimensional structure (pp. 254–259).

Cutting the lattice along the directions of highest atomic density (lowest Miller indices) exposes the most uniform crystal surface, exemplified by the flat (111) face [1] (Fig. 6-1). Cleaving the lattice along planes of lower atomic density (higher Miller indices) gives more complex crystal surfaces exhibiting

Top view   Oblique view

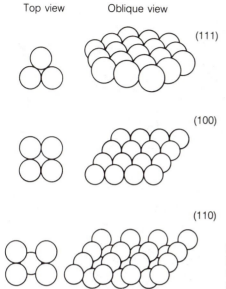

(111)

(100)

(110)

Figure 6-1
Low-Miller-index planes of a cubic close-packed (face-centered cubic) metal. Top views at the left show the symmetries of sites on the surfaces.

ordered terrace, step, and kink structures [2] (Fig. 6-2). Changes in the cleavage angle lead to changes in the terrace width and step density. The terrace, step, and kink structures are quite stable; but sometimes, even under vacuum, surface reconstruction may take place, affecting just the top few layers. The surfaces with steps and kinks more nearly resemble the structures of typical catalyst surfaces (with their nonuniformities and imperfections) than do the simpler structures.

These single-crystal surfaces are ideal models of more typical surfaces. The step and kink structures illustrate how surfaces are nonuniform (heterogeneous) on an atomic scale. There are also other kinds of nonuniformities (or defects) in surfaces. Point defects are either adsorbed atoms or vacancies (where single atoms are missing from the lattice). Surface defects on a larger scale almost always exist, and some can be detected by microscopy. There may be dislocations in a crystal due to a mismatch of atomic planes; these show up as line defects. Dislocation densities of the order of 10^6 to 10^8 per square centimeter are typical of metal or ionic single-crystal surfaces, and densities of 10^4 to 10^6 per square centimeter are typical of metal oxides.

6.2.2 Experimental Methods of Determining Surface Structures of Single Crystals

Details of surface structure can be discerned with optical and electron microscopy. The scanning tunneling microscope provides evidence of the terraces and steps on a gold surface, as shown in Fig. 6-3 [3]. The transmission electron microscope provides structural information down to dimensions of a few tenths of nanometers under the most favorable circumstances. But for catalysis, it is

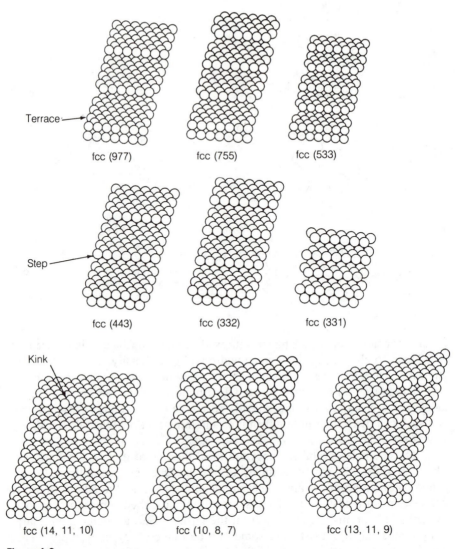

Terrace

fcc (977) fcc (755) fcc (533)

Step

fcc (443) fcc (332) fcc (331)

Kink

fcc (14, 11, 10) fcc (10, 8, 7) fcc (13, 11, 9)

Figure 6-2
Structures of several high-Miller-index stepped surfaces of a face-centered
cubic (fcc) metal with different terrace widths and step orientations and several
with different kink concentrations in the steps [2].

the atomic-scale structure that is the most important, and the structural detail
available from microscopy is usually not fine enough to provide information
about catalytic sites on surfaces, that is, the specific arrangements of atoms
where reactant molecules form chemical bonds with the surface and are converted into products.

A powerful method for determining structures of *ordered* surfaces on an
atomic scale is an electron diffraction technique that is similar in principle to

the X-ray diffraction technique for determining three-dimensional structures of single crystals or powders (used, for example, in zeolite structure determinations). The surface-specific technique is low-energy electron diffraction (LEED) [4]. In LEED, the surface of a single crystal is bombarded with a beam of electrons having so little energy that they barely penetrate the solid beneath the surface. The electrons, which have a wavelength less than the interatomic spacing on the surface, are diffracted. The back-diffracted beam of elastically scattered electrons is characteristic of the geometric arrangement of the surface atoms. These electrons impinge on a fluorescent screen where the diffraction pattern is displayed. The structure of the surface (provided that it is not too complex) can be inferred from the diffraction pattern. Well-prepared single crystals often have surface structures very nearly the same as those of ideal surfaces.

There are important limitations to the LEED technique; the sample must be investigated in an ultrahigh vacuum (pressure less than about 10^{-10} Torr, approximately 10^{-13} atm) to ensure that the surface is rigorously clean and that the electrons reach it unimpeded. Experiments carried out to investigate surfaces of single crystals under ultrahigh vacuum are commonly called surface science experiments. Some of the other surface science techniques (Table 6-2) include the electron spectroscopies: Auger electron spectroscopy, X-ray photoelectron spectroscopy, ultraviolet photoelectron spectroscopy, and high-

Figure 6-3
Image of the surface of a gold single crystal determined by scanning tunneling microscopy [3].

Table 6-2

ULTRAHIGH VACUUM TECHNIQUES FOR CHARACTERIZATION OF STRUCTURES AND COMPOSITIONS OF SOLID SURFACES

Method	Acronym	Nature of Experiment	Information Obtainable
Auger electron spectroscopy	AES	Electron emission from near-surface atoms excited by impinging photons	Near-surface composition
Secondary ion mass spectrometry	SIMS	Ion beam induced ejection of surface atoms as ions	Surface composition
X-ray photoelectron spectroscopy (electron spectroscopy for chemical analysis)	XPS (ESCA)	Electron emission from near-surface atoms excited by impinging X-rays	Near-surface composition; approximate oxidation states of near-surface atoms
Ultraviolet photoelectron spectroscopy	UPS	Electron emission from near-surface atoms excited by impinging ultraviolet radiation	Bonding of surface species
Low-energy electron diffraction	LEED	Elastic backscattering of low-energy electrons	Atomic structure of surface
Ion scattering spectroscopy	ISS	Elastic reflection of impinging inert gas ions	Atomic structure and composition of surface
Electron energy loss spectroscopy	EELS	Vibrational excitation of surface atoms by inelastic reflection of low-energy electrons	Structure and bonding of surface species

resolution electron energy loss spectroscopy. Often, many of the surface science experiments can be carried out in a single vacuum chamber, holding a sample with a surface area of only about 1 cm^2 and a series of detectors to characterize the radiation emitted from the sample bombarded in turn with beams of X-rays, electrons, and small ions. One of the challenges is to find relations between the catalytic properties and the structures of clean surfaces determined by surface science techniques. This is often difficult because a surface functioning as a catalyst (even if it is a single crystal) is usually far different from a clean surface—it is covered with a menagerie of reactants, products, and other species, including impurities, which may cause substantial restructuring.

6.2.3 High-Surface-Area Amorphous Solids

MACROSCOPIC STRUCTURES OF POROUS SOLIDS

Surfaces of typical catalysts are quite different from those of single crystals. A catalyst particle usually consists of a jumble of small crystallites aggregated into a high-surface-area porous structure. These primary particles are usually too small to yield to structural determination by X-ray powder diffraction, and the bulk solid is therefore referred to as amorphous. The small crystallites expose a variety of crystal planes with various compositions and many defects. Little more can be said to generalize about structures of real catalyst surfaces; specific examples are considered in some detail in the following parts of the chapter.

The need for high surface areas of solid catalysts is easy to understand: When a catalytic reaction takes place on a surface, the rate of the reaction increases in proportion to the surface area, provided that transport restrictions are negligible. Therefore, the optimum form of a catalyst is usually a porous solid that has a high internal surface area. The only other way to produce a high surface area per unit volume of reactor is to use minute particles (extremely fine powders). Practical concerns make this option unrealistic. For example, the pressure drop across a fixed-bed reactor would be intolerably high, and the fine particles would be easily entrained in the product stream and plug downstream pumps and lines.

EXAMPLE 6-1   Surface Areas of Practical Catalysts [5]
Problem

According to a rule of thumb, stated in Chapter 1, a catalytic reactor, to be practically useful, must transform on the order of 10^{-6} mol of reactant/ (s·cm^3 volume in the reactor). What does this imply about the minimum surface area of a catalyst for conversion of a reactant with a molecular weight of 200 at 500°C and atmospheric pressure?

Solution

To get a rough indication, assume that the collision theory will allow an estimate of the upper limit of the rate of a surface-catalyzed reaction. From the kinetic theory of gases, the rate N of surface bombardment by reactant is about $\frac{1}{4}vC$, where v is the average velocity of the molecule and C is its concentration; this works out to be about 0.1 mol/(s·cm^2). Weisz [5] has estimated that only a fraction $f = e^{E_{act}/RT}$ of the bombarding molecules can react and only a fraction α of the surface will be catalytically active. Therefore, a criterion for estimating the minimum surface area is $N\alpha f S > 10^{-6}$ mol/s, where S is the surface area (cm^2/cm^3). The calculation shows, for example, that for values of α of 0.01 and $E_{act} = 30$ kcal/mol, S is about 10 m^2/cm^3; for values of α of 0.01 and $E_{act} = 40$ kcal/mol, S is about 10^3 m^2/cm^3. ∎

Industrial catalysts typically have internal surface areas of hundreds of square meters per gram. Many metal oxides are readily prepared in the form of high-surface-area solids. Some high-surface-area metals are also known; Raney nickel, used to catalyze hydrogenation of fats and other organic compounds, is prepared by extracting Al from an AlNi alloy, the void spaces (pores) being left where the Al had been present in the solid.

Because many catalytic materials cannot easily be prepared or used in the form of small particles or suitably robust high-surface-area particles, they may be dispersed as minute particles on a sturdy high-surface-area support such as γ-Al_2O_3 or carbon. Many industrial catalysts consist of such supported metals, metal oxides, and metal sulfides; examples are described later in this chapter.

The physical properties are important in determining the effectiveness of a catalyst particle, which is determined by the interplay of chemical reaction and transport processes (Section 4.8). Instruments are available for the routine and even automated measurement of catalyst physical properties. Surface areas are determined by the measurement of the amount of N_2 adsorbed. Adsorption of N_2 on most surfaces is not chemical in nature, but physical—it has some of the character of condensation. Evaluation of data that characterize adsorption of N_2 (and many other adsorbates) is based on the Brunauer–Emmett–Teller (BET) model of adsorption, presented in Section 6.3.1.

Pore size distributions are measured by mercury penetration, which is used in conjunction with other techniques such as those characterizing N_2 adsorption. With an instrument called a mercury porosimeter, liquid mercury is forced into the pores of the catalyst at various pressures. Mercury is an appropriate liquid because it has a high surface tension and does not wet the surface. The effect of internal surface tension is to resist the entry of the mercury into the pores. The force opposing the entry into a narrow cylindrical pore of radius r is $2\pi r\sigma \cos \phi$, where σ is the interfacial tension and ϕ is the angle of contact between the liquid and solid. The force driving the mercury into a pore is $\pi r^2 P$, where P is the applied pressure. At equilibrium, the forces are equal and

$$P = - \frac{2\sigma\pi \cos \phi}{r} \qquad (6\text{-}1)$$

A pressure of about 700 atm is required to fill a pore with a diameter of 10 nm; the porosimeter is not used routinely for smaller pores.

The sizes of the smaller pores are measured in complementary experiments characterizing desorption of N_2 from the catalyst. Adsorption experiments can provide the same information, but often hysteresis effects in adsorption–desorption cycles complicate matters. The basis of the measurement is the capillary condensation that occurs in small pores at pressures less than the saturation vapor pressure of the adsorbate. The smaller the diameter of the pore, the greater is the lowering of the vapor pressure of the liquid in it. Measurements of the amount of N_2 desorbed as a function of pressure determine a pore size distribution in the range of about 1 to 20 nm; this range conveniently overlaps that measured with the mercury porosimeter. Some data obtained by these methods are shown in Fig. 6-4 [6].

Pores with diameters <2.0 nm are called *micropores*. Those with diameters >50 nm are called *macropores*. Pores of intermediate size are called *mesopores*. The pores of catalysts are often highly irregular in size and shape. Analysis of the pore size distribution when there are complex pore shapes (e.g., ink bottles) is too complicated to be considered here [6].

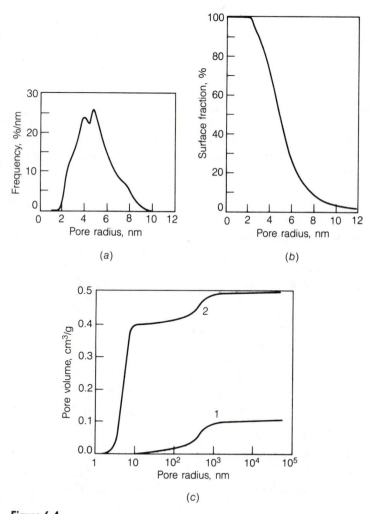

Figure 6-4
Pore size distribution of γ-Al₂O₃. (*a*), (*b*) Distribution determined by analysis of N₂ adsorption and desorption, as explained in ref. 6. (*c*) Curve 1, distribution of macropore sizes as determined by mercury penetration; curve 2, complete size distribution determined by a combination of the data of both kinds [6].

Other important physical properties of catalyst particles include size, shape, and strength, both to resist crushing and to resist abrasion in handling. The optimum size, shape, internal surface area, and pore size distribution may be inferred from the rates of the reaction and transport processes described in Chapter 4. There are some obvious trade-offs: Larger pores (which allow more ready access of reactants to the particle interior) can be gained only at the expense of physical strength and of internal surface area (hence, catalytic activity, which is proportional to surface area in the absence of intraparticle transport influence). Often catalysts contain binders, such as clays, which at the high temperatures of preparation may undergo physical changes comparable to melting so that they wet the primary particles of the catalyst and glue them together. Catalysts may also contain stabilizers, materials that do not play a direct catalytic role but impart improved physical properties.

SURFACE STRUCTURES OF METAL OXIDES

Most catalytic materials are amorphous and cannot even be prepared as single crystals that would allow exact determinations of surface structure. Many metals and a few metal oxides (e.g., TiO_2 and MgO) have been prepared as single crystals and have been investigated with ultrahigh vacuum techniques, but the most commonly applied catalytic materials have not been characterized in such detail.

The best available characterizations of most catalytic materials are based on inferences from the bulk structure (which by themselves can often be misleading), combined with evidence of surface composition and identification of functional groups determined spectroscopically (e.g., —OH groups on metal oxide surfaces). The most valuable characterizations of surface functional groups are obtained with organic adsorbates as probe molecules [7].

MgO

Consider several examples of metal oxides commonly used as catalyst supports. Magnesia, MgO, can be made by heating $Mg(OH)_2$ or $MgCO_3$ in air. MgO has a simple bulk structure, that of rock salt; the Mg and the O are octahedrally coordinated. The solid and the surface have a strongly ionic character, well represented as $Mg^{2+}O^{2-}$. By cleaving the bulk to expose the (100) face (Fig. 6-5), one obtains the surface that has been observed by microscopy to be the most common on MgO powders. The Mg^{2+} and O^{2-} ions on this ideal surface are fivefold coordinated. This surface is neutral, and there is no need to postulate the presence of additional ions to account for the necessary charge balance. Nonetheless, protons are usually present in surface —OH groups, observed by infrared spectroscopy and formed by dissociative adsorption of water.

A schematic cross section of this surface is shown in Fig. 6-6 [8], where the process of dehydroxylation is depicted. Hydroxylation, the reverse, is the reaction with water to give surface —OH groups and to protonate O^{2-} ions of the surface; analogous chemistry was illustrated for zeolites in Chapter 5. In

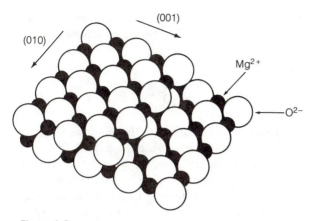

Figure 6-5
Model of the MgO (100) plane, showing a unit atomic step in the [001] direction.

this simplified picture of the surface, there are two kinds of —OH groups, one providing the sixth group coordinating a Mg^{2+} ion (making it coordinatively saturated) and the other resulting from protonation of an O^{2-} ion terminating the bulk. These different —OH groups are distinguished by their frequencies in the infrared spectrum.

Mg^{2+} ions are exposed at the surfaces of the crystalline primary particles as a result of dehydroxylation. Treatment at high temperatures (say, 600°C) gives a highly dehydroxylated surface that exposes many of these Lewis acid sites.

The O^{2-} ions and —OH groups on the MgO surface are basic, and the

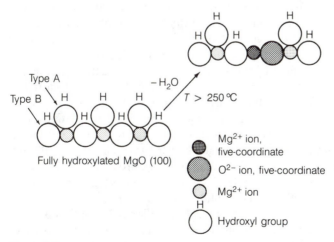

Figure 6-6
Simplified representation of the dehydroxylation of MgO.

chemistry of the surface of MgO is dominated by its basicity. Brønsted acids chemisorb on MgO to form carbanions and —OH groups. The adsorbates react dissociatively with surface ion pairs:

$$\{Mg^{2+}O^{2-}\} + XH \rightarrow \{Mg^{2+}O^{2-}\} \overset{\displaystyle X^- \quad H^+}{} \tag{6-2}$$

Increasing dehydroxylation increases the base strength of the surface; highly dehydroxylated MgO is such a strong base that it deprotonates weak Brønsted acids such as NH_3 ($pK_a = 36$) and propylene ($pK_a = 35$); even the heterolytic dissociation of H_2 occurs on MgO. Surface sites with such strong basicity are more highly unsaturated than the fivefold coordinate ions shown in Fig. 6-5. Three- and four-coordinate Mg^{2+} and O^{2-} ions have been detected by ultraviolet–visible and photoluminescence spectroscopy on the surfaces of polycrystalline MgO prepared from $Mg(OH)_2$, following heating to 800°C. The locations of these highly unsaturated sites are not resolved, but they may be at edges and corners of (100) planes.

Besides being a strong base, the highly dehydroxylated MgO surface is a good reducing agent. The reducing sites are apparently defects, possibly surface cation vacancies; the dissociative chemisorption of Brønsted acids blocks the reactivity of the reducing sites.

SiO_2

Silica, SiO_2, is an amorphous solid that can be made by heating a silica hydrogel, which can be made by acidifying a dilute solution of sodium silicate and allowing sufficient time for the silica in solution to polymerize. The hydrogel is an aggregate of primary particles about 5 nm in diameter, made up of a network of interconnected covalently bonded SiO_4 tetrahedra (already mentioned in Chapter 5). Drying of the hydrogel by heating leaves open spaces constituting a pore network and an internal surface area of perhaps 500 m^2/g.

The surface is that of the aggregated primary particles, which are noncrystalline (hence the term silica gel). In comparison with surfaces of most metal oxides, the silica surface is nearly inert. The most reactive groups are the —OH groups (called silanol groups) that terminate the primary particles; these are weakly acidic, comparable to alcohols. The bulk may be terminated entirely by —OH groups, which can be removed by dehydroxylation. Two types of silanol groups are usually distinguished, isolated groups and neighboring (vicinal) groups that may be hydrogen bonded to each other. Fully hydrated samples, heated in air (i.e., calcined) at temperatures <200°C, also contain geminal groups, $Si(OH)_2$. Complete removal of silanol groups requires temperatures greater than about 700°C and results in significant changes in surface morphology. The aprotic sites present after dehydroxylation at 600–800°C have been suggested to be primarily highly strained Si–O–Si linkages.

Commercial samples of silica may contain impurities such as iron. Even very small amounts of impurities may greatly affect the surface and catalytic properties.

γ-Al₂O₃

The most widely applied catalyst support is γ-alumina, γ-Al_2O_3, one of the family of transition aluminas. These are defective, metastable solids formed from $Al(OH)_3$ as it is heated through temperatures of some hundreds of degrees Celsius (Fig. 6-7). As the solid is heated in air, it is decomposed into an oxide with a micropore system and a surface area of hundreds of square meters per gram. Raising the temperature to about 1100°C leads to further transformation of the solid, with changes in structure of the primary particles and collapse of the pore structure, leading to loss of almost all of the internal surface area and ultimately giving the stable, extremely hard, crystalline α-Al_2O_3 (corundum), made up of linked AlO_6 octahedra [9]; this has a melting point of about 2100°C.

Transition aluminas are the most widely used support materials for catalysts, for several reasons. They are inexpensive, stable at relatively high temperatures (even under hydrothermal conditions, i.e., in the presence of steam), mechanically stable, easily formed in processes such as extrusion into various shapes (typically, cylinders) having good physical strength, and easily formed with a variety of pore structures.

One way to build macroporosity into a transition alumina is to add a removable material such as carbon black or sawdust to the green body (a term used in ceramics to denote the solid before firing, i.e., heating in air as described above). The metal oxide forms a stable structure around the removable material, and when the latter is removed, for example, by burning, it leaves macropores with dimensions determined by the removable material.

The dimensions of the micropores are determined by the packing of the primary particles (crystallites); the micropores and some mesopores are the void spaces between the primary particles and have dimensions of the order of 10 nm. The dimensions of the macropores may be thought of as related to

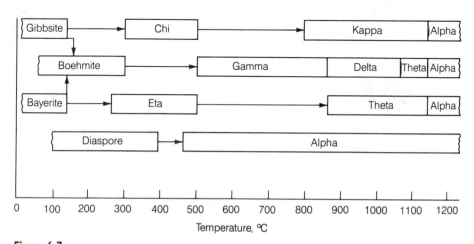

Figure 6-7
Decomposition sequence illustrating the conversion of aluminum hydroxides to give transition aluminas and, ultimately, α-Al₂O₃ [10].

the dimensions of agglomerates of the primary particles. Some representative dimensions are as follows: micropore plus mesopore volume, 0.5 cm^3/g; average diameter of micropores and mesopores, 10 nm; surface area of micropores and mesopores, 200 m^2/g; macropore volume 0.5 cm^3/g; and average macropore diameter, 12,000 nm (the microparticle density is about 1.2 g/cm^3) [10].

γ-Al$_2$O$_3$ is an amorphous solid having a defect spinel structure (spinel is the mineral MgAl$_2$O$_4$) with layers of O^{2-} ions in a cubic close packed arrangement and layers of Al^{3+} ions, some having tetrahedral and some having octahedral coordination in the oxygen lattice. The bonding is intermediate between the ionic bonding of MgO and the covalent bonding of SiO$_2$.

The primary particles of a transition alumina are crystallites terminated by layers of oxygen anions; and, for charge neutrality, the surfaces must incorporate cations, which are typically H$^+$, present in —OH groups. Heating the fully hydroxylated γ-Al$_2$O$_3$ under vacuum to temperatures exceeding 200°C leads to dehydroxylation, as depicted schematically in Fig. 6-8 [7]. The degree of dehydroxylation of the surface can be regulated by the temperature, as shown in Fig. 6-9 [11]. This process leads to the formation of coordinatively unsaturated O^{2-} ions and adjacent surface anion vacancies. If the surface layer is a close-packed (111) surface, then the anion vacancy is expected to expose either two five-coordinate Al^{3+} ions or one three-coordinate Al^{3+} ion. Therefore, the dehydroxylation creates Lewis acidic as well as Lewis basic sites on the surface.

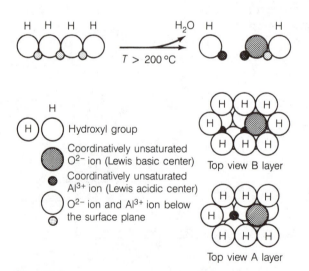

Figure 6-8
Simplified representation of the dehydroxylation of the (111) surface of γ-Al$_2$O$_3$ showing two structures of the resulting Lewis acid–base sites, depending on the placement of an underlying Al^{3+} ion, which may be in either a tetrahedral or an octahedral surrounding [7, 8a]. In the B layer there are Al^{3+} ions located in octahedral positions only. Formally, the other layer (the A layer) can be obtained from this by transferring two-thirds of the Al^{3+} ions from octahedral to tetrahedral positions.

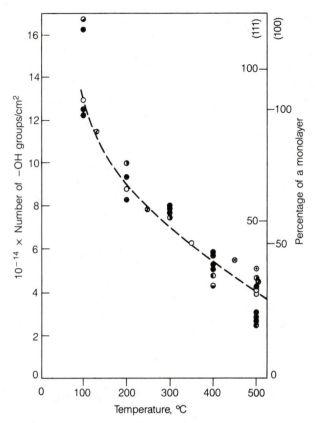

Figure 6-9
Density of —OH groups on γ-Al₂O₃ determined by the temperature of treatment. Vertical scale on the right depicts coverages as a percentage of a monolayer for both the (111) and (100) planes [11].

The acid–base properties of the surface have been investigated by titration and by adsorption of organic probe molecules and measurement of their infrared and NMR spectra. The surface —OH groups act as hydrogen bond donors to benzene, but the interaction is weaker than that for SiO_2, indicating a very weak Brønsted acidity of γ-Al_2O_3. Infrared spectra of adsorbed pyridine and other bases give evidence of Lewis acid sites, and shifts in the N—H stretching frequency of adsorbed pyrrole show the presence of strongly basic sites, which may be the O^{2-} ions and/or the —OH groups. Bifunctional adsorbates such as alcohols, with proton donor and proton acceptor sites, interact with the acid–base pair sites on the surface.

Models of the structure of solid γ-Al_2O_3 (which are still subject to refinement owing to the complexity of the structure) have been used as a basis for inference of models of structures of sites on the (111) faces [7, 8a], as shown in Fig. 6-8. The models of these surfaces are also based on infrared spectra providing evidence of surface —OH groups, infrared spectra of adsorbed bases, and other chemical evidence of Lewis acid groups (exposed Al^{3+} ions, as in zeolites), and various kinds of chemical evidence of the basic character of O^{2-} ions and of —OH groups.

It is emphasized that the structural models shown in Fig. 6-8 rest on a much less firm foundation than the structural models mentioned for clean surfaces of single crystals of metal, which have been determined in LEED experiments. The structural models available for surfaces of any amorphous solid are tentative and subject to continuing revision. The models shown for γ-Al_2O_3 are for particular crystal faces, but the γ-Al_2O_3 used in the form of an amorphous material exposes a variety of crystal faces in various and unknown proportions. Furthermore, defects and impurities complicate the structures of the actual catalytic materials.

Mixed Metal Oxides

There are many catalytically important metal oxides in addition to those mentioned above, and some are combinations, such as silica–alumina, which was used extensively as a hydrocarbon cracking catalyst prior to the advent of zeolites. It is still used in large quantities as an amorphous matrix for zeolites in cracking catalysts. Silica–alumina hydrogel can be prepared by coprecipitation of the two components. The solid with a low alumina content may consist of corner-sharing SiO_4 and AlO_4 tetrahedra (like those mentioned in the description of zeolites), almost randomly distributed in the solid. At higher alumina contents, the structure is evidently more complicated. Charge neutrality requires cations at the surface, and silica–alumina has both Lewis and Brønsted acid sites, some of the latter being very strong.

Some other mixed metal oxides are also strong acids; some are even superacids. Some of the strongest solid superacids are formed by adsorption of Lewis acids like SbF_5 on solid Brønsted acids, such as silica–alumina; others are formed by sulfating solids such as zirconia.

Mixed metal oxides with redox properties are important catalysts for selective oxidation of hydrocarbons, as described in Section 6.4.6.

6.3
ADSORPTION

When a molecule or atom is brought in contact with a surface, it may combine with the surface through the formation of a chemical bond (or bonds). This is called **chemisorption,** a first step in surface catalysis. Alternatively, the adsorbing molecule may combine with the surface in a physical process resembling simple condensation, with the interaction forces being weak van der Waals forces. The latter is called **physical adsorption,** or **physisorption.** Not surprisingly, the distinction between the two kinds of adsorption is not always clear cut; usually, if the magnitude of the enthalpy of adsorption is about the same as that for condensation (say, 5 kcal/mol), the process is regarded as physisorption.

6.3.1 Adsorption Isotherms

It is helpful to begin with a discussion of adsorption equilibria before considering the structure and bonding of chemisorbed species. The simplest model of adsorption (Langmuir adsorption) was introduced in Chapter 4. Although the model is sometimes a good approximation for adsorption on solids with nearly uniform surfaces, it usually fails to provide an accurate representation when the adsorbent is an inorganic solid of the kinds used as catalysts.

The reasons for the inadequacy of the simple model are found by examining the underlying assumptions (pp. 195–196). The primary assumption of uniformity of the surface sites is usually invalid for inorganic solids. Evidence consistent with the nonuniformity of sites on a surface is shown in Fig. 6-10 [12]. Adsorption up to about one-third of a monolayer appears to be ideal, but the enthalpy of adsorption decreases markedly at higher coverages. Evidently, the sites of greatest coordinative unsaturation are energetically most favored and occupied first; then, others of less coordinative unsaturation are occupied. Ultimately, steric and/or ligand effects caused by the adsorbate may come into play, and then another assumption underlying the Langmuir model (the lack of interaction of adsorbed species) will also be invalid. But the nonuniformity of the surface sites is usually the major reason for the model's inadequacy.

Recognition of the fact that the simple Langmuir isotherm is usually not a good approximation of adsorption equilibrium raises the question of what mathematical formulations do a better job of representing experimental results. Many empirical equations have been found to be useful but to have limited applicability; the most important are summarized in Table 6-3. For example, the mathematically simple Freundlich isotherm for some compound A has been

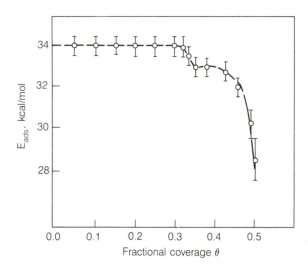

Figure 6-10
Dependence of the heat of adsorption of CO on coverage of the palladium (111) surface [12].

Table 6-3
ADSORPTION ISOTHERMS[a]

Langmuir	$\dfrac{V}{V_m} = \theta = \dfrac{KP}{1 + KP}$	Chemisorption on ideal surfaces
Freundlich	$V = kP^{1/n} \ (n > 1)$	Chemisorption and physical adsorption
Frumkin–Temkin	$\dfrac{V}{V_m} = \theta = \dfrac{1}{a} \ln KP$	Chemisorption
Brunauer–Emmett–Teller (BET)	$\dfrac{V}{V_m} = \dfrac{cP}{(P_o - P)\left(1 + \dfrac{(c-1)P}{P_o}\right)}$	Multilayer physical adsorption

[a] V is the volume adsorbed (usually expressed in cm^3 at standard temperature); V_m is the volume required to cover the adsorbent surface with a monolayer of adsorbate; θ is the fraction of the monolayer covered at an equilibrium pressure P; P_o is the normal vapor pressure of the adsorbate; all other symbols represent temperature-dependent constants.

found to fit many sets of data successfully:

$$\theta_A = kP_A^{1/n} \tag{6-3}$$

(where $n > 1$). It has been shown that this is the form of equation that one would expect for low (submonolayer) coverages on a nonuniform surface having adsorption sites characterized by an exponential distribution of enthalpies of adsorption.

One adsorption isotherm is of great value for determining surface areas of catalysts: the BET isotherm [13]. In contrast to the Langmuir isotherm, it accounts for multilayer adsorption and therefore gives a much better representation of physisorption than the Langmuir isotherm.

The BET equation is based on the same assumptions made by Langmuir, but with the complication that multilayers of adsorbate are allowed to exist on top of the monolayer. Each adsorbed species in the first layer is assumed to be a site available for adsorption of a species in the second layer; species in the second layer provide sites for those in the third layer, and so forth. To derive the equation (Table 6-3), Brunauer, Emmett, and Teller [13] assumed that the rate of adsorption on the bare surface was equal to the rate of desorption from the monolayer; the development is similar to Langmuir's in this respect. They further assumed that the rate of adsorption onto the monolayer is equal to the rate of desorption from the second layer, and so on. And in a further simplification, they assumed that the enthalpy of adsorption to give the second and higher layers is equal to the enthalpy of condensation of the adsorbing gas; the enthalpy of adsorption giving the first layer was assumed to be different (and greater in magnitude). These assumptions imply that the adsorption on the layers beyond the first is characterized by a purely physical process just like condensation. The model is still an oversimplification, but it represents many experimental results very well and is widely used.

The BET development leads to the isotherm

$$V = \frac{V_m cP}{(P_o - P)\left(1 + \frac{(c - 1)P}{P_o}\right)}$$

(6-4)

where c is a temperature-dependent constant related to the enthalpies of adsorption of the first and higher layers; P_o is the normal (saturation) vapor pressure of the adsorbing gas at the temperature of the experiment; V is the volume of adsorbed gas; and V_m is the volume adsorbed to give a monolayer. The equation reduces to the Langmuir form in the limiting case of a low partial pressure of adsorbing gas and a large value of c. When the BET equation is rearranged into the form

$$\frac{\frac{P}{P_o}}{V_m\left(1 - \frac{P}{P_o}\right)} = \frac{1}{V_m c} + \frac{(c - 1)\left(\frac{P}{P_o}\right)}{V_m c}$$

(6-5)

it becomes apparent that a plot of the left-hand side of the equation vs. P/P_o gives a straight line with intercept $1/V_m c$. (Other linearized forms are also used.) Experimental results often conform to the straight-line plots for low values of P/P_o; capillary condensation is to be avoided. This linear plot allows a simple determination of V_m, and the assumption of the cross-sectional area of an adsorbed N_2 molecule (taken to be 0.162 nm^2) allows an estimate of the surface area of the solid.

EXAMPLE 6-2 Determination of an Adsorption Isotherm from Experimental Results

Problem

Consider the adsorption equilibrium data for N_2 and other gases on silica gel, shown in Fig. 6-11 [14]. Propose a suitable representation in terms of an adsorption isotherm.

Solution

The data are not consistent with the Langmuir isotherm; some of the curves are S-shaped, suggesting multilayer adsorption and therefore the BET isotherm. The data are plotted according to the linearized form of the BET equation used by Brunauer et al. [13] (Fig. 6-12). There is a good linear fit, confirming the applicability of the equation. Can you confirm that the linearity of the plots is consistent with the BET equation? What

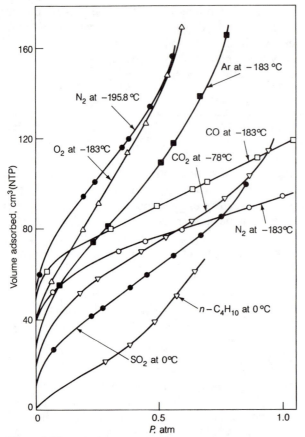

Figure 6-11
Adsorption isotherms for N_2 and other gases on 0.606 g of silica gel [14].

is the surface area of the catalyst determined by the nitrogen adsorption data?

It would be difficult to overemphasize the importance of the BET equation, which was published in 1938. It allows routine, exact determination of surface areas of porous solids, so that catalytic reaction rates can be normalized to unit surface area. The BET model is inexact, and the surface areas are to some degree arbitrary; but as the BET equation is almost universally used, the internal consistency of results obtained in many laboratories is assured. Before the BET development, quantitative comparisons of activities of high-surface-area catalysts were not feasible. Since it is often difficult to reproduce a catalyst preparation exactly and since surface areas vary from batch to batch, the implication is that before the BET method was available, it was hardly possible to make meaningful quantitative comparisons of activities of various catalysts.

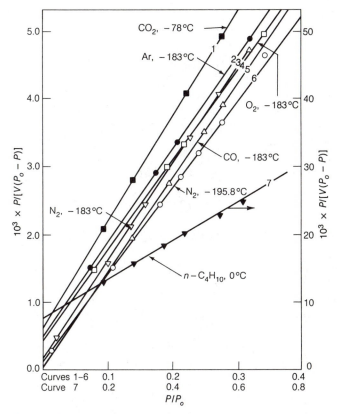

Figure 6-12
Adsorption isotherm for N_2 and other gases on silica gel [13]. Data
are plotted in a linearized BET form.

6.3.2 Structures of Adsorbed Species on Single-Crystal Metal Surfaces

The well-defined cycles of catalysis in solution (Chapter 2) all rest on the identification of reaction intermediates by spectroscopic methods and X-ray crystallography. Methods for identification of adsorbed species include a wide range
of spectroscopies, but the definitive determinations of crystallography are not
applicable. Therefore, only a few structures of adsorbates are firmly based;
most are structures of simple species on single crystals of metal. For the most
part, the structures have been inferred by comparison of surface spectra with
spectra of well-known molecular species used as models. Metal clusters are
valuable models, since they resemble metal surfaces with adsorbed ligands.

Information about molecular configurations of species adsorbed on metals
can be determined by vibrational spectroscopies, for example, infrared or a
more sensitive technique requiring ultrahigh vacuum and applicable to single-
crystal surfaces, electron energy loss spectroscopy (EELS), often used in com-

bination with ultraviolet photoelectron spectroscopy (UPS) to give information about electronic structure as well as molecular configuration (Table 6-2).

Sometimes the adsorbed species form ordered structures on metal surfaces; these can be characterized by LEED. One of the simplest and most important adsorbates is hydrogen, a reactant in many catalytic processes. When H_2 reacts with a metal surface, it dissociates as a result of reactions that may be compared with oxidative additions, giving structures analogous to metal hydride complexes. For example, hydrogen chemisorbed on the nickel (111) surface has been found by LEED to give the structure shown in Fig. 6-13, provided that the surface coverage is one-half a monolayer [15]. In this ordered structure, the hydrogen atoms are bridging ligands at threefold sites, with a Ni—H distance of about 0.18 nm. Analogously, in the metal cluster $H_3Ni_4(C_5H_5O)_4$, the hydride ligands are triply bridging, with about the same Ni—H distance (0.169 nm).

The most thoroughly investigated adsorbate is CO; CO on metal surfaces can be bonded in all the modes evident in metal clusters (Fig. 2-52). The chemisorption of CO on the clean (100) surface of palladium occurs by a ligand association reaction, and the resulting structure has been determined by LEED to be that shown in Fig. 6-14, when the surface coverage is half a monolayer; the figure gives an indication of the bond distances [17]. The CO bridges pairs of Pd atoms.

Ordered structures of CO on the (111) face of Pd are shown in Fig. 6-15 [18, 19, 20]. At surface coverages less than one-third of a monolayer, the energy of adsorption is constant (Fig. 6-10), but the structure changes with increasing coverage beyond one-third of a monolayer, consistent with the expected steric effects, and the bonding modes evident in metal clusters appear. The spectroscopic data of Fig. 6-16 [18] are explained (at least in part) by the changes in energetics of the CO adsorption with coverage.

The bonding of some simple hydrocarbon ligands also seems to be understood now, again on the basis of comparisons of vibrational spectra with those of metal clusters [21]. The structures shown in Fig. 6-17 [22], ethylidyne, propylidyne, and butylidyne, are formed by room-temperature reaction of the respective olefins with Pt; a C—H bond is broken as the olefin interacts with the surface (the adsorption is dissociative), and three strong metal–carbon

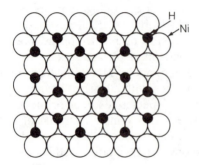

Figure 6-13
Structure of hydrogen chemisorbed on the nickel (111) surface at a coverage of half a monolayer.

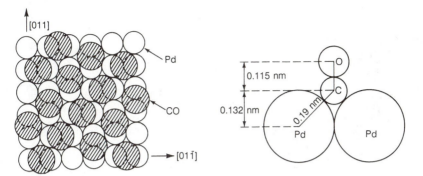

Figure 6-14
Structure of CO chemisorbed on the (100) face of palladium at a coverage
of half a monolayer [17].

bonds are formed [18]. The remainder of the hydrocarbon chain is structurally
similar to that in a gas phase paraffin.

Structures like those described above form on the surfaces of many metals. There are differences in reactivity from one metal to another, and there
are differences in reactivity from one crystal face to another of a given metal.

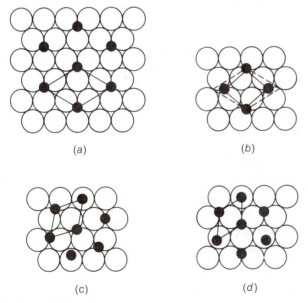

Figure 6-15
Models for the adsorption of CO (closed circles) on the (111)
surface of Pd [18, 19, 20]. At low surface coverages (a), CO
occupies threefold sites (the structure is analogous to that of
a face-bridging CO ligand in a metal cluster (Fig. 2-52). At
higher surface coverages, other structures form, including
those analogous to edge-bridging (b, c) and terminal (c, d)
CO ligands in metal clusters.

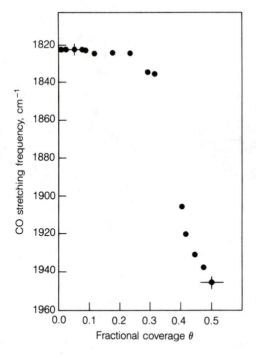

Figure 6-16
Variation with surface coverage of the C—O stretching frequency of CO adsorbed on a Pd (111) surface [18].

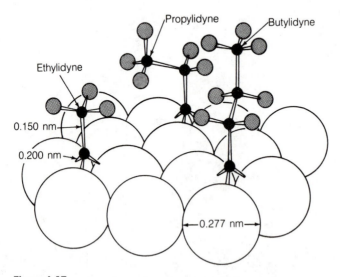

Figure 6-17
Structures of akylidyne species on the (111) face of Pt, formed by adsorption of olefins at room temperature. Large open circles represent Pt, small full circles represent carbon, and hatched circles represent hydrogen [22].

For example [22], the hexagonal array of metal atoms on the (111) terrace face of Pt produces an ethylidyne layer upon adsorption of ethylene at room temperature. The more open (100) surface of Pt is apparently more reactive; the square face of metal atoms on this surface dehydrogenates ethylene to give acetylene at temperatures less than room temperature, and the ethylidyne layer appears to form only at temperatures less than $-70°C$. The stepped and kinked surfaces are even more reactive, since the atoms exposed at ledges and especially at kinks have higher degrees of coordinative unsaturation than the terrace atoms.

When the adsorbed alkylidynes are brought to relatively high temperatures, they react; the reactions include breaking of C–H bonds, among others. The products include ill-defined carbonaceous (hydrocarbon) overlayers of various compositions and structures, not greatly different from the coke formed in zeolites; even graphitelike structures can form [23]. A schematic representation of a Pt surface partially covered with such an overlayer is shown in Fig. 6-18 [23]. The sketch shows some regions of exposed metal surface and different structures of the overlayer. The surface of a metal catalyst for hydrocarbon conversion in the working state may resemble this picture.

Metal atoms can be adsorbed on metal surfaces. Cu atoms deposited from the gas phase on a Ru surface first form a monatomic layer (characterized by LEED); increasing amounts of Cu lead to three-dimensional crystal growth [24]. When Cu atoms are removed from the surface in a temperature-programmed desorption experiment (one in which the temperature of the sample in vacuum is increased at a linear rate and the evolved gases analyzed by a mass spectrometer), the results show that the chemisorption energy of Cu on Ru is about 3 kcal/mol higher than the sublimation energy of bulk Cu. The Cu–Ru interaction in the bulk, on the other hand, is less favored energetically than the Ru–Ru interaction, so that no bulk alloy formation occurs. Therefore, the stable monolayer structure is formed [24]. Small amounts of Cu on the Ru surface sharply reduce the amount of strongly chemisorbed hydrogen [25]; this result gives an indication of the possibilities for modifying metal surface reactivity by incorporation of a second metal component. An application of this idea arises in the catalytic reforming of hydrocarbons (Section 6.4.5).

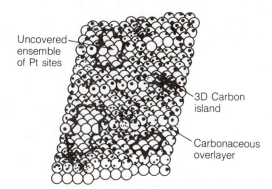

Uncovered ensemble of Pt sites

3D Carbon island

Carbonaceous overlayer

Figure 6-18
Schematic representation of carbonaceous overlayers on the surface of Pt [23].

The simple picture of chemical bonding of an adsorbate presented in the preceding pages is sometimes not observed. The adsorbate may exert such a strong influence on the surface that it causes the bulk to be restructured; the symmetry of the crystal plane may be changed by addition of the adsorbate, or the surface may be locally restructured (the process sometimes being called corrosive chemisorption). The adsorbate may also become incorporated in the bulk; for example, hydrogen has such a high solubility in palladium that diffusion through palladium thimbles is used to purify hydrogen.

6.3.3 Adsorption on Complex Surfaces

The typical catalyst surface is far more complex than that of a clean single crystal of metal in a vacuum. The models of surfaces of metal oxides presented in Section 6.2.3 just begin to illustrate the complexity of bare catalyst surfaces. A typical catalytic material exposes a number of different crystal faces, and impurities are almost always present and greatly complicate matters. Consequently, one must be satisfied with sometimes vague models of the structures of adsorbed species. These models are usually inferred from the chemistry of the surface and the adsorbate combined with spectroscopic data characterizing the adsorbed species. The vibrational spectroscopies, especially infrared, are widely applied.

For example, surfaces of oxides such as γ-Al_2O_3 have some acidic character (as well as basic character), and adsorption of many compounds is interpreted in terms of acid–base chemistry. Adsorption of the base pyridine (a frequently used probe molecule) leads to the formation of several species, some of them distinguishable by infrared spectroscopy: Pyridinium ions are formed in association with proton donor —OH groups on some metal oxides (e.g., silica–alumina, but not on γ-Al_2O_3 [26]), as inferred from the infrared spectra. Pyridine also combines with the exposed Al^{3+} sites (Lewis acid sites) on surfaces, giving distinctive spectra, but detailed structural models of the adsorbed species are lacking [7]. Infrared spectroscopy is also used to observe changes in the bands indicative of the surface —OH groups.

CO is often used as a probe of surface structure, since relatively intense infrared spectra of the adsorbed ligand give evidence of the sites to which it is bonded. CO on metals was mentioned above. CO on metal oxides gives evidence of the nature and number of coordinatively unsaturated cationic centers, and the frequencies of the vibrations provide evidence of the degree of backbonding. Experiments at very low temperatures are often the most informative. NO is similarly a useful probe molecule, but it is more reactive than CO, and interpretation of its adsorption is often complicated by its reactions on the surface.

Hydrogen bonds to surfaces in a multitude of ways [27]. Surface —OH groups on metal oxides have been mentioned; they have acidic and basic character, being both proton donors and acceptors. Metal–hydrogen bonds also form, giving hydrides on some metal oxides, for example, ZnO, as on metals. More specific structures are presented later.

There is often a question of the oxidation states of metals on surfaces, especially surfaces of oxides of metals that are stable in more than one oxidation state. Examples are Cr_2O_3, V_2O_5, and TiO_2 (these are sometimes called reducible oxides, in contrast to oxides such as Al_2O_3 and SiO_2). The oxidation state of a metal on the surface is often not the same as that of the metal in the bulk solid. The metal oxidation state may affect the reactivity and catalytic activity in important ways; and in some surface reactions, for example, redox reactions and those involving oxidative additions and reductive eliminations, changes in oxidation state are required for catalytic cycles. Probe molecules adsorbed on these surfaces may provide some information about the metal oxidation state—but they may also change it! Some physical methods such as electron spin resonance spectroscopy and XPS can also provide information about oxidation states of metals on surfaces, but often this information is not definitive. A relatively new technique that gives measurements of X-ray absorption near edge structure (XANES), offers good prospects.

6.3.4 Functionalized Surfaces

Catalytically active groups can be anchored to surfaces of supports by chemisorption. In Chapter 4, acidic, basic, redox, and metal complex groups attached to solid polymers were mentioned. These same groups can be attached to inorganic supports, and metal oxides such as silica and alumina are the most commonly used. Silica, being relatively simple and unreactive, can be thought of as a "pegboard" [28]. The functional groups that terminate silica include —OH groups, believed to occur in the following [7]:

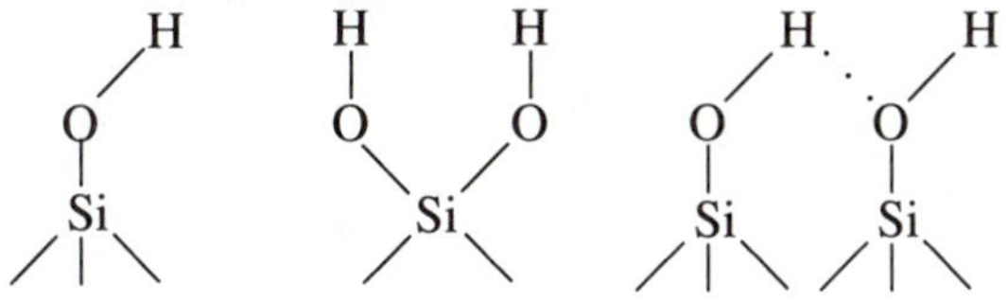

When the surface is heated, water is driven off; and if the temperature is raised to about 450°C, stable and relatively unreactive "siloxane bonds" are formed:

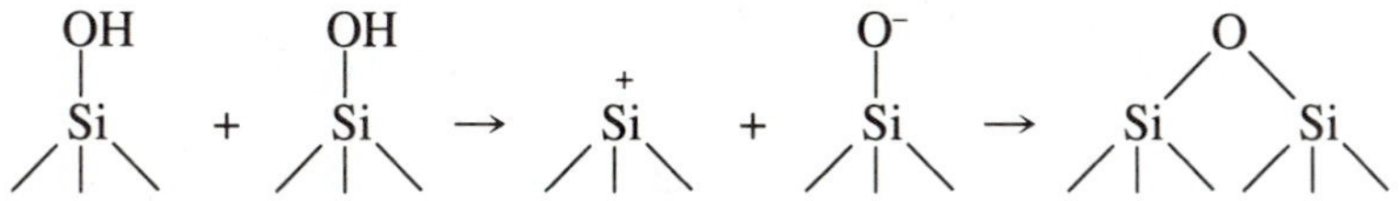

$$(6\text{-}6)$$

Taking advantage of the reactivity of these groups, which are weakly acidic ($pK_a \cong 7$), one can attach many kinds of catalytic groups to the surface and expect that a reaction will be catalyzed by the anchored groups, with the support playing a negligible role. Typical reactions used to form catalytically active surface-bound groups are esterifications and other condensations.

For example, strongly acidic groups can be anchored to the SiO_2 surface by the incorporation of phenyl groups by reaction with $PhSiCl_3$ followed by sulfonation:

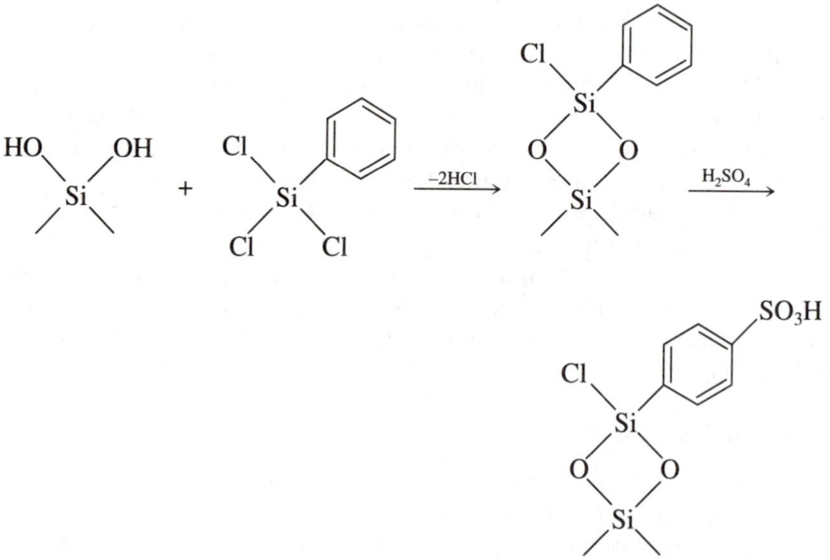

$$(6\text{-}7)$$

This solid acid is an ion exchanger, a close analogue of the cross-linked poly(styrenesulfonic acid) described in Chapter 4.

Mononuclear complexes of many metals (e.g., Cr, Mo, Ni, Rh, Pd, W, Re, and Pt) can be anchored to a silica surface by reaction of the π-allyl complex with surface —OH groups, as illustrated below for Mo [29]:

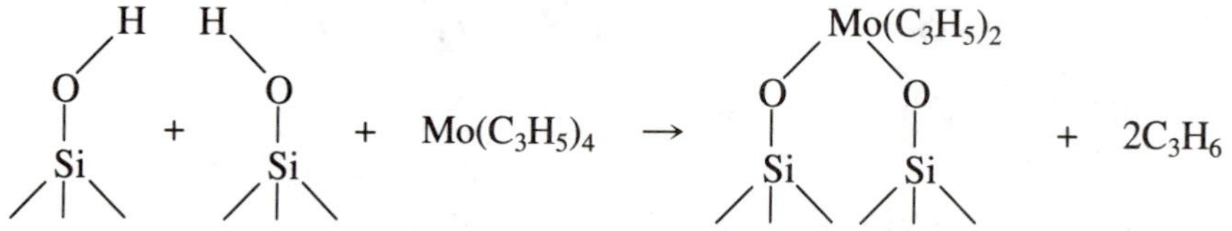

$$(6\text{-}8)$$

Catalytic activities can be predicted from those of the soluble analogues. Groups like these containing Mo (or Re or W) are catalytically active for metathesis and a number of other reactions of olefins, for example, which are discussed below in Section 6.4.1.

Surface metal complexes like these are easily prepared, and the metal–oxygen bonds may be quite strong for oxophilic metals such as those of Groups 6 and 7 (and lower) of the periodic table, but they are unstable in air (being rapidly oxidized) and under conditions of some catalytic reactions. The Group 8 metals in surface structures like that shown in Eq. 6-8 are readily reduced by hydrogen under mild conditions; when the metal is reduced to the zero-valent state, it migrates over the surface and forms aggregates of metal atoms, thereby giving the most common form of a supported metal catalyst, discussed below in Section 6.4.5.

Anchored transition metal complexes have also been synthesized to have ligands closely analogous to those of soluble catalysts used in industrial processes. For example, rhodium complexes for olefin hydrogenation and hydroformylation have been prepared by first attaching amine or phosphine ligands to SiO_2 and then incorporating the metal complex; the structures are analogous to those of the polymer-supported rhodium complexes described in Chapter 4. The following structure has been inferred and suggested to be a catalyst precursor (here, the hatched structure represents the SiO_2) [30]:

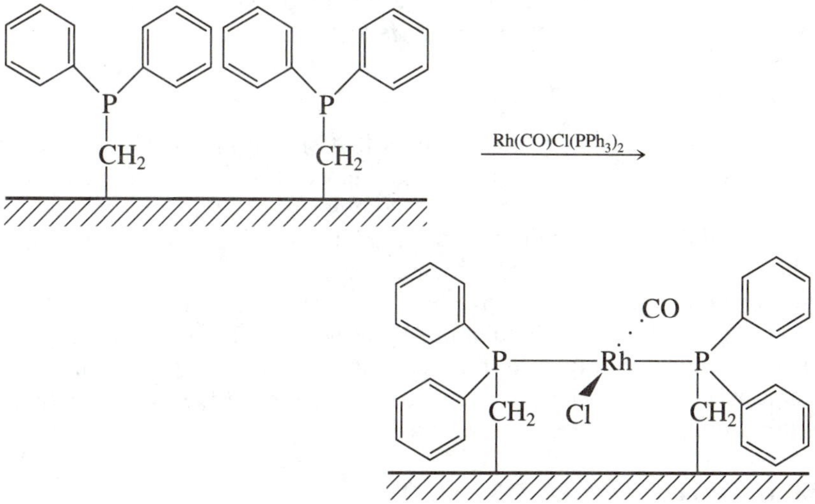

$$(6-9)$$

The silica-supported acids, bases, and metal complexes take their place in a near-continuum beginning with the solution chemistry of the same groups described in Chapter 2. Attachment of these groups to a solid support provides new opportunities in catalysis. The stabilization of coordinative unsaturation, for example, is a characteristic of the groups on inorganic supports as it is of those on rigid, porous polymers, as illustrated in Chapter 4. The interactions of groups on the rigid inorganic solids, however, are much more restricted than those of the groups in all but the most inflexible polymers.

The ideas are illustrated here with SiO_2 as the support. It has been used the most because it is more nearly inert than most other metal oxide supports. But SiO_2 is usually more reactive than poly(styrene-divinylbenzene), and not all the syntheses possible with this hydrocarbon support are possible with SiO_2.

A few metal-oxide-supported metal complexes are applied as industrial catalysts, for example, supported complexes of Cr and of Zr for olefin polymerization (discussed in Section 6.4.1), but these are exceptions. The structures anchored to SiO_2 through organic ligands are not applied industrially, since they suffer from the same disadvantage of lack of stability as their polymer-supported analogues discussed in Chapter 4.

6.4
SURFACE CATALYSIS

6.4.1 Catalysis on Functionalized Surfaces: Connections to Molecular Catalysis

When a catalyst consists of functional groups such as acids, bases, and metal complexes grafted to a nearly inert support, then the catalysis may be very similar to that occurring in a solution of analogous functional groups, and the catalytic sites are easily identified as the functional groups. This point has been illustrated for polymer-supported groups, and little elaboration is needed. The silica-supported metal complex groups, for example, provide some of the most direct connections between solution catalysis and surface catalysis.

CATALYSIS BY SILICA-SUPPORTED Rh COMPLEXES

Silica-supported rhodium complexes such as that shown in Eq. 6-9 are analogous to soluble and polymer-supported rhodium complexes mentioned previously. As expected, the silica-supported complex is catalytically active for olefin hydrogenation and hydroformylation, among other reactions. Catalysts such as this have been investigated as candidates for commercial operation in olefin hydroformylation, but they have not been successful because of a slow loss of rhodium into the reactant solution. What is important about these supported metal complex catalysts is the conceptual bridge they form, linking solution and surface catalysis; the chemistry is in essence the same whether the catalytic group is dissolved or supported.

WACKER OXIDATION OF ETHYLENE ON V_2O_5 SURFACES FUNCTIONALIZED WITH Pd

The Wacker oxidation of ethylene to give acetaldehyde was introduced in Chapter 2, and the design of a polymer-supported catalyst was introduced in Chapter 4. The design has been extended straightforwardly to inorganic surfaces; as before, the components are Pd^{2+} complexes and redox groups. The redox groups are provided by the surface of V_2O_5, a solid having vanadium cations that can be present on the surface in various oxidation states. A schematic representation of a catalytic cycle is shown in Fig. 6-19 [31].

OLEFIN METATHESIS CATALYZED BY METAL-OXIDE-SUPPORTED METAL COMPLEXES

Organometallic complexes such as the molybdenum allyl mentioned above react with —OH groups on the surfaces of metal oxides to give structures that are catalytically active for olefin metathesis. The metals are those known from solution chemistry to be active for this reaction (Mo, W, and Re), and the extension to surface catalysis is relatively straightforward. However, there are complications, expected because the solution catalysis involves an intricate cycle and the reaction is promoted by Lewis acids. Similarly, metal-oxide-

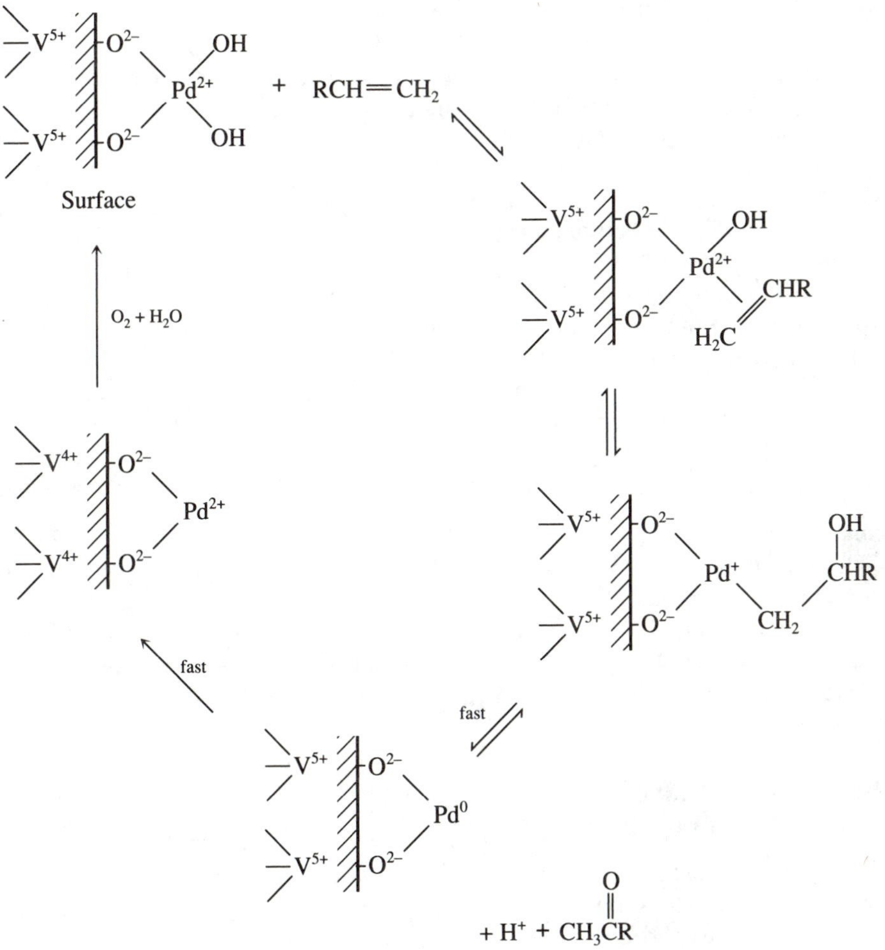

Figure 6-19
Schematic cycle proposed for the Wacker oxidation of ethylene catalyzed by a Pd-doped V_2O_5 catalyst [31]. The hatching represents the V_2O_5 surface.

supported catalysts are promoted by Lewis acids, and γ-Al_2O_3-supported Mo complexes are more active than SiO_2-supported analogues because of the Lewis acid sites on the former [32]. Catalysts prepared from allyl complexes are active, whereas those prepared from ethoxide complexes are not, possibly because the former provide the hydride ligands needed for the catalysis and the latter do not. Physical techniques such as XPS have been used to characterize the catalysts, leading to suggestions that the catalytically active forms incorporate metal ions in relatively high oxidation states [33].

The patterns of the catalysis by surface complexes are similar to those in the solution catalysis, but a stronger statement is not warranted because details of the surface-catalyzed reactions are lacking. The complexity of the situation approaches that for more typical surface catalysts. In practice, the catalysts used for olefin metathesis are not the supported mononuclear complexes described above. Instead, they are metal-oxide-supported metal oxides, consisting of layers resembling Re_2O_7 on alumina; these are structurally complicated and not fully characterized.

EXAMPLE 6-3 Determination of Number of Catalytic Sites in Poisoning Experiments

Problem

Kuznetsov and co-workers [32] have used O_2 to poison silica-supported propylene metathesis catalysts prepared from the reaction of tungsten allyl with the —OH groups of silica. They observed that oxygen poisoned the catalyst. The activity, measured at 100°C with a propylene partial pressure of 0.16 atm, decreased linearly from 6 molecules/(W^{4+} ion·s) to zero as the atomic ratio of oxygen to surface tungsten increased from 0 to 2. What are the implications of the observation?

Solution

The fact that the activity decreased linearly with increasing O/W ratio implies that the catalytic species were equivalent in activity, suggesting that they were nearly "molecular" in character, being tungsten complexes anchored to the silica surface. The data suggest that one O_2 molecule reacts with one tungsten complex to form a catalytically inactive species.

∎

OLEFIN POLYMERIZATION ON SURFACES FUNCTIONALIZED WITH Cr COMPLEXES

Silica-supported complexes of Cr and Zr are industrial catalysts for polymerization of olefins. The supported Cr catalysts are formed from organochromium complexes such as chromium allyl or chromocene or from inorganic precursors such as CrO_3 deposited on the support from aqueous solution and then heated in air (calcined). The precursors react with surface oxygen or —OH groups, forming Cr–O bonds. A variety of surface structures can be formed, with Cr existing in a range of oxidation states, from 0 to +6 [34]. A number of these different Cr complexes are catalytically active for the olefin polymerization, but the evidence suggests that Cr^{2+} complexes may be the active species in the commercial catalysts, although this point is disputed [34], and some Cr^{3+} complexes appear to be active in solution, whereas similar Cr^{2+} complexes are not.

The exact catalytic sites are unidentified, and little is known about the reaction mechanism, but there is, at least, an imperfect analogy to the solution-catalyzed olefin polymerizations discussed in Chapter 2. These and related organochromium [35] and organozirconium [36] catalysts are the only supported metal complex catalysts that are of great technological importance. They are a family of catalysts used for the manufacture of polyethylene and other polymers.

Silica is the preferred support, but others, such as aluminas, can also be used. In commercial processes, the exothermic polymerization of ethylene takes place in the presence of gas phase reactants in a fluidized-bed reactor or, alternatively, in the presence of liquid phase reactants, giving either dissolved or solid product, the latter present in a slurry. The well-mixed gas in the fluidized bed or the stirred liquid facilitates heat transfer. Pressures are typically low and temperatures are roughly 100°C, depending on the catalyst and reactor type.

The catalysts give yields as high as thousands of grams of nearly linear polyethylene per gram of catalyst. The catalyst is a high-area porous solid, and the massive amount of polymer that is formed on the interior surface as monomer molecules are inserted between the catalytic sites and the growing polymer chains exerts a strong force that causes fragmentation of the porous catalyst particles. Ultimately, the resulting fine particles of inorganic catalyst are enveloped by the mass of organic polymer, and the catalyst remains as a minor impurity in the product.

Since olefin polymerization processes are so important in technology, there has been extensive research on the catalyst and reactions, but the lack of detailed understanding is representative of the complexity of industrial catalysis. Notwithstanding the lack of understanding, the surface-complex-catalyzed olefin polymerization is important to the discussion of this text because of the scale of the industrial processes and because it forms a central part of the near-continuum, illustrating the general pattern of transition-metal-complex-catalyzed reactions, whether they occur in solution or on surfaces. The development is continued below in Section 6.4.2, where catalysis by $TiCl_3$ surfaces is described in terms of the molecular concepts stated in Chapter 2.

CATALYSIS BY MULTICENTER SURFACE SITES

The supported metal complex catalysts described in the preceding sections consist of mononuclear metal complexes acting more or less independently of each other on the support. When neighboring metal centers are present, the catalytic properties may be markedly changed. A few examples of multicenter surface sites provide a transition to the most typical metal catalysts, which are extended surfaces of metal atoms.

Surface sites that have a specified number of metal centers can be prepared by using a molecular precursor with that number of metal atoms and allowing it to react with a support surface. Such preparations have been illustrated with organometallic precursors that are dimers and with metal clusters containing three metal atoms.

A family of surface-bound dimers and metal pair sites (formed by fragmenting the dimers) has been prepared from the allyl complexes $Cr_2(\eta^3\text{-}C_3H_5)_4$ and $Mo_2(\eta^3\text{-}C_3H_5)_4$ on silica supports [29] (Fig. 6-20). The catalytic activities and selectivities of the molybdenum species for propylene oxidation depend strongly on the surface structure. The Mo(IV) monomers and dimers on SiO_2 have a selectivity for acrolein formation of only about 30 percent, whereas the

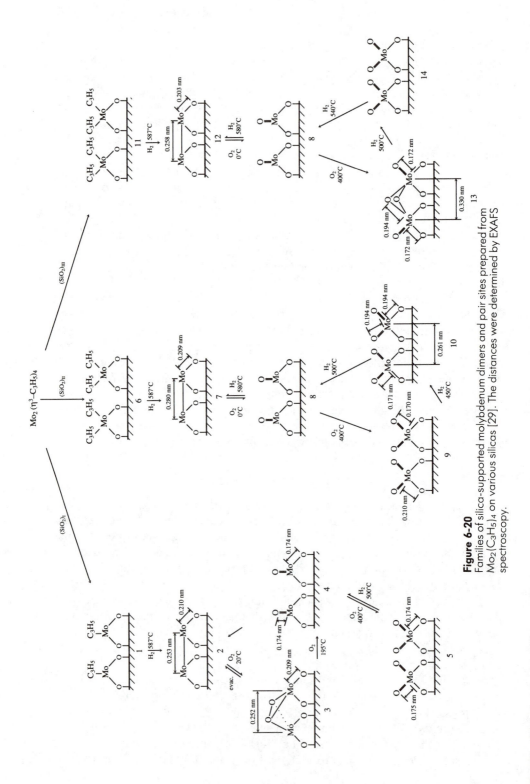

Figure 6-20
Families of silica-supported molybdenum dimers and pair sites prepared from $Mo_2(C_3H_5)_4$ on various silicas [29]. The distances were determined by EXAFS spectroscopy.

oxygen-bridged molybdenum dimers shown as structure 14 in Fig. 6-20 have an acrolein selectivity of about 80 percent [29]. Structures of these simple surface species have been determined by EXAFS spectroscopy and other techniques. The structural results suggest that the high selectivity of structure 14 is associated with the steric crowding in the Mo dimer.

Rhenium is a sufficiently oxophilic metal that it, like chromium, titanium, zirconium, and molybdenum, is able to form strong bonds to oxygen ligands of a metal oxide surface. Mononuclear, dinuclear, and trinuclear rhenium carbonyl complexes have been used to prepare catalysts supported on MgO. The mononuclear complex $HRe(CO)_5$, a very weak acid, was chemisorbed without dissociation of a proton on the strongly basic surface of largely dehydroxylated MgO [37]. When the surface complex was oxidized, it gave a new surface species formulated as $Re(CO)_3\{O\text{-}Mg\}_2\{HO\text{-}Mg\}$ (where the braces { } enclose groups terminating the MgO support), and these were presumably isolated from each other on the support surface [37]. The structure has been determined by EXAFS spectroscopy (Fig. 6-21) [37b]. The dinuclear complex $Re_2(CO)_{10}$ is inferred to give pair sites, and the trinuclear complex $H_3Re_3(CO)_{12}$ to give ensembles consisting of three Re centers on the surface; EXAFS spectroscopy gave evidence of the neighboring Re centers [37a]. The catalytic properties depend on the number of Re atoms in the surface sites. The catalyst with isolated Re centers is active for olefin hydrogenation but not for cyclopropane hydrogenolysis. The catalyst incorporating ensembles of three of these Re centers is active for both reactions, having nearly the same activity per Re atom for the olefin hydrogenation reaction as the former catalyst [38]. These results indicate that isolated Re centers are active for propylene hydrogenation but not for cyclopropane hydrogenolysis. This result is consistent with the known ability of mononuclear metal complexes in solution (e.g., the Wilkinson complex) to catalyze the hydrogenation but not the hydrogenolysis.

The details of how these multicenter Mo and Re catalytic sites operate are not known, but the strong dependence of the catalyst performance on the presence of neighboring metal centers and on small changes in the structure indicates many opportunities for designing surface catalytic sites with subtly different structures and compositions to give catalysts with markedly different

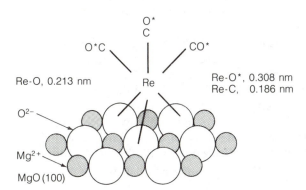

Figure 6-21
Model of the structure of rhenium subcarbonyl on MgO [37b]. The distances were determined by EXAFS spectroscopy. O* represents the oxygen of carbonyl groups.

properties. With stable structures like these on metal oxide surfaces, there are opportunities for preparation of stable catalysts; in contrast, soluble compounds with neighboring metal centers (metal clusters) are for the most part too unstable to offer comparable opportunities. The robust surface species dominate in practical catalysis.

6.4.2 Olefin Polymerization Catalysis on Titanium Trichloride Surfaces: Further Links Between Molecular and Surface Catalysis

Since the catalysts described in the foregoing pages are well represented as molecular analogues, the connections between solution and surface catalysis are straightforward. The analogy has been extended to include catalysts that are bulk solids, exemplified by α-TiCl$_3$, which, when combined with aluminum alkyls, is a Ziegler catalyst, used for olefin polymerization. The reasoning used to explain the performance of this catalyst is a direct extension of the reasoning used in Chapter 2 to account for the performance of soluble transition metal complex catalysts. This example completes a transition to the surfaces of solid catalysts, demonstrating the strength of the concepts of coordination chemistry and organometallic chemistry as an important part of the foundation of catalytic chemistry.

Soluble olefin polymerization catalysts were mentioned in Chapter 2. Many metal complexes and combinations of metal complexes in solution are catalytically active for this reaction, and a few are stereoselective. The first stereoselective catalysts for olefin polymerization were solids, discovered serendipitously in 1953 by K. Ziegler, who added TiCl$_4$ to an aluminum alkyl and found it to be active for the polymerization of ethylene to give high-molecular-weight polyethylene. This catalyst was found by G. Natta to be active for polymerization of propylene to give rubberlike solids, and it has proved to be the forerunner of a large family of catalysts, typified by α-TiCl$_3$ combined with aluminum alkyls such as Al(C$_2$H$_5$)$_2$Cl, that play a major role in polymerization technology [39]. These catalysts are used primarily for polymerization of propylene, whereas catalysts exemplified by the supported chromium complexes described in Section 6.4.1 are used primarily for polymerization of ethylene. Each is also used with other olefin monomers and combinations of monomers to make copolymers.

Ziegler's discovery led to research by Natta et al. [40], who elucidated details of the stereochemistry of the polymerization, which provided evidence of the stereoregularity of the polymer products and crucial clues about the nature of the catalytic surface. Ziegler and Natta shared the Nobel prize for their work on these catalysts, which are used on an enormous scale.

The uniqueness of the Ziegler catalysts lies in their ability to catalyze propylene polymerization to give high yields of isotactic polypropylene (p. 105). Atactic polypropylene is an undesired by-product of the Ziegler–Natta polymerization, having less desirable physical properties than the stereoregular form.

The essence of the polymerization mechanism suggested by Cossee and Arlman [41] is its ability to account for the stereospecificity of the propylene polymerization. The mechanism is incomplete in detail and fails to account for all the observations, but it accounts for the essential observations and is especially important conceptually, for it is built on the concepts of molecular coordination chemistry, extended to the surface of the α-TiCl$_3$. To understand this mechanism, it is necessary to understand the structure of the bulk of the catalyst, and from that to infer some important characteristics of the surface. Recall that inferences about surface structures based on knowledge of the bulk structure are often not correct; the example presented here should not be overgeneralized.

There are several structures of bulk TiCl$_3$; the α and γ forms are stereospecific catalysts, but the β form gives atactic polypropylene. The α form has the bulk structure depicted in Fig. 6-22. This is a layer structure; layers containing Ti^{3+} ions intersperse close-packed layers of Cl$^-$ ions. Since there is only one Ti cation for every three Cl anions, the cation layers are only partially filled, and some cation layers are missing. The structure is built up of alternating layers in the pattern Cl layer, Ti layer, Cl layer, empty layer, and so on. Each Ti^{3+} ion in the bulk is octahedrally surrounded by six Cl$^-$ ions, and two-thirds of the cation positions in the layer are filled.

Extrapolating this structural information to the surface, Cossee and Arlman assumed that the octahedral coordination of the Ti was retained. To meet the criterion of charge neutrality of the solid, they assumed that there were anion vacancies at the crystal edges, that is, that some of the Cl$^-$ ions were missing from the positions where they would be expected in a structure formed

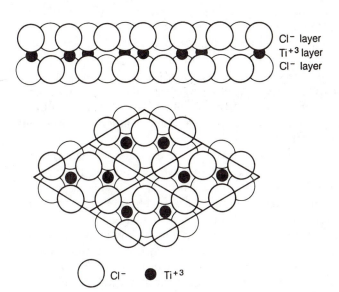

Cl$^-$ layer
Ti^{+3} layer
Cl$^-$ layer

Cl$^-$ Ti^{+3}

Figure 6-22
Structure of α-TiCl$_3$.

by simple cleavage of the bulk. This assumption implies that Ti^{3+} ions are exposed at the surface at these anion vacancies and that there are different surroundings of the exposed cations, depending on their relationship to the anion vacancies.

The challenge was to explain the mechanism of the polymerization and to reconcile it with the surface structure and bonding and account for the stereospecificity of the reaction. Cossee and Arlman reasoned that the coordinative unsaturation was a key to the surface reactivity, assuming that propylene could bond to the Ti cations where they were five- rather than six-coordinate by virtue of the missing Cl anions. Furthermore, they inferred that steric constraints imposed by the neighboring surface anions (analogous to ligands in transition metal complexes) would restrict the reactants in such a way as to give only stereoregular polypropylene on the surface. They assumed that propylene first coordinates to the exposed Ti cation and is then inserted into the Ti–alkyl bond, where the alkyl ligand is the growing polymer chain. The assumption of an insertion step is consistent with the early work of Ziegler showing that with aluminum alkyls, polyolefins were built up by insertion of monomers between mononuclear metal centers and the neighboring ligands which are growing polymer chains. This insertion step (an alkyl migration) increases the length of the growing polymer chain by one monomer unit, and the process can be repeated *ad infinitum*, as in the solution polymerization discussed in Chapter 2.

According to this simplified picture, the catalytic sites are present on the crystal edges of α-TiCl$_3$, shown in Fig. 6-23 [42]. The Ti cations exposed at the surface have neighboring Cl anions, some of them part of the bulk solid and some seeming to dangle at the surface. The octahedral surrounding of each Ti cation is evident in the figure. The anion vacancies (sites of coordinative unsaturation of the Ti cations) present opportunities for bonding of more than one organic ligand, presumed to be the growing polymer (alkyl group) and the olefin monomer, which is assumed to be π-bonded, although there is no direct experimental evidence of the bonding. It is straightforward to postulate that the olefin is inserted between the Ti cation and the growing polymer chain, with the result being the lengthening of the attached chain by one monomer unit and the formation of another site of coordinative unsaturation where another monomer can be coordinated.

In the catalytic site depicted in Fig. 6-23, the growing polymer chain is represented as a large solid sphere; it can adopt two different positions, each giving an octahedral surrounding to the neighboring Ti cation, with one coordination site remaining unoccupied, ready for adsorption of propylene in a ligand association reaction. The catalytic site is a molecular analogue; only the immediate neighbors around the Ti cation are assumed to influence the reactivity, by virtue of steric crowding. As shown in Fig. 6-23, the rigid surface geometry establishes two positions for the alkyl chain, and the one shown in structure (*a*) is inferred to be much more stable than the other, leaving only the other site for the bonding of propylene. As the insertion step takes place, the growing polymer chain switches to the less favored position. Presumably,

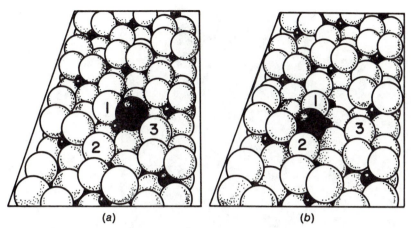

(a) (b)

Figure 6-23
Catalytic sites for stereospecific polymerization of propylene on α-TiCl$_3$ crystal edges [42]. Small black spheres represent Ti^{3+} ions, large black spheres alkyl groups (growing polymer chains), and white spheres Cl$^-$ ions. Structure (b) is converted to structure (a) as a result of migration of the alkyl group, the driving force being the greater steric crowding of the alkyl group in (b). Propylene chemisorbs to give structure (a) in one sterically preferred configuration, with the CH$_2$ end downward and the CH$_3$ group on the side of the Cl$^-$ ion marked 1 [42].

it rapidly switches back to the more favored position, leaving the other position for bonding of the monomer. Since this open coordination site is sterically crowded, it is assumed that the propylene can bond to it in only one configuration, with the small —CH$_2$ group pointed downward and the bulkier —CH$_2$—CH$_3$ group pointed outward. Consequently, the monomer is evidently always aligned in the same way with respect to the growing polymer chain, which leads to its incorporation in the chain to give the stereoregularity of isotactic polypropylene.

There is a close parallel to the picture of Fig. 2-42. Consistent with this model, Ti cations at the surface having a higher degree of coordinative unsaturation or less steric crowding give atactic and not stereoregular polymer. Such unselective sites include a small fraction of the sites on α-TiCl$_3$ and virtually all of the sites on β-TiCl$_3$.

To repeat, this model is based on many assumptions and must be regarded as oversimplified. The details of the organometallic surface structures and the reaction mechanisms are unknown, and the model fails to address the role of the cocatalyst (the aluminum alkyl), which may provide a ligand that participates in the catalytic site. Nonetheless, the model is plausible and powerful in its ability to account for the essential characteristics of a remarkably selective and important reaction. One of the demonstrations of its essential correctness is shown in the electron micrograph of Fig. 6-24; this shows a hexagonal crystal of α-TiCl$_3$, with dots believed to be attached chains of polypropylene located on crystal edges present on a crystal growth spiral [43]. The Cossee–Arlman

Figure 6-24
Electron micrograph of hexagonal crystals of α-TiCl$_3$. Pattern of dots indicates growing chains of polypropylene on sites located along a crystal growth spiral [43].

model is a strong conceptual link between molecular chemistry and surface catalysis.

The Ziegler catalysts used today in industry are more complicated structurally than those described above. The modern catalysts are supported, being formed, for example, from TiCl$_4$ on MgCl$_2$, a support chosen to have a cation (Mg^{2+}) with nearly the same ionic radius as Ti^{4+}, which somehow gives an active, stable surface structure in registry with the support. The catalyst surfaces also incorporate aluminum alkyls [often Al(C$_2$H$_5$)$_2$Cl] and electron donors such as ethyl benzoate, which improves the catalyst selectivity for isotactic polypropylene, possibly by selectively poisoning the unselective sites. The catalytic sites are thought to be similar to those described above for the TiCl$_3$ catalysts, but little is known about their structures.

Many methods are described in patents for preparing supported catalysts [44]. For example, a catalyst may be prepared by grinding the MgCl$_2$ support in the presence of ethyl benzoate or another electron donor to expose new surface and to make fine particles. Then, the support may be brought in contact with hot TiCl$_4$ liquid, washed with hydrocarbon, and used in the presence of a cocatalyst such as Al(C$_2$H$_5$)$_2$Cl and an electron donor compound. These supported catalysts are exceedingly complex in composition and structure, and continuing industrial research is leading to better catalysts and processes. The reactions are usually carried out in slurry reactors at moderate temperatures (about 100°C) and pressures (about 10 atm). Selectivities to isotactic polypropylene of 95 percent and higher have been reported, with yields exceeding 1000 g of polypropylene/g of Ti in the catalyst. The catalyst remains as a minor impurity in the product. Hydrogen is used as a reactant to regulate the molecular weight distribution; it causes chain termination by reacting not with propylene but with the growing polymer chain and forming a polymer detached from the surface.

6.4.3 Catalysis on Metal Surfaces

Molecules are activated by metal atoms in metal surfaces much in the way that they are activated by metal atoms in metal complexes. For a reactant to be activated, it must be bonded to the metal, which requires coordinative unsaturation (in the catalysis literature, a coordinatively unsaturated surface site is often abbreviated as *cus* [45]). Metal surfaces offer a marked advantage over metal complexes in this respect; they can be made to be coordinatively unsaturated (e.g., by treatment under vacuum to remove adsorbates) without loss of their structures. Metal complexes and clusters, on the other hand, are usually unstable when they have even low degrees of coordinative unsaturation. The compounds require stabilizing ligands. So do solid metals; the ligands that stabilize a surface metal atom are the neighboring metal atoms in the bulk and surface.

Therefore, reactivity of metal surfaces is distinct from that of complexes and clusters of the same metal and not easily predicted. Many metals have catalytic activity for many reactions. In practice, the catalytically important metals include most of the transition metals that are important in metal complex catalysis; the platinum group metals find especially wide use, even though they are expensive.

Metal surfaces offer many opportunities for catalysis that are not offered by metal complexes. Because the metals are robust, they can be used efficiently at temperatures that are too high to be practical for solutions and too high for the stability of most metal complexes in solution or anchored to supports. The metal surfaces can be used with varying degrees of coverage and can be applied as catalysts with feeds providing a wide variety of reactants; the surface may be able to accommodate and facilitate the reactions of them all. As many as hundreds of different compounds may be converted simultaneously on surfaces of metal catalysts for conversion of petroleum fractions, as discussed in Section 6.4.5.

One of the important concepts articulated about metal catalysis (and surface catalysis more generally) was H. S. Taylor's 1925 postulate of active centers [46]. Taylor inferred that surfaces are nonuniform and that only a minority of specific surface sites may be active for a particular catalytic reaction under a particular set of conditions. Experiments with selective poisons have shown that typically only a small fraction of the accessible surface is catalytically active. Experiments with well-defined single-crystal surfaces having various crystal faces and various densities of steps and kinks have now shown unequivocally that catalytic activity for some reactions depends strongly on the coordination of a metal center and that, in some instances, only a very small fraction of surface metal atoms, for example, those with high degrees of coordinative unsaturation, account for the activity of a metal surface. Usually the catalytic sites remain unidentified.

The reactive intermediates on the surfaces of most catalysts are present in such low concentrations that their identities also remain largely unknown. The more stable adsorbate structures that have been characterized spectroscopically are not reactive enough to play a role in a reasonably fast catalytic reaction; instead, they may have nothing to do with catalysis and just be "spectators" present on the surface.

Ligands are varied to control the catalytic properties of metal complexes and, similarly, ligands on metal surfaces can be varied to control the catalytic properties. When atoms of a second metal are mixed into the bulk of a metal to form an alloy, the surface composition and often the structure are also changed, with modification of the catalytic properties. Similarly, the presence of components referred to as promoters, activators, stabilizers, and poisons can all lead to modifications of the surface structure and reactivity and therefore the catalytic properties of a metal surface. An example of an alloylike catalyst, discussed below in Section 6.4.5, is the combination of iridium with platinum used for naphtha reforming, whereby the iridium modifies the reactivity of nearby platinum. This catalyst is poisoned by strongly adsorbed compounds of sulfur or phosphorus, which bond strongly to the catalytic sites and render them inactive. A fraction of a monolayer of sulfur on a metal surface can drastically reduce its catalytic activity; consequently, feeds to processes using this catalyst must be pretreated catalytically to remove the sulfur-containing compounds before they come in contact with the metal catalyst.

EXAMPLE 6-4 Poisoning of Metal Catalysts by Sulfur

Problem

Nickel is an active catalyst for the hydrogenation of CO to give methane and water. An investigation of the poisoning of the catalyst by H_2S was carried out in a gradientless reactor (a perfectly mixed flow reactor operated at steady state) with the nickel present as a low-area polycrystalline film [47]. The data determined rates of the catalytic reaction, showing, for example, that 15 ppb of H_2S in the reactant stream reversibly decreased the activity of the catalyst by more than two orders of magnitude, and 50 ppb decreased it by an additional order of magnitude. Product analyses and mass balance calculations determined the results plotted in Fig. 6-25. Kinetics data also showed that the activation energy for methane formation did not depend on the H_2S concentration in the reactant. What are the implications of these results?

Solution

The first conclusion is that most of the catalytic activity was lost when only tens of parts per billion of H_2S were present in the feed. Extreme efficiencies of purification of a feed stream would be required to allow operation of the catalyst with nearly its maximum activity. Under almost any practical conditions, the surface would be nearly saturated with sulfur. The results also imply a high strength of adsorption of species formed from H_2S on the nickel surface; the enthalpy of adsorption of H_2S on nickel has been estimated to be about twice the enthalpy of formation of the bulk Ni_3S_2 [47]. The linearity of the plot also provides some information about the surface intermediate in methane formation. The plot shows that the rate of methane formation is proportional to the square of the fraction of unpoisoned surface, which implies that the rate-determining step requires two Ni sites. These results and the lack of dependence of the activation energy on the H_2S partial pressure imply that the poisoning by sulfur is primarily a geometric effect whereby one sulfur atom

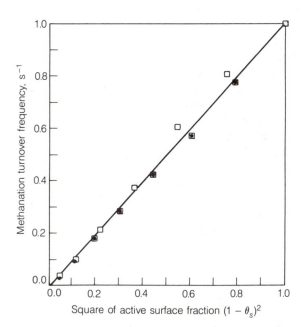

Figure 6-25
Poisoning by H_2S of a nickel catalyst used for conversion of $CO + H_2$ into methane. Data show the dependence of the catalytic activity on the fraction of the nickel surface not covered with sulfur [47a].

(presumably formed by dissociative adsorption of H_2S) blocks a surface catalytic site. (What unstated assumption underlies these inferences?)

The phenomena are more complicated than they are represented to be here [48]. ∎

For years, researchers have attempted to rationalize the effects of additives such as promoters and poisons in terms of their *geometric* and *electronic effects* on surface reactivity and catalytic activity. The implication is that added components, such as sulfur adsorbed on a metal surface or a second kind of metal atom in a metal surface, might block adsorption or reaction at a particular set of surface metal atoms (exerting a geometric influence, as in the preceding example) or, alternatively, that they might influence the reactivity of a particular metal site by electron donation or withdrawal. Geometric and electronic effects are not easily separated, and many of the earlier interpretations in these terms are now recognized as oversimplifications.

Metal clusters have been referred to as models of metal surfaces, and it is helpful to examine the limitations of the concept. The structure and bonding of metal clusters often mimic well the structure and bonding of adsorbates on metal surfaces, but the analogy does not extend to reactivity and catalytic activity. Moreover, cluster reactivity is not a good basis for prediction of the reactivity of ligands adsorbed on a surface [49]. Nonetheless, the concept of isolated reactive sites on surfaces as molecular analogues is extremely valuable and is borne out by a mass of experimental results. The neighboring groups in a surface do influence the reactivity of an adsorbate (as the groups on a metal cluster influence the reactivity of ligands on a neighboring metal center), and the influence is largely localized; for the most part, the bulk properties of the solid are not of much significance in determining the surface reactivity and catalytic activity.

For years, researchers had worked from the opposite hypothesis and sought to relate catalysis to physical properties of bulk solids (such as electrical properties reflecting the band character of electrons in metals and the semiconductor properties of metal oxides), but the search was largely fruitless. Today, the reasoning is based much more on concepts of local chemical bonding. Clusters are regarded as important, although imperfect, models because theoretical chemistry and computers are sufficiently advanced to allow accurate calculations of chemical states and reactivities of groups of as many as of the order of 10 atoms.

One of the examples illustrating connections between metal surface catalysis and molecular chemistry is the metal-catalyzed dehydrogenation of formic acid:

$$HCOOH \rightarrow CO_2 + H_2 \qquad (6\text{-}10)$$

Many metals are active for this reaction, and Fig. 6-26 shows a correlation between the activity of each metal catalyst and the standard enthalpy of for-

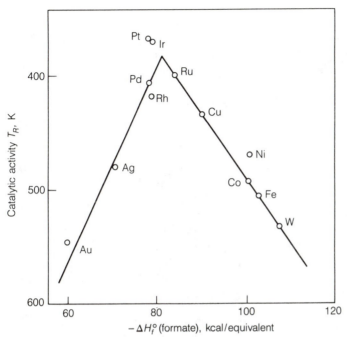

Figure 6-26
Volcano plot: Catalytic activities of metals for formic acid dehydrogenation correlated with the enthalpy of formation of the metal formates.
T_R is the activity given as the temperature required to achieve 50 percent
conversion with all other conditions constant [50].

mation of the respective metal formate [50]. Activity in this plot is represented
by the temperature required for a particular conversion under otherwise fixed
conditions. The volcano-shaped plot is consistent with the hypothesis that the
surface reaction intermediates closely resemble metal formates. When formic
acid adsorbs on metals, it gives surface species with infrared spectra resembling
those of metal formates. The surface formates are assumed to be similar to
bulk metal formates. The shape of the plot illustrates the principle of Sabatier,
who played a central role in the discovery of practical catalytic routes for
organic synthesis and was awarded the Nobel prize in 1912. According to Sabatier's principle, there is an optimum strength of bonding between the reactant
and catalyst. The metals represented on the left-hand side of the plot (Au and
Ag) interact so weakly with formic acid that the rate of formation of the surface
intermediate is low; it may be rate determining. In contrast, the metals on the
right-hand side of the plot (Fe and W) form such strong bonds with formic acid
that the decomposition of the surface intermediate becomes rate determining.
The intermediate metals (Pt, Ir, and Ru) are the most active catalysts. The
principle of an optimum strength of bonding between the reactants and catalyst
was enunciated in Chapter 2. It is quite general.

CO OXIDATION ON Pd

A few catalytic reactions on metal surfaces are relatively well understood, and it is helpful to examine what is known in some detail and to extract some general lessons. A simple oxidation reaction occurs on platinum group metals:

$$CO + \tfrac{1}{2}O_2 \rightarrow CO_2 \qquad (6\text{-}11)$$

This is technologically important, as it occurs in tens of millions of catalytic reactors in automobiles, helping to reduce the air pollution resulting from exhaust. Most exhaust abatement catalysts contain platinum and rhodium supported on ceramic honeycomb reactors that facilitate the heat and mass transport. These metals catalyze a number of reactions in addition to the oxidation of CO, also converting unburned hydrocarbons and nitrogen oxides [51].

The CO oxidation catalyzed by palladium has been investigated in many experiments, including some carried out with single crystals of catalyst in ultrahigh vacuum apparatus. The results obtained in these surface science experiments are directly pertinent to the practical catalysis taking place at atmospheric pressures for the following reasons [52]:

1. Surface contaminants such as carbon or sulfur are burned off the catalyst, so that the surface is clean, except for adsorbed reactants.
2. The catalytic activity does not depend on the crystallographic structure of the metal surface, as illustrated by the rate data of Fig. 6-27.
3. The rate of CO oxidation is independent of pressure, depending only on a ratio of partial pressures, at least over a restricted range of conditions.

Thus the situation is almost as favorable as one could wish for elucidation of the details of the surface catalysis with a full arsenal of experimental methods. A deep understanding of the surface catalysis has evolved from characterization of the adsorption of CO and oxygen combined with a series of kinetics measurements. The interpretation is that developed by Engel, Ertl, and coworkers [20, 52].

Adsorption of CO on palladium has already been discussed (Figs. 6-14 and 6-15). CO_2 is negligibly adsorbed under catalytic conditions; when formed on the surface, it is rapidly desorbed. Oxygen is dissociatively adsorbed, and LEED has been used to determine the structure shown in Fig. 6-28. Under reaction conditions, the surface diffusion of chemisorbed CO is evidently much faster than that of chemisorbed O atoms, so that the surface may be approximated as a fixed lattice of O atoms and a mobile layer of CO.

Investigations of the coadsorption of CO and oxygen, mostly by LEED at low temperatures, have given details of the interactions:

1. At very low surface coverages, both CO and O are chemisorbed and randomly distributed over the Pd (111) surface.
2. When the coverage of the surface by CO exceeds one-third of a

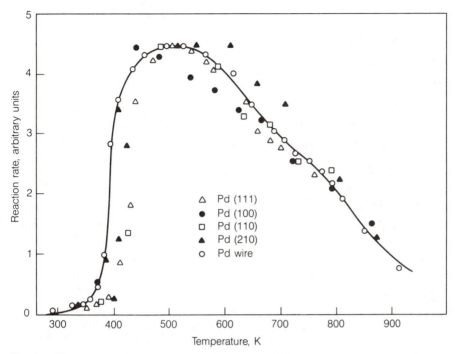

Figure 6-27
Rates of CO oxidation catalyzed by various Pd surfaces [52].

monolayer, chemisorption of oxygen is blocked. When CO is pread-sorbed with a lower coverage than one-third of a monolayer, then oxygen chemisorption occurs, but the CO and O are adsorbed in separate domains, as depicted schematically in Fig. 6-29. Then, reaction between chemisorbed CO and chemisorbed O is expected to take place only at the boundaries of these domains.

3. When oxygen is adsorbed first, saturating the clean palladium (111) surface, adsorption of CO does take place, and the uptake of CO leads to a compression of the adsorbed oxygen layer into domains with a local coverage of one-third of a monolayer, with the CO forming separate islands (Fig. 6-29c). Further addition of CO leads to formation of mixed domains with both CO and O, with the coverage by each being one-half of a monolayer. Measurements with UPS indicate that the electron donor oxygen atoms on the palladium surface in the close neighborhood of CO reduce the net electron transfer from the metal to the CO ligand and thereby weaken the metal—CO bond, which facilitates the catalytic reaction. Consequently, the mixed phase is highly reactive, giving CO_2 at temperatures far below ambient.

Characterization of the individual surface species has led to the recognition of the following steps as possible elementary steps in the catalytic cycle.

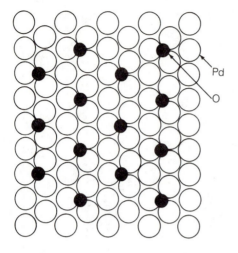

Figure 6-28
Structure of oxygen adsorbed on the (111) face of Pd at a coverage of one-fourth of a monolayer, as determined by LEED [52].

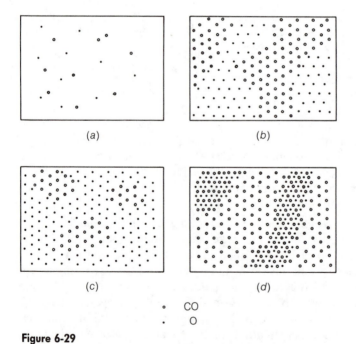

Figure 6-29
Schematic representation of domains of CO and O on the (111) face of Pd [19]. (a) Random distribution observed at low coverages; (b) island formation resulting when CO is preadsorbed and the surface is subsequently saturated with oxygen; (c) compression of the layer of adsorbed oxygen (oxygen preadsorbed); and (d) mixed CO + O phase and islands of CO when oxygen is preadsorbed and the surface subsequently saturated with CO.

Here, the subscript ads refers to adsorbed species, and S represents surface sites, which are not specified more precisely because they are not known more precisely; the sites for one adsorbate are not meant to be the same as the sites for another adsorbate in this representation:

$$O_2 + 2S \rightarrow O_{2\ ads} \rightarrow 2O_{ads} \tag{6-12}$$

$$CO + S \rightarrow CO_{ads} \tag{6-13}$$

$$CO_{ads} + O_{ads} \rightarrow CO_2 + 2S \tag{6-14}$$

$$CO + O_{ads} \rightarrow CO_2 + S \tag{6-15}$$

$$2CO_{ads} + O_2 \rightarrow CO_2 + 2S \tag{6-16}$$

Some apparently plausible steps are omitted from this sequence, on the basis of the following experimental results: No oxygen was dissolved in the bulk palladium, and the reaction of two adsorbed oxygen atoms to give gas phase O_2 has been observed to be kinetically insignificant at temperatures less than about 430°C.

The postulated step (an Eley–Rideal reaction) shown in the final equation has been ruled out on the basis of the observation that exposure of a surface saturated with adsorbed CO to gas phase O_2 does not lead to CO_2 formation. Evidently, dissociative chemisorption of O_2 is a necessary prerequisite to catalysis, and the surface plays the crucial role of facilitating it. As this first step in the sequence is inhibited by adsorbed CO, the CO coverage must be below a certain limit for it to take place.

Experiments to distinguish the third and fourth steps in the sequence were carried out with molecular beams impinging on the palladium surface; these were non-steady-state kinetics experiments. If the latter steps prevailed, then the rate of reaction would be expected to be greatest when a beam of CO molecules impinged on the surface with the highest coverage of adsorbed oxygen. But in an experiment with the initial oxygen coverage on the surface being one-fourth of a monolayer, the CO (impinging at a constant rate) initially gave CO_2 at a very low rate, with the rate increasing to a maximum after a few seconds (Fig. 6-30). During the first second, the concentration of adsorbed oxygen remained constant and the CO coverage increased. During this period, the oxygen layer was compressed as the rate of the catalytic reaction increased. The reaction rate reached a maximum when the concentrations of surface CO and surface O were approximately equal (Fig. 6-30). These data can only be reconciled with the Langmuir–Hinshelwood mechanism whereby adsorbed oxygen atoms react with adsorbed CO.

Quantitative experiments have led to a schematic potential energy diagram characterizing the elementary steps on the Pd (111) surface (Fig. 6-31). The nature of the transition state on the surface continues to be debated, but the depth of understanding of the surface reaction is noteworthy. Still, the steps

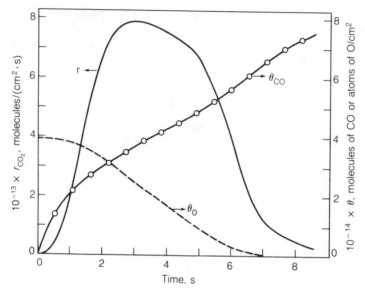

Figure 6-30
Transient data showing the rate of CO_2 formation from CO and the coverages of the Pd (111) surface by oxygen and CO [20]. The temperature was 647°C. At time zero, the surface, which had been precovered with oxygen, was exposed to a constant flux of CO at a pressure of 7.9×10^{-11} atm.

are not depicted in a catalytic cycle because the exact nature of the structure and bonding of the catalytic intermediates on the surface is not known. (Can you construct a cycle, assuming that the CO and O ligands are bonded as stated above?) This potential energy diagram emphasizes the role of the surface in providing an efficient pathway for the reaction. Most of the energy is liberated upon adsorption of the reactants, and the activation energy barrier for com-

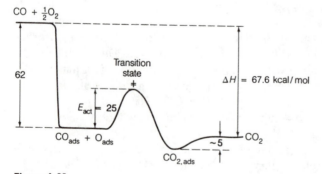

Figure 6-31
Schematic potential energy diagram illustrating the changes associated with the individual reaction steps on the Pd (111) surface at low coverages [52].

bination of the adsorbed CO and the adsorbed O is relatively small; this step is only weakly exothermic.

Although the sequence of elementary steps is quite simple, the overall kinetics of the reaction is not. The nonuniformity of the surface and the segregation of the reactants in surface domains rules out the simple Langmuir–Hinshelwood treatment of the kinetics, except in the special case of low surface coverages by CO and O, when they are randomly distributed and might, to a first approximation, be considered to be part of an ideal surface.

EXAMPLE 6-5 Kinetics of CO Oxidation on Pd (111)

Problem

The rate equation describing CO oxidation on this surface has been found to be the following when surface coverages are relatively low:

$$r = \frac{kP_{O_2}}{P_{CO}} \tag{6-17}$$

Reconcile this with the sequence of elementary steps stated above.

Solution

On the basis of the discussion above, the sequence of kinetically significant elementary steps is as follows:

$$O_2 + 2S \rightarrow 2O_{ads} \tag{6-18}$$

$$CO + S \rightarrow CO_{ads} \tag{6-19}$$

$$CO_{ads} + O_{ads} \rightarrow CO_2 \tag{6-20}$$

Presume that the surface concentrations of CO and O are low enough that they are randomly distributed on the surface and that ideal kinetics, based on the assumption of Langmuir adsorption, will be a satisfactory approximation. By inspection, it is evident that if the surface reaction between CO and O were rate determining, the observed kinetics could not be accounted for. Instead, consider the possibility that adsorption of oxygen is rate determining:

$$r = k_{21}P_{O_2}\theta_v \tag{6-21}$$

And if θ_{CO} is large in comparison with θ_O (as has been demonstrated experimentally) and the adsorption–desorption of CO is so fast that it achieves virtual equilibrium, then

$$\theta_v = \frac{1}{1 + K_{CO}P_{CO}} \tag{6-22}$$

and, if the surface coverage is large, then $K_{CO}P_{CO} \gg 1$, and

$$r = \frac{k_{21}P_{O_2}}{K_{CO}P_{CO}} \tag{6-23}$$

in agreement with the observations. Can you reconcile the data of Fig. 6-29 with this interpretation? Over what range of temperature is the interpretation consistent with the data? What might the rate-determining step be in another temperature range? ∎

The oxidation of CO on palladium is one of the most satisfying examples in the science of surface catalysis. Results from a number of experiments have been combined to give a firm identification of the reactants on the surface (although details of the structure and bonding are missing) and a detailed, quantitative interpretation of the reaction. This example shows the value of ideal kinetics in representing some data, and it also shows its limitations; it is quite clear that the ideal models are not sufficient to account for the kinetics when the adsorbed reactants are segregated in domains on the surface. Understanding of the surface catalysis rests firmly on results obtained in surface science experiments; without these techniques, this depth of understanding would have been unachievable.

NH₃ SYNTHESIS ON Fe

Modern catalytic technology was born in Germany early in this century with the development by Haber, Bosch, and Mittasch of the process for synthesis of ammonia, needed for nitrogen "fixation" and the production of nitrates for fertilizers [53, 54]. This technology allows the human population to feed itself, but it was also crucial for the production of conventional explosives. Haber was awarded a Nobel prize for his invention of the catalytic process, and Bosch was awarded a Nobel prize for his contributions to high-pressure technology. In experiments directed by Mittasch, the modified iron catalysts were identified in laboratory tests of some 20,000 candidate catalysts with high-pressure flow reactors. Today's industrial research and development leading to new and improved catalysts is done with virtually the same methodology, but now the work is more efficient because of the greatly improved automation of the test reactors and the computer processing of the data.

The ammonia synthesis reaction

$$N_2 + 3H_2 \rightarrow 2NH_3 \tag{6-24}$$

is exothermic and accompanied by a decrease in the number of moles. At equilibrium the reaction is therefore favored by high pressures and low temperatures. In practice, pressures of hundreds of atmospheres are required, even for low conversions. Temperatures of about 400°C and higher are used, representing a compromise between kinetics (rates being higher at higher temperatures) and thermodynamics. A successful process requires a catalyst with sufficient activity to allow operation at a temperature low enough to give an economical conversion within the thermodynamic constraint.

Because of its great industrial importance, the ammonia synthesis reaction has been the subject of extensive continuing investigation, and it is now one of the best understood surface-catalyzed reactions and the one of greatest historical significance. The insights have arisen from investigations of adsorption of the reactants and product, kinetics and surface science experiments, and characterizations of the catalyst structure. The modern catalysts [55] are not much different from what Haber and colleagues discovered in the beginning, consisting of iron with chemical and textural promoters.

The catalyst in the working state contains metallic iron, and therefore extensive surface science experiments have been done with single crystals of iron. The Fe (111) surface is highly active, about 25 times as active as the Fe (100) surface and about 400 times as active as the Fe (110) surface at 20 atm and 525°C [56]. These results show that the activity depends strongly on the structure of the iron surface.

Evidence of the adsorbed species formed from H_2 and N_2 on iron has been obtained in surface science experiments [57]. The adsorption of N_2 is slow and is characterized by a very low sticking coefficient (about 10^{-6}) and a substantial activation energy. Adsorption of N_2 leads to reconstruction of the surface. The N_2 is dissociated, forming complicated structures that have been investigated with LEED and other techniques and inferred to be "surface nitrides" with depths of several atomic layers and compositions of roughly Fe_4N, a bulk compound that is, however, thermodynamically unstable under conditions for which the surface structure is stable.

The rate of adsorption of N_2 is dependent on the structure of the iron surface [58, 59]. The ratios of initial rates of adsorption on the (111), (100), and (110) surfaces at about 275°C are 60:3:1. The activation energy for adsorption varies from one iron surface to another and depends on the surface coverage. The activation energy is least and the rate of adsorption greatest for the Fe (111) surface, the one found to be the most active catalytically. This observation is consistent with conclusions drawn years earlier (on the basis of measurements of rates of the catalytic reaction and rates of adsorption of N_2) that the adsorption of N_2 is an activated process and that it is rate determining in the catalytic cycle [60].

In contrast, the adsorption of hydrogen on iron is fast and is characterized by a high sticking probability (of order 10^{-1}) and little or no activation energy. The adsorption is dissociative, giving hydrogen atoms covalently bonded but highly mobile on the surface. It is inferred that hydrogen attains virtual adsorption–desorption equilibrium under conditions of the catalytic ammonia synthesis reaction.

Adsorption of ammonia is rapid and is characterized by a low activation energy. Ammonia decomposes to give H_2 and N_2, and desorption of N_2 is rate determining under conditions of many investigations, consistent with what might be inferred from the statement above about the adsorption of N_2 being rate determining for the ammonia synthesis reaction. Numerous researchers have chosen to investigate the catalytic ammonia decomposition reaction at relatively low pressures to learn more about the important reverse reaction, which is difficult to investigate at low pressures because of the equilibrium limitation. One may properly draw inferences about the mechanism of the ammonia synthesis reaction from the observations of the ammonia decomposition reaction, but it is important to recall the limitation of the principle of microscopic reversibility, which states that the forward and reverse reaction proceed through the same transition state *at equilibrium*.

Engel and Ertl [52, 57] used surface science techniques to investigate catalysts that had been used for ammonia synthesis at atmospheric pressure. They found that one very stable surface species, adsorbed N atoms, was predominant on the Fe (111) surface following evacuation. All the other adsorbed species desorbed under vacuum at temperatures less than 200°C, but the adsorbed nitrogen was stable at temperatures up to about 450°C. The stable atomic nitrogen species on the surface was inferred to be an intermediate in the catalytic reaction, and the possible involvement of an adsorbed dinitrogen species was ruled out on the basis of an analysis of the kinetics showing that the concentration of N_{ads} would depend on the H_2 partial pressure if this surface species were involved in the rate-determining step and would not if it were simply formed as an unreactive species from adsorbed dinitrogen that was instead involved in the rate-determining step. It was observed that the surface concentration of N_{ads} decreased with increasing H_2 partial pressure at a given N_2 partial pressure, and thus it was concluded that the former was involved in the rate-determining step [52, 57]. Other evidence from kinetics experiments supports this conclusion [61].

Other, less stable surface intermediates identified spectroscopically include NH and NH_2. Consistent with all these observations, the sequence of elementary steps has been formulated as follows, where, as before, S refers to unspecified surface sites (not necessarily all the same) on the catalyst surface:

$$N_2 + 2S \rightarrow 2N_{ads} \qquad (6\text{-}25)$$

$$H_2 + 2S \rightarrow 2H_{ads} \qquad (6\text{-}26)$$

$$N_{ads} + H_{ads} \rightarrow NH_{ads} + S \qquad (6\text{-}27)$$

$$NH_{ads} + H_{ads} \rightarrow NH_{2ads} + S \qquad (6\text{-}28)$$

$$NH_{2ads} + H_{ads} \rightarrow NH_3 + 2S \qquad (6\text{-}29)$$

The exact nature of the surface site(s) S remains to be determined, but the high activity of the Fe (111) surface suggests that geometric arrangements of Fe atoms on this surface are uniquely favorable for the reaction.

A thermochemical profile for the adsorption of nitrogen on a Fe (100) surface has been formulated and is shown in Fig. 6-32 [62]. This Lennard–

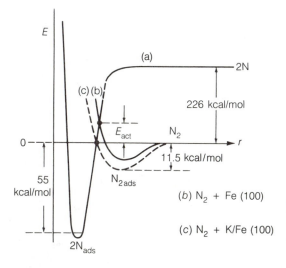

(b) N$_2$ + Fe (100)

(c) N$_2$ + K/Fe (100)

Figure 6-32
Potential energy diagram for adsorption of atomic and molecular nitrogen on clean and potassium-covered Fe (111) surfaces [62].

Jones diagram indicates the activated nature of the exothermic dissociative adsorption and the formation of strong metal-nitrogen bonds.

An approximate potential energy diagram for the sequence of steps in the catalytic reaction is shown in Fig. 6-33 [57]. The values are based on some approximations and depend on the surface coverage, but they constitute a strong statement of the nature of the catalytic process and indicate how the iron greatly facilitates the reaction, which is orders of magnitude (can you estimate how many?) faster than the gas phase reaction proceeding through atomic intermediates. The gas phase reaction is energetically unfavorable because the dissociation energies in the first steps are so great. An advantage of the surface is that it facilitates these bond-breaking reactions, and the energy gain associated with the formation of the strong metal–nitrogen and metal–hydrogen bonds at the surface actually makes the first steps exothermic. Nonetheless, dissociative adsorption of the nitrogen is still rate determining, not because of a high activation barrier, but because the preexponential factor in the rate constant is so small. The subsequent hydrogenation steps on the surface are slightly endothermic.

A simplified version of the reaction kinetics of the ammonia synthesis can be derived on the basis of the assumptions that (1) the surface is ideal, (2) the adsorption of nitrogen, the first step in the sequence shown above, Eq. 6-25, is rate determining, and (3) the conversion is so low that the rate of the reverse reaction can be neglected:

$$r = kP_{N_2}\theta_v \tag{6-30}$$

The steps after the rate-determining step may be summed and an equilibrium expression written to determine an equation for θ_N, with N_{ads} being the predominant surface intermediate; θ_v is then $1 - \theta_N$ according to the Lang-

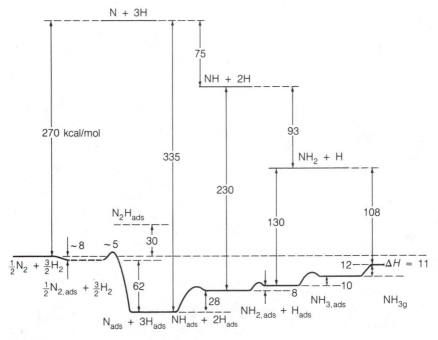

Figure 6-33
Schematic potential energy diagram for NH₃ synthesis and decomposition on iron. Energies are given in kcal/mol [57].

muir formalism. Substituting in the equation above gives the result

$$r = \frac{kP_{N_2}}{1 + K'P_{NH_3}^2/P_{H_2}^3} \tag{6-31}$$

where K' incorporates the reaction equilibrium constant (can you identify K' more explicitly?). This equation is approximated well by the following form:

$$r = k'P_{N_2}\left(\frac{P_{H_2}^3}{P_{NH_3}^2}\right)^n \tag{6-32}$$

where $0 < n < 1$ [63].

A very similar form of equation accounting for the nonuniformity of the catalyst surface was derived by Temkin and Pyzhev [64], as summarized in Boudart and Djéga-Mariadassou's book [63]. The full Temkin–Pyzhev equation accounting for the reverse reaction is the following:

$$r = kP_{N_2}\left(\frac{P_{H_2}^3}{P_{NH_3}^2}\right)^m - k'\left(\frac{P_{NH_3}^2}{P_{H_2}^3}\right)^{1-m} \tag{6-33}$$

This equation was presented almost 50 years ago, but it is still the standard and is used successfully to predict the performance of industrial catalysts.

The simplified form of the rate equation, based on the Langmuir assumptions of ideal kinetics, is remarkably similar to the more rigorous Temkin–Pyzhev equation, which accounts for the nonuniformity of the catalyst surface. The nonuniformity of the iron surface is well demonstrated, and it might seem paradoxical that the ideal kinetics is so successful. But ideal kinetics is often entirely sufficient for reactions on highly complex surfaces. To a first approximation, one may regard the (often small) subset of surface sites on a nonuniform surface that actually participates in a catalytic reaction as virtually uniform and therefore fairly well represented by Langmuir's assumptions [65]. More often than not, kinetics of the form indicated by Langmuir's model is appropriate; seldom are sufficiently detailed kinetics data available as would be needed to justify a more rigorous treatment and a more complex form of equation.

Stoltze and Norskov [66], beginning with the experimental result that the dissociation of adsorbed nitrogen is the rate-determining step, calculated the overall rate of the ammonia synthesis reaction for technological catalysts under realistic conditions from the rate of this step and the equilibrium constants of all the other steps. The adsorption–desorption equilibria were computed on the basis of an assumed Langmuir adsorption and by evaluation of the partition functions of the gaseous and adsorbed species. Experimental results for the potassium-promoted Fe (111) surface were used for the rate of dissociative adsorption of nitrogen. The active area of the commercial catalyst was estimated from CO chemisorption data. Remarkably, the calculations based on the results of the single-crystal experiments at very low pressures gave results that matched well the results for commercial catalysts. Rates of the catalytic reaction were within a factor of 2 of the observed values. The predictions also account well for the observed effects of temperature, pressure, and potassium coverage, as well as the effect of water, a catalyst poison.

The work constitutes the most rigorous prediction of the kinetics of a technologically important surface-catalyzed reaction and a highly satisfying consolidation of decades of scientific and technological effort on the ammonia synthesis reaction.

The industrial catalysts used for ammonia synthesis differ from the original Mittasch catalyst only slightly, by containing more promoters. The original and still-used catalysts are made from magnetite (Fe_3O_4), which is largely reduced to metallic iron under conditions of catalytic reaction. The catalyst is "doubly promoted," the promoters being potassium (about 1 wt% of the catalyst and possibly present as K_2O), a chemical promoter, and alumina (about 3 wt% of the catalyst), a textural promoter [55]. An additional textural promoter, CaO, is used in most industrial catalysts.

Reduction of the oxide catalyst leads to formation of pores, and the pore structure is optimized to reduce the resistance to intraparticle mass transport and give maximum rates of reaction per unit of reactor volume. An industrial catalyst may have many pores with diameters of roughly 8 nm and a number with diameters of roughly 20 nm. The internal surface area is about 15 m^2/g, and the primary iron particles are of the order of 30 nm in size [55]. Promoters are concentrated at the surfaces of the primary iron particles, and only a few percent of the surface is exposed iron. A fraction of the textural promoter is dispersed within the iron crystallites. The textural promoters stabilize the catalyst by maintaining the dispersion of the iron, preventing it from sintering into a nonporous mass, and the alumina apparently stabilizes the active (111) surface of iron [67]. There is a trade-off here: If there were too much textural promoter, the activity of the catalyst would be reduced by blocking of the iron surface; whereas if there were too little textural promoter, the activity could not be satisfactorily maintained because of the loss of iron surface area. Good industrial catalysts can last for years.

The chemical promoter K$_2$O, which may be present in part in distinct phases incorporating Al$_2$O$_3$ or other metal oxide components, increases the catalytic activity of the iron. This promoter is believed to be present on the iron surface and to donate electrons to the iron, increasing its reactivity for dissociation of N$_2$. This effect has also been noted with metallic potassium on iron single crystals, and the result is shown in Fig. 6-32, where a lowered activation energy for dissociative adsorption of N$_2$ is evident for the promoted surface.

The modern process for ammonia synthesis is essentially the same as that conceived by Haber [53, 68]. High-pressure (typically 200 atm) hydrogen and nitrogen are fed in near-stoichiometric proportions to a flow system incorporating a catalytic reactor, a compressor for recycle of unconverted reactants, and a condenser for recovery of ammonia from the loop. The space velocity is typically in the range 8,000–60,000 h^{-1}. The reactor may be a bundle of small-diameter tubes or a multibed tube with a much larger diameter. Heat removal is crucial, and cooling between the stages of the multibed reactor is necessary. Since the reaction is exothermic, the hot product gas flows through heat exchangers to heat the incoming reactant stream. Compression costs are high, and most of the process improvements in recent years have resulted from more efficient engineering and not from catalyst improvements.

The catalyst performance and lifetime are dependent on the purity of the feed, and a large part of the technology (and much of the cost) is related to production of the hydrogen, usually from natural gas, but alternatively from petroleum or coal [68].

The ammonia synthesis is one of the best understood catalytic reactions, and it is remarkable how well the understanding based on the results of modern surface science experiments confirms that developed by an earlier generation of scientists, primarily P. H. Emmett, whose research is summarized in a review [68b]. Emmett's interpretation was based on adsorption, kinetics, and catalyst structure data combined with excellent chemical sense. It is instructive to read the early work, not only to appreciate the inventiveness and insight of the researchers, but also, by contrast, to understand how much more incisive are the modern experimental tools.

HYDROCARBON CONVERSION ON Pt

Countless reactions of organic compounds take place on surfaces of metal catalysts, ranging from those used on a small scale in preparative organic chemistry [69] to those used on much larger scales in conversion of commodity chemicals, petroleum, petrochemicals, fats and oils, fine chemicals, and pharmaceuticals. The platinum group metals find the greatest application, with platinum itself being the most widely applied metal catalyst because it is active for so many reactions. A few of these reactions are introduced here, and others are considered in Section 6.4.5.

The reactions of hydrocarbons on Pt surfaces include hydrogenation and dehydrogenation, isomerization, hydrogenolysis (hydrogenation with breaking of bonds, e.g., C—C, C—S, and C—N bonds), and a number of cyclization reactions, including dehydrocyclization, whereby a paraffin such as *n*-heptane is converted into hydrogen and a cyclic compound such as toluene. Rates of these reactions have been measured with single-crystal platinum catalysts, and Somorjai [2] has summarized a large body of data in a semiquantitative way in Fig. 6-34. Some metal-catalyzed reactions that are related to the hydrocarbon conversions result in hydrocarbon formation from other organic reactants, for example, the hydrogenation of CO (the Fischer–Tropsch reaction) to give paraffins and olefins [70, 71].

Olefin hydrogenation, especially ethylene hydrogenation, has been investigated intensively for decades, having been regarded for many years as potentially one of the best test reactions to provide detailed information about the catalytic nature of metal surfaces. However, the simple stoichiometry of ethylene hydrogenation belies a mechanistic complexity, and even today the

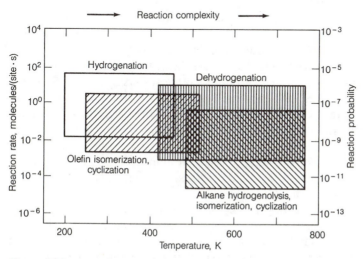

Figure 6-34
Approximate rates of hydrocarbon conversion reactions on platinum catalysts [2].

reaction intermediates are not known with certainty. The reaction occurs readily on many metal surfaces. At high temperatures, the working surface of platinum or palladium is nearly clean; but at low temperatures, it is largely covered with carbonaceous residues. It is possible that there are catalytic sites for olefin hydrogenation in addition to those provided by the bare metal surface, such as parts of overlayers. The question of the role of surface-bound organometallic species such as ethylidyne remains open. Ethylidyne ligands have been observed on the surface of a functioning Pd catalyst by infrared spectroscopy at room temperature and found to be spectators during ethylene hydrogenation [72]. Experiments with a pulse reactor containing Pt catalyst operated at −31°C showed that carbonaceous deposits were formed rapidly but were not involved in the ethylene hydrogenation [73]. One might be tempted to assume an analogy between the reaction on the nearly clean metal surface and the reaction catalyzed by Wilkinson's complex in solution, but the strength of the analogy remains to be assessed.

Olefin hydrogenation and paraffin dehydrogenation reactions are considered again in later sections that are concerned with metal oxide and supported metal catalysts.

A number of hydrocarbon rearrangement reactions have been investigated with platinum single crystals cleaved to expose various densities of terraces, steps, and kinks [74]. Experiments were done to measure rates of catalytic reactions such as the dehydrocyclization of *n*-hexane to give benzene, the simultaneous dehydrocyclization to give methylcyclopentane, isomerization to give 2-methylpentane, and hydrogenolysis to give propane. The apparatus allowed preparation of a clean single crystal in ultrahigh vacuum, transfer to an attached reactor (operated at atmospheric pressure) and, after catalysis, transfer back to the ultrahigh vacuum chamber for experiments to demonstrate the lack of surface contamination and restructuring.

The results show that the activity of platinum for the hydrocarbon conversions varies from one crystal face to another [74]. For example, the conversion of *n*-hexane into benzene and hydrogen occurs most readily on platinum surfaces with the flat hexagonal (111) structure; these surfaces are several times more active than the (100) surface [75]. Many other results of this kind provide evidence of the dependence of catalytic activity for hydrocarbon conversion on the structure of the platinum surface. Just what the catalytic sites are for each of these reactions is still not known, but extensive research with isotopic tracers has been done to elucidate details of the reaction pathways and allow inferences of the nature of the surface reaction intermediates, which include cyclic structures (analogues of metallacycles), some bonded to the metal surface at more than one position [76]. The working catalysts are typically largely covered by hydrocarbon overlayers like that sketched in Fig. 6-18.

These hydrocarbon conversions are important in processes for converting naphtha to high-octane-number gasoline and are considered further in Section 6.4.5, where the process and the industrial platinum-containing catalysts are introduced.

6.4.4 Catalysis on Metal Oxide Surfaces

Extensive research with single crystals of metals, almost all of it done in the last 15 years and made possible by efficient ultrahigh vacuum apparatus and complementary spectroscopic methods, has helped to create a foundation for understanding of catalysis on metal surfaces. In contrast, research with single crystals of metal oxides has developed much less rapidly, in part because the metal oxides are poor electrical conductors and build up charge when bombarded with electrons, which hinders ultrahigh-vacuum electron spectroscopy experiments. Consequently, there is a paucity of examples to illustrate general lessons of catalysis by metal oxide surfaces.

The next sections introduce some concepts of catalysis on metal oxides from the basis of relatively well understood examples. These have been chosen in part to make connections to earlier chapters, illustrating catalysis that is essentially acid–base chemistry and catalysis that involves organometallic chemistry; catalysis that involves redox chemistry is exemplified by the Wacker oxidation mentioned in Section 6.4.1.

ALCOHOL DEHYDRATION ON γ-Al$_2$O$_3$

Many reactions on surfaces of metal oxides are explained by the concepts of acid–base chemistry, as is illustrated in Chapter 5 by hydrocarbon conversions catalyzed by zeolites. Metal oxides exhibit wide ranges of acid and base strength. The proton donor groups are usually —OH groups, and the Lewis acid sites are exposed metal ions. The basic groups are usually oxygen ions and —OH groups. As in solution acid–base catalysis, the important elementary steps on the surface involve transfer of protons and hydride ions, and the catalytic activity may be simply related to the proton donor strength, although sometimes the chemistry is more subtle, involving the concerted action of proton donor and proton acceptor groups.

These ideas are illustrated by alcohol dehydrations, which are among the most thoroughly investigated reactions catalyzed by metal oxides [77]. The catalysts for alcohol dehydration include the sulfonated polymers mentioned in Chapter 4 and the acidic zeolites and clays mentioned in Chapter 5, but metal oxides such as γ-Al$_2$O$_3$ have been used industrially for these reactions for decades. Today, they are important only in countries such as Brazil where there is a large supply of ethanol from biomass; this is catalytically dehydrated to give ethylene. (In most parts of the world, ethylene is manufactured by steam cracking of hydrocarbons.)

In contrast to the sulfonic acid polymers, which are strong acids, γ-Al$_2$O$_3$ is neither a strong acid nor a strong base, and it therefore lacks catalytic activity for reactions such as paraffin cracking. Its activity for alcohol dehydration is associated with both its proton donor and proton acceptor groups; the same conclusion was drawn in Chapter 4 for the polymer-supported catalysts. Understanding of the mechanism of alcohol dehydration on γ-Al$_2$O$_3$ is based on investigations of chemisorption of alcohols and other molecules, kinetics of the catalytic reactions, and spectroscopic investigations of the adsorbed species [78].

Pyridine, a base that adsorbs on Lewis acid sites, inhibits olefin isomerization on the surface but does not affect the rate of alcohol dehydration; this result indicates that the Lewis acid sites (Al^{3+} ions) are not involved in the dehydration reaction. A fully hydroxylated surface [that of $Al(OH)_3$] is catalytically inactive, indicating the need for surface oxygen ions. The patterns of reactivity parallel those observed in acidic solutions, which implicates the proton donor groups. Infrared spectra show that alcohols bond to the alumina surface through hydrogen bonds; the alcohol and the surface have complementary functional groups, each being a proton donor and a proton acceptor. The following structure accounts for the observations:

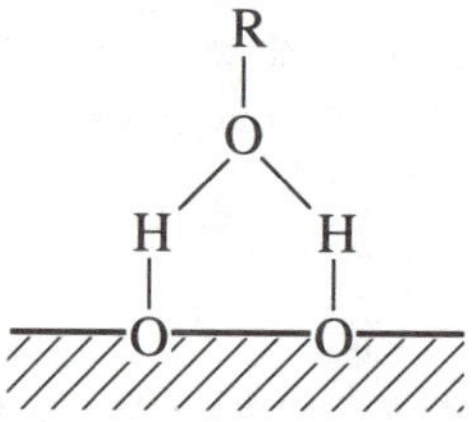

The catalytic reaction is accounted for as shown in Fig. 6-35 [79]. A proton is transferred from a surface —OH group to the oxygen of the alcohol, facilitating the transfer of a proton on a neighboring carbon atom to a basic oxygen atom of the surface. Olefin product can then desorb. The surface is left partially hydrated, and product water is a strong reaction inhibitor.

This is a simplified model. Experiments done with various alcohols as reactants have given details of the mechanism, including the stereochemistry [77, 78]. There is evidence for carbenium ion intermediates in some situations and for concerted reactions and nonionic intermediates in others. The parallels to alcohol dehydration in acidic solutions and in ion exchange resins are strong.

The alcohol dehydrations to give ethers also occur on the γ-Al_2O_3 surface. The reaction is believed to proceed through a Langmuir–Hinshelwood mechanism as two neighboring adsorbed alcohol molecules (one possibly an alkoxide, RO, and the other a hydrogen-bonded intermediate) react.

These interpretations are still somewhat uncertain, but the essential pat-

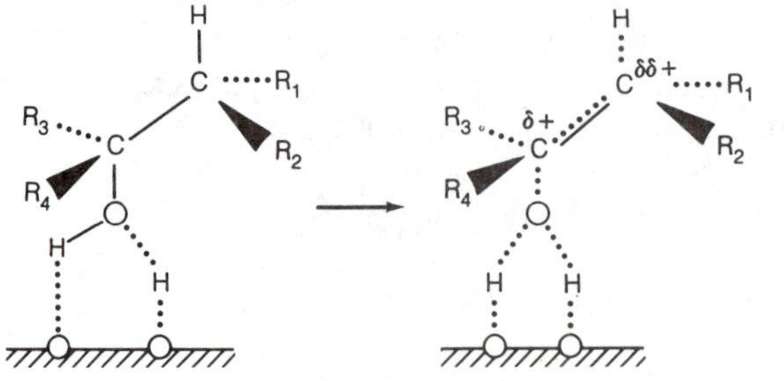

Figure 6-35
Proposed step in alcohol dehydration catalyzed by γ-Al_2O_3 [79].

tern is well understood. The results are not easily generalized to other metal oxide catalysts. For example, the intermediates formed from alcohols on thoria are evidently bonded to Lewis acid sites.

REACTIONS OF OLEFINS ON ZnO

Most useful catalytic reactions proceed so rapidly that it is almost impossible to observe the reactive intermediates on the surface, let alone identify them. Not surprisingly, some of the best-understood surface-catalyzed reactions are those that proceed so slowly that the reactive intermediates can be observed spectroscopically. When the spectroscopic experiments are done in conjunction with measurements of chemisorption and reaction kinetics, they may lead to a detailed model of the surface catalytic chemistry. This is the situation with reactions of olefins on zinc oxide (ZnO).

Zinc oxide has the virtue of excellent optical transparency that facilitates investigation of the surface species by infrared spectroscopy. And ZnO has such a low catalytic activity for a number of reactions of olefins that the surface reaction intermediates are present in high enough concentrations to be observed directly by this technique during catalysis. Consequently, it becomes possible to observe relationships between the concentration of a surface intermediate and the rate of a catalytic reaction. This information is crucial for identification of a reaction intermediate and to distinguish it from a mere spectator on the surface.

The following section is a summary of the work of Kokes and Dent [80] with reactions of olefins on ZnO; it shows the importance of organometallic chemistry in catalysis on metal oxide surfaces. As before, there are strong parallels between the chemistry of catalysis in solution and that occurring on surfaces.

Hydrogen is chemisorbed dissociatively on ZnO, and the surface species have been identified by their infrared spectra [81]. One hydrogen atom becomes bonded as a proton to a surface oxygen atom (identified by the O–H stretching frequency), and the other becomes bonded to an exposed zinc atom as a hydride ligand (identified by the Zn–H stretching frequency). Consequently, the dissociative adsorption is described as heterolytic.

Measurements of the rates of adsorption and desorption of hydrogen demonstrate the nonuniformity of the surface. Adsorption occurs on two distinct kinds of sites on the ZnO; adsorption on the first (type I) is rapid and easily reversible, whereas adsorption on the second (type II) is initially rapid but falls off in rate after a fraction of a monolayer has been adsorbed. Adsorption on the one type of site is virtually independent of adsorption on the other.

Adsorption of ethylene on the surface is rapid and reversible [80]. The isotherms are nonlinear and show some evidence of saturation at a value about five times the saturation coverage by type I hydrogen. Evidently the ethylene is chemisorbed, and some may be physisorbed as well. The initial enthalpy of adsorption is -14 kcal/mol (implying chemisorption, since the enthalpy of liquefaction is about a fourth of this value). Analysis of the infrared spectrum and comparison with the spectra of molecular metal complexes having π-bonded olefin ligands lead to the identification of the chemisorbed olefin as π-bonded, presumably to a Zn atom at the surface [80]. Evidently, the chemisorption is a ligand association reaction requiring a coordinatively unsaturated Zn atom at the surface.

Water adsorption on ZnO occurs dissociatively, as described above for water adsorption on MgO and γ-Al$_2$O$_3$. This acid–base chemistry gives —OH groups bonded to zinc at the surface and protons bonded to oxygen at the surface. Consequently, the sites for adsorption of H$_2$ are blocked.

When hydrogen and ethylene are simultaneously present on the surface of ZnO, they are catalytically converted into ethane. The chemistry is complicated at high temperatures, but at temperatures of about 100°C, Kokes and Dent [80] found that the surface and catalytic chemistry are so simple that they could use infrared spectroscopy to investigate the catalyst in the working state and to elucidate many details of the catalysis.

When H$_2$ was brought in contact with chemisorbed ethylene on ZnO, a new surface species was formed and observed in the spectrum. When experiments were done with tracers (D$_2$ reacting with adsorbed C$_2$H$_4$, and H$_2$ reacting with adsorbed C$_2$D$_4$), the resulting new surface species could be identified by analysis of the spectrum. This species is an ethyl ligand σ bonded to Zn, similar to that in diethyl zinc. It is formed by reaction of ethylene with hydrogen and is an obvious candidate for an intermediate in the catalytic cycle.

To determine whether it is a reactive intermediate, the researchers monitored its infrared spectrum during catalysis at steady state in a recirculation flow reactor. Then, they replaced the flowing mixture of hydrogen and ethylene with a mixture of hydrogen and inert helium, observing the intensity of a band in the spectrum characteristic of the surface-bound ethyl ligand, which was the predominant form of hydrocarbon on the surface. The band intensity decayed with time, and from the rate of decay an initial rate of disappearance of the ethyl ligand was determined. This was approximately the same as the steady-state rate of the catalytic hydrogenation reaction observed in the earlier part of the experiment. These observations confirm the identification of the surface-bound ethyl ligand as an intermediate in the catalytic cycle.

It follows that the catalytic sites may be pairs of Zn and O sites, and the low coverage of the surface by reactants, determined in the adsorption experiments, implies that the pairs are widely separated. The sequence of steps stated schematically below is consistent with the observations [80] (here, the hatched structures represent the surface):

$$\text{H}_2 + \overset{}{\underset{/\!/\!/\!/\!/\!/\!/}{\text{Zn}\!-\!\text{O}}} \quad \rightarrow \quad \overset{\overset{\text{H}\quad\text{H}}{\mid\quad\mid}}{\underset{/\!/\!/\!/\!/\!/\!/}{\text{Zn}\!-\!\text{O}}} \tag{6-34}$$

$$\overset{\overset{\text{H}\quad\text{H}}{\mid\quad\mid}}{\underset{/\!/\!/\!/\!/\!/\!/}{\text{Zn}\!-\!\text{O}}} + \underset{/\!/\!/\!/}{\text{O}} \quad \rightarrow \quad \overset{\overset{\text{H}}{\mid}}{\underset{/\!/\!/\!/\!/\!/\!/}{\text{Zn}\!-\!\text{O}}} + \overset{\overset{\text{H}}{\mid}}{\underset{/\!/\!/\!/}{\text{O}}} \tag{6-35}$$

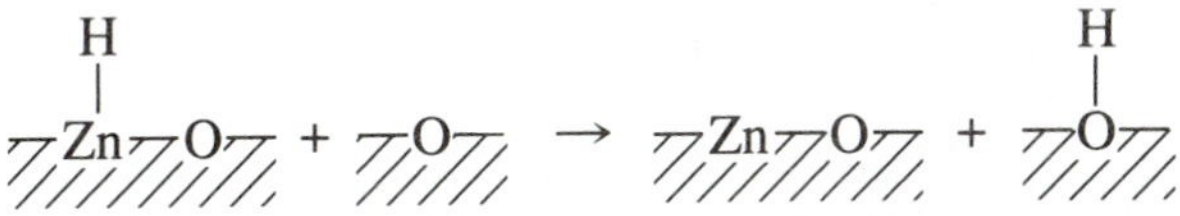

$$\text{H} - Zn - O + - O \rightarrow \text{H} - Zn - O + - O \tag{6-36}$$

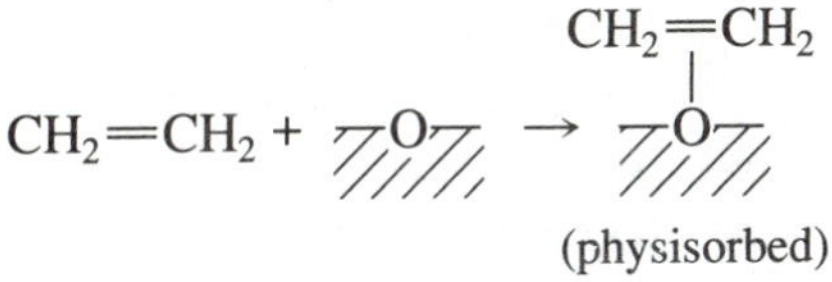

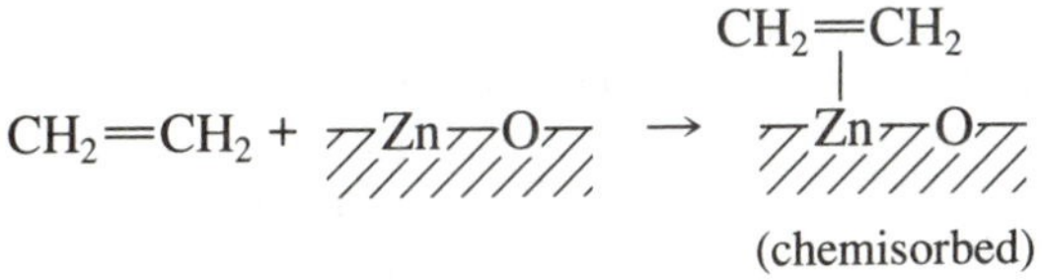

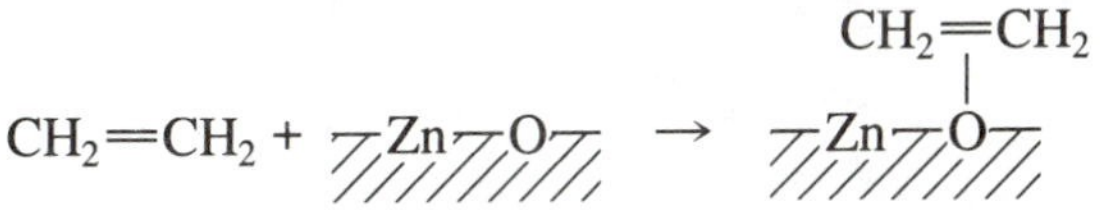

$$CH_2{=}CH_2 + -O- \rightarrow CH_2{=}CH_2\text{-}O- \tag{6-37}$$
(physisorbed)

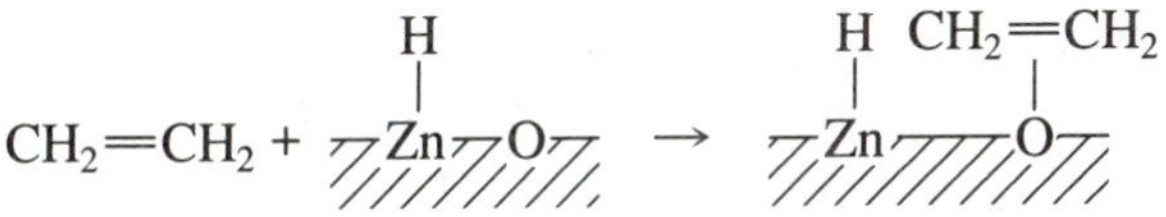

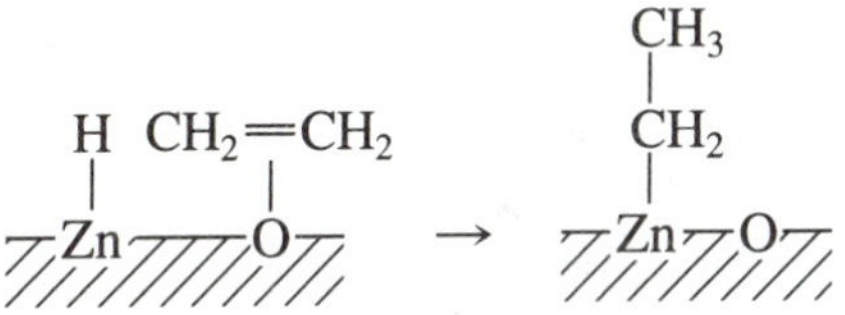

$$CH_2{=}CH_2 + -Zn-O- \rightarrow CH_2{=}CH_2\text{-}Zn-O- \tag{6-38}$$
(chemisorbed)

$$CH_2{=}CH_2 + -Zn-O- \rightarrow -Zn-O\text{-}CH_2{=}CH_2- \tag{6-39}$$

$$CH_2{=}CH_2 + \text{H}-Zn-O- \rightarrow \text{H}-Zn-O\text{-}CH_2{=}CH_2- \tag{6-40}$$

$$\text{H } CH_2{=}CH_2 \text{ on } -Zn---O- \rightarrow CH_3\text{-}CH_2\text{-}Zn-O- \tag{6-41}$$

$$CH_3\text{-}CH_2\text{-}Zn-O- + \text{H}-O- \rightarrow C_2H_6 + -Zn-O- + -O- \tag{6-42}$$

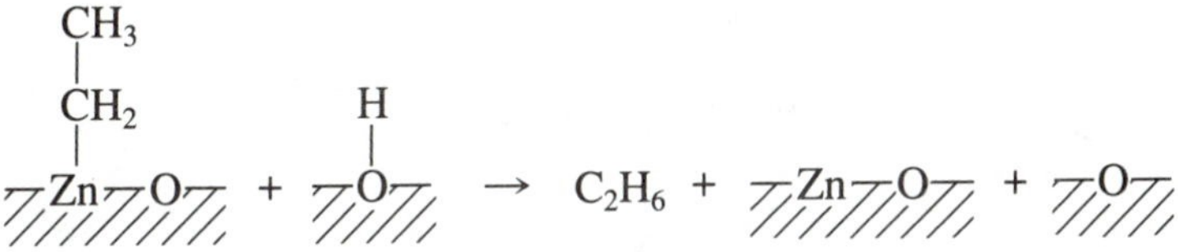

Many of the details are not known. For example, the nature of the π-bonded ethylene remains to be elucidated, and the reaction of this or another adsorbed ethylene species with hydrogen is not characterized. Experiments with D_2 tracer showed that type II hydrogen was only a spectator on the surface. Evidently, type I hydrogen on the surface reacts with the adsorbed ethylene in one step and with the ethyl ligand in another; the hydrogen must migrate to the Zn–O pair sites where the ethyl ligand is adsorbed.

The proposed sequence illustrates reactions of ligands adsorbed on neighboring surface sites, not the same surface atom. Roughly speaking, then, the elementary steps (which are not well known) are believed to be different from those illustrated in Chapter 2 for mononuclear metal complexes (for which the reactant ligands were bonded to the same metal atom) and more akin to those suggested for metal clusters.

The Langmuir–Hinshelwood mechanism stated above involves chemisorbed reactants. An Eley–Rideal mechanism whereby hydrogen from the gas phase reacts with chemisorbed hydrocarbon was ruled out on the basis of poisoning experiments with water. Water strongly poisons the catalyst, blocking the adsorption of hydrogen but not the adsorption of ethylene, which implies the necessity of chemisorbed hydrogen for the catalysis.

EXAMPLE 6-6 Kinetics of Ethylene Hydrogenation on ZnO
Problem

Examine the kinetics data of Kokes and Dent [80] shown in Figs. 6-36 and 6-37 and demonstrate their consistency with the mechanism presented above.

Solution

Figure 6-36 shows that the rate is proportional to the square root of the H_2 partial pressure, and Fig. 6-37 shows a Langmuir dependence on the

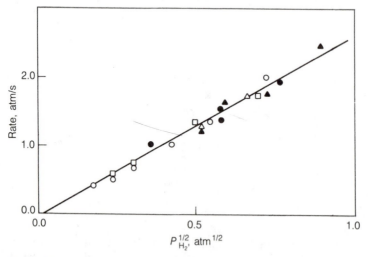

Figure 6-36
Kinetics of ethylene hydrogenation catalyzed by ZnO at room temperature and various H_2 and ethylene partial pressures [80]. The data obtained at different ethylene partial pressures have been "scaled" to bring them near the same line; the different symbols represent different ethylene partial pressures. The data determine the order of reaction in H_2.

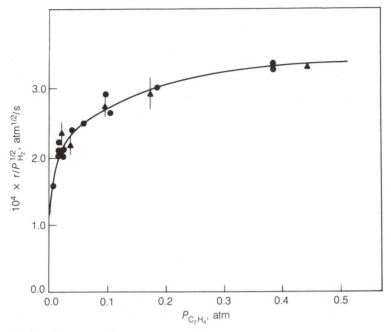

Figure 6-37
Kinetics of ethylene hydrogenation catalyzed by ZnO at room temperature
[80]. Data determine the dependence of rate on the ethylene partial pressure.

ethylene partial pressure. A rate equation of the following form is indicated:

$$r = \frac{kP_{H_2}^{1/2}P_{C_2H_4}}{1 + K_{C_2H_4}P_{C_2H_4}} \tag{6-43}$$

If this is interpreted in terms of ideal kinetics, it implies that

$$r = k\theta_H\theta_{HC}' \tag{6-44}$$

where θ_H is the fraction (presumed to be small) of one kind of surface site covered with hydrogen (which has been dissociated), and θ_{HC}' is the fraction of another (independent) kind of surface site covered with hydrocarbon formed from ethylene. This equation is explained straightforwardly if the rate-determining step is the reaction of adsorbed hydrogen with the adsorbed hydrocarbon, for example, the ethyl ligand, which is the predominant hydrocarbon surface species indicated by the infrared spectra. ∎

Olefins besides ethylene also react on the ZnO surface, and the patterns are more complex [80]. For example, propylene reacts dissociatively with the surface acid–base pair sites, giving surface —OH groups and allyl ligands coordinated to zinc at the surface. Butenes adsorb similarly, and the π-allyl species are intermediates in catalytic isomerization. This isomerization route is different from that occurring on more strongly acidic metal oxides such as silica–alumina, and the parallels between solution catalysis and surface catalysis are again evident.

The examples of metal oxide catalysis illustrated with alcohol dehydration and olefin hydrogenation and isomerization indicate only a small fraction of the catalytic reactions that take place on metal oxide surfaces. But the broad patterns of acid-base chemistry and organometallic chemistry are consistent with most of the known catalytic chemistry, and these concepts provide a good starting point for an understanding of the more complicated chemistry, some of which is addressed later in this chapter.

6.4.5 Catalysis by Supported Metals

Many industrial catalysts are metals, and the structurally simple single crystals described above provide an introduction to the more typical metal structures used in industrial catalysts. Some metal catalysts are used in the form of powders or screens (gauzes) in industrial processes, but most are used in the form of highly dispersed aggregates on supports. One reason why highly dispersed metals are preferred is that many of the industrial catalytic metals are very expensive, and it is efficient to have most of the metal exposed at a surface and accessible to reactants. Typical supports are robust porous solids, including metal oxides, such as alumina and silica, and carbon. The following sections include descriptions of the structure and properties of supported metals and examples of processes using these catalysts.

THE NATURE OF SUPPORTED METALS

The supported metal catalysts used widely in technology consist of aggregates (sometimes called clusters or crystallites or particles) of metal of various sizes and shapes dispersed on a support [82], as illustrated by the electron micrograph of Fig. 6-38. A catalyst (e.g., rhodium on γ-Al_2O_3, abbreviated Rh/γ-Al_2O_3) is often prepared by bringing an excess of an aqueous solution of a metal salt (e.g., $RhCl_3$) in contact with a porous support (e.g., γ-Al_2O_3) in a process called impregnation. The solution fills the pores, and some of the metal salt may be adsorbed, with the adsorption depending on the polarization of the support surface, which is influenced by the solution pH. When the solution is decanted and the remainder of the solvent is evaporated, the metal salt is left dispersed on the surface. Alternatively, in the incipient wetness technique, only enough solution is added to fill the pores of the solid. To give a higher initial dispersion of the metal species on the support, an ion exchange process may be used instead of impregnation; for example, $[Pt(NH_3)_4]^{2+}$ in aqueous solution may exchange with protons on the metal oxide support.

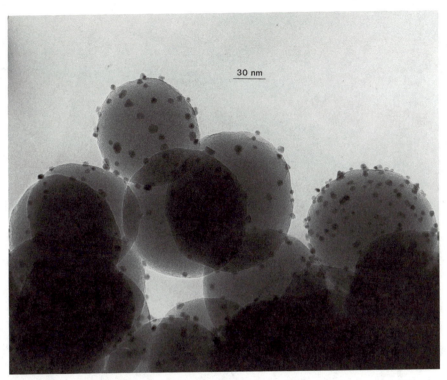

Figure 6-38
Electron micrograph of a supported metal catalyst, Rh/SiO₂. The metal crystallites are present on the surfaces of primary particles of SiO₂ [83]. [Courtesy of Professor A. Datye.]

In the next step of a typical preparation, the impregnated or ion-exchanged sample is calcined (heated to a high temperature, usually in air) and then treated with hydrogen at high temperature to reduce the metal. Metals readily reduced to the zero-valent state include the platinum group metals as well as silver, copper, and others. During the reduction, the metal migrates and forms aggregates dispersed on the support. Anions, for example, Cl^-, from the precursor may also be left on the support surface. The metal aggregates may be extremely small clusters, consisting of only a few atoms, but some are particles consisting of hundreds, thousands, or more atoms. Since the synthesis chemistry is imprecise and the support surfaces nonuniform, the metal aggregates are nonuniform in size and shape. The distribution of sizes is strongly dependent on the details of the preparation and not easily predicted, since the surface phenomena occurring during the preparation are not well understood.

The larger metal particles in supported metal catalysts are three-dimensional and have been characterized in detail by transmission electron microscopy [83] and other techniques. These particles may be considered small chunks of bulk metal. Their surfaces present a number of different crystal faces (Fig. 6-39) [84], and it appears that, to a good approximation for particles larger than

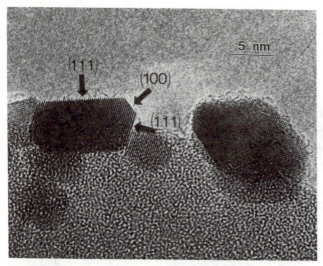

Figure 6-39
Electron micrograph of a supported metal catalyst, Rh/TiO₂ [84]. The
resolution is so high that different crystal faces of the metal particle
can be discerned. [Courtesy of Professor A. Datye.]

roughly 5 nm, they present more or less the same distribution of crystal faces,
independent of the particle size or shape.

Aggregates smaller than about 1 nm may be the most important catalytically because they have a large fraction of the metal exposed at the surface
and therefore accessible to reactants (this fraction is referred to as the *dispersion*). The very small aggregates (clusters) are also of most interest structurally,
since they may resemble molecular species more than bulk metals. These clusters are difficult to characterize because they are not easily accessible to direct
observation and may interact strongly with the support through most of their
metal atoms. The support may be considered to be a set of rigid ligands.

The simplest supported metal structures have been derived from organometallic clusters bonded to metal oxide surfaces. Preparation from these molecular precursors requires control of the chemistry on the support surface,
and since most organometallic clusters are fragile, the method is of interest in
research and not large-scale applications [8]. The anion $[Os_{10}C(CO)_{24}]^{2-}$ has
been formed on the surface of MgO. This cluster has a robust metal framework,
and electron micrographs and EXAFS spectra suggest that it can be denuded
of its carbonyl ligands to give Os_{10} units on the support, which are catalytically
active for typical hydrocarbon conversion reactions; these are among the supported metals with the best defined and most uniform structures [8b]. EXAFS
spectroscopy is a method that takes advantage of backscattering of photo-
electrons emitted from atoms absorbing X-radiation to give precise structural
information about the atoms in the immediate neighborhood of the absorbing
atom [85].

Other highly dispersed supported metals having some regularity of structure are two-dimensional, consisting of monolayer rafts on metal oxide supports. The raft structure is consistent with results of transmission electron microscopy of $Rh/\gamma-Al_2O_3$, for example [86], but there is still a question about whether the monolayer structure pertains to the catalyst with metal in the zerovalent state or only to oxidized forms. It seems likely that the raft structures are most stable when the metals are in positive oxidation states and strongly bonded to the metal oxide support. The oxophilic Group 6 and 7 metals are therefore considered better candidates for raft structures than the Group 8 metals, at least, when they are used in a reducing atmosphere such as H_2. A highly dispersed catalyst such as $Re/\gamma-Al_2O_3$ may well consist of cationic Re rafts on the support even at temperatures of several hundred degrees Celsius in the presence of a reducing atmosphere, as indicated by X-ray photoelectron and EXAFS spectra [87]. A schematic representation of the structure of such a catalyst is shown in Fig. 6-40. A second metal that is easily reduced, such as Pt, may facilitate the reduction of the Re by hydrogen spillover, a process whereby hydrogen is dissociatively adsorbed on metal aggregates, forming atomic hydrogen that is transported readily over the surface to other locations.

Rhodium supported on $\gamma-Al_2O_3$ is stable in a highly dispersed form and has been thoroughly investigated as a prototype supported metal catalyst. Many structural details have been obtained by the application of electron microscopy, XPS, and EXAFS spectroscopy. A structural model of a fully reduced $Rh/\gamma-Al_2O_3$ sample (Fig. 6-41) depicts clusters of about seven Rh atoms present in a bilayer with an fcc structure on the (111) surface of the $\gamma-Al_2O_3$; each Rh atom has, on average, two oxygen neighbors. This structure represents an

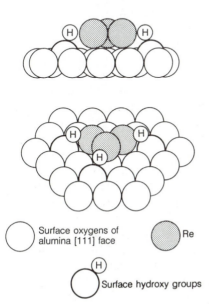

Surface oxygens of alumina [111] face

Re

Surface hydroxy groups

Figure 6-40
Structural model of a highly dispersed supported rhenium raft ($Re_3/\gamma-Al_2O_3$) determined with EXAFS spectroscopy. The metal entities consist of three Re atoms; the catalyst was prepared from the trinuclear cluster $H_3Re_3(CO)_{12}$ [87].

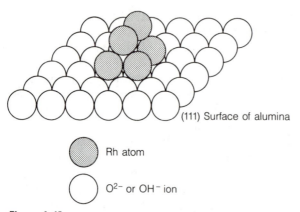

(111) Surface of alumina

Rh atom

O^{2-} or OH^- ion

Figure 6-41
Structural model of a highly dispersed supported metal catalyst, Rh/γ-Al$_2$O$_3$, inferred from EXAFS spectra and other data
[90]. The structure represents an average of the nonuniform
structures of this conventionally prepared catalyst.

average; a variety of related structures exist simultaneously. The catalyst is
therefore more complex structurally than the supported Os$_{10}$ catalyst mentioned above, but it is still almost molecular in character.

The metal–metal bonding in the larger, three-dimensional particles of supported metals is very similar to that in bulk metals; the metal–metal distances
are the same. The smallest entities of supported metals, however, may have
contracted metal–metal distances. When rafts of metal atoms in a positive
oxidation state are present, as in Re/γ-Al$_2$O$_3$ formed from H$_3$Re$_3$(CO)$_{12}$ (Fig.
6-40) [87], the unusually short metal–metal distances suggest that there may
be metal–metal multiple bonds [88]. Much remains to be learned about the
bonding in these surface structures.

Researchers have paid careful attention to the complex effects of different
supports and the nature of metal–support interactions [89], but there is still
only an incomplete understanding of the interfaces between dispersed metals
and metal oxide surfaces. The most helpful physical characterization method
is EXAFS spectroscopy, which is element specific and can provide information
about the atoms at the interface and their surroundings. This technique has
been used to characterize the structure and metal–oxygen distances of metal/
metal oxide interfaces in highly dispersed supported metals. Metal–oxygen
distances of about 0.25–0.27 nm have been found for metal-oxide-supported
metals such as Rh/γ-Al$_2$O$_3$ in the presence of hydrogen [90]. These distances
are too long to be evidence of bonds between metal ions and oxygen ions, and
they have been suggested to be indicative of metal atoms interacting with O^{2-}
ions of the surface, with hydrogen at the interface. The metal–oxygen distance
decreases to typically 0.21 nm when the samples are evacuated at high temperature. This is the distance in the cationic Re complex of Fig. 6-21. As yet,
no direct connection between these changes in the structure of the metal–

support interface and the surface and catalytic chemistry of the supported metal particles has been established.

The metal-support interactions are sometimes more complicated than has been stated above [89, 91, 92]. For example, some metals have an affinity for interstitial positions in metal oxides; nickel may be present in the bulk of γ-Al_2O_3 and simultaneously present as zero-valent metal particles on the surface of this support. Some reducible metal oxide supports such as TiO_2 form structures that migrate and partially envelope supported metal particles (Fig. 6-42) [84], drastically affecting their properties for chemisorption and catalysis.

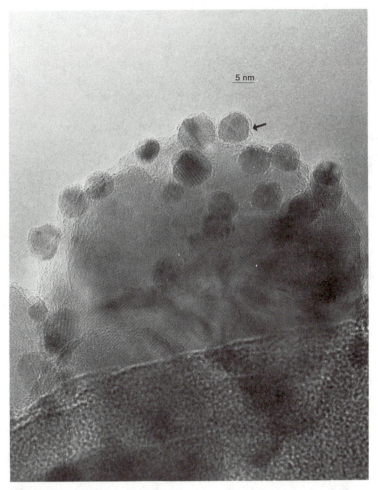

Figure 6-42
Electron micrograph of a supported metal catalyst, Rh/TiO_2 [84]. There is a layer of titanium oxide surrounding the metal particle, indicated by the arrow. This is evidence of one of the strongest metal-support effects. The overlayer strongly affects the catalytic properties of the metal. [Courtesy of Professor A. Datye.]

Most reactions catalyzed by supported metals take place on the metal surface, with the support usually being inert. (Sometimes the support also plays a direct role, as discussed below in this section.) It is therefore logical to represent the rate of a catalytic reaction per unit surface area of exposed metal. An experimental challenge is to determine the surface area of metal or, equivalently, the dispersion. If the metal aggregates are relatively large (between about 5 and 50 nm in diameter), the average size can be estimated from broadening of the X-ray diffraction lines; smaller aggregates hardly show up in the X-ray diffraction pattern. Electron microscopy can also be used to estimate the sizes of aggregates large enough to observe (which may not be all of them!); and because the aggregates and the porous solid are nonuniform, many measurements are usually needed to ensure good statistics. The dimensions of the scattering centers observed in the electron micrographs can be used to estimate metal dispersions, but some aggregate shape must be assumed. Often, particles are nearly hemispherical, but small aggregates, especially those on strongly interacting supports, may have much different shapes. EXAFS spectroscopy is also valuable for estimating the sizes and shapes of small metal aggregates on supports.

The most reliable methods for determining metal surface areas in supported metal catalysts involve selective chemisorption of H_2, O_2, or CO under conditions chosen to give coverage of the exposed metal surface but minimal adsorption on the support. These methods are essentially titrations, and they give good, reproducible results for a number of supported metals (e.g., Pt/Al_2O_3). The accuracy of the determination of metal surface area depends on knowledge of the stoichiometry of the adsorption, and the method as it is commonly used is limited because the stoichiometry is not well known for the smallest aggregates. In H_2 chemisorption, it is usually assumed that one H atom is adsorbed for each exposed metal atom, corresponding to results obtained for (111) faces of single crystals of fcc metals [93]. The approximation is good for most platinum group metals (not including Pd, which dissolves hydrogen), provided that the aggregates are larger than about 5 nm. The application of hydrogen chemisorption to determine the dispersion of platinum on γ-Al_2O_3 is illustrated in Fig. 6-43 [94]. The raw gas uptake data shown in the figure are not sufficient to determine the metal dispersion. Another set of measurements is made to determine how much hydrogen is removed in a short evacuation, and the amount of this weakly held hydrogen is subtracted from the raw uptake data, with the difference representing the strongly held hydrogen, which is used as a measure of the metal dispersion.

When the metal aggregates are smaller than about 1 nm in diameter, the stoichiometry of adsorption may differ significantly from one H atom per surface metal atom. Chemisorption measurements have been made for such highly dispersed clusters and have been interpreted on the basis of EXAFS data obtained to provide an independent measure of the average sizes [95]. Assuming that the metal clusters were hemispherical, the researchers determined the number of H atoms adsorbed per metal atom in the surface, finding that this ratio can approach 3 for Ir clusters of the highest dispersion (Fig. 6-44). The

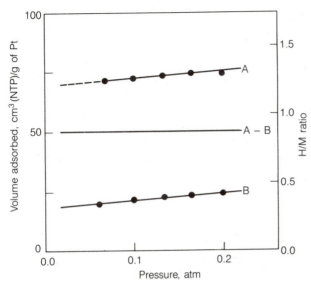

Figure 6-43
H_2 chemisorption on Pt/γ-Al$_2$O$_3$ for determination of the metal
surface area [94]. The isotherms were determined for chemisorption of hydrogen at room temperature. Curve A represents
the total adsorption on the initially evacuated catalyst. Curve B
is a second isotherm, determined by readmission of H_2 after the
sample had been evacuated for 10 min subsequent to the completion of the experiment to determine isotherm A. The difference
isotherm A − B was obtained by subtracting isotherm B from
isotherm A.

limiting chemisorption stoichiometries for Pt and Rh clusters are lower, approximately one and two H atoms per metal atom, respectively. (Are the data
of Fig. 6-43 consistent with the data of Fig. 6-44?) The significantly lower
hydrogen adsorption stoichiometry for Pt clusters than for Rh and Ir clusters
is evidently an electronic effect associated with the presence of one additional
valence electron for Pt. The higher limiting H/metal ratio for Ir clusters than
for Rh clusters can be understood by considering the generally greater metal–
H bond strengths and stabilities of metal hydride complexes of the 5d metals
as compared to their 4d congeners. There appear to be strong parallels between
the structural chemistry of molecular polyhydride clusters and the species
formed by hydrogen chemisorption on small supported metal clusters. The
nonuniform structures of the supported metal clusters in these samples ultimately limit the precision with which the surface polyhydride structures can
be defined.

Rates of catalytic reactions on supported metals are represented in units
of, for example, molecules/(cm^2 of exposed metal·s) or, better, if the number
of exposed metal sites has been measured by a technique such as hydrogen
chemisorption, in units of molecules/(site·s). Rates in the latter units are referred to as *turnover frequencies* (or sometimes turnover numbers, the term

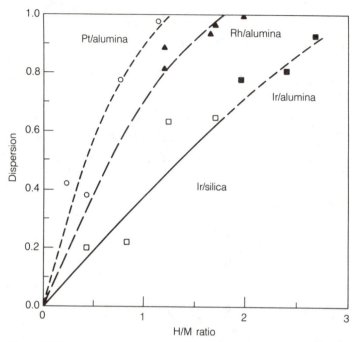

Figure 6-44

Chemisorption stoichiometry of hydrogen on highly dispersed supported metals [95].

introduced in Chapter 3). Strictly, the term should apply only when the reaction is zero-order in the reactant (when the surface is saturated with reactant), but this limitation is rarely recognized in the literature of surface catalysis. The distinction between catalytically active surface sites and surface sites measured by chemisorption is usually beyond the reach of the experimenter and is ignored; hence, in catalysis by surfaces of metals (and most other solids) the term turnover frequency usually lacks the fundamental meaning attributed to it earlier in this book.

Supported metals are often stable in highly dispersed forms and may work for years in a commercial catalyst, but their stabilities are limited and the dispersions subject to change. The process called *sintering* leads to a decrease in dispersion as aggregates or particles are combined to give larger ones [96]. Sintering can occur as metal particles literally migrate on metal oxide surfaces, with rates being greatest for the smallest aggregates. Sintering can also occur as individual metal atoms or metal complexes enter the gas phase to be transported to neighboring particles. Metal particles may also spread out and wet a support surface, and so facilitate the transport that leads to sintering. Sintering rates are dependent on the nature of the support surface and metal, on the reactive atmosphere, and on the nature of surface impurities.

As metals sinter during operation in a catalytic reactor, the activity per unit volume of catalyst usually decreases because the exposed metal area decreases. Consequently, sintering is one cause of catalyst deactivation. Sometimes the loss of metal surface area is reversed in a process called regeneration as catalysts are subjected to conditions that lead to *redispersion* of the metals [96]. For example, when Pt/Al$_2$O$_3$ is subjected to an atmosphere of oxygen at high temperature, the metal is redispersed; the process probably involves formation of a film of platinum oxide, which spreads on the support surface. Breakup of the thin film of this metal oxide leads to highly dispersed particulate structures, which upon treatment in hydrogen are reduced to give small metal aggregates. Other methods of redispersion may involve formation of volatile compounds, such as metal chlorides; these are transported through the gas phase to locations on the support where they form aggregates, which upon reduction become metal.

STRUCTURE-INSENSITIVE AND STRUCTURE-SENSITIVE REACTIONS

Experiments described above with single crystals of metal have shown that for some catalytic reactions (e.g., the oxidation of CO), the rate is the same on various crystal faces of a given metal, whereas for other reactions (e.g., ammonia synthesis), it can vary by orders of magnitude from one crystal face to another. It is not surprising, then, that rates of some catalytic reactions are the same for aggregates of supported metals of different sizes and shapes and that rates of other catalytic reactions depend on the aggregate size and shape. Experiments with supported metal catalysts having aggregates in various size ranges have been done to elucidate the patterns of catalytic activity, and the results have provided many insights into the nature of catalysis by metals [97].

Metal-catalyzed reactions are classified as *structure insensitive* when the rate (expressed per unit surface area of exposed metal) is almost independent of the average aggregate or particle size of dispersed metal. On the other hand, when there is a substantial difference in rate from one average metal aggregate size to another, the reaction is classified as *structure sensitive*. These definitions reflect the fact that systematic variation of the size of a small metal aggregate leads to systematic changes in the surface structure, although just what these changes are is largely unknown and surely varies from one metal to another. These concepts arose because one of the best ways to gain some understanding of the properties of supported metal catalysts through variation of their structures is to vary the average metal aggregate size, for example, by preparing a highly dispersed catalyst and then increasing the average size in steps by controlled sintering or by preparing catalysts with different metal loadings. The size range that is most appropriate is about 1 to 5 nm. Smaller aggregates (clusters) are more nearly molecular in character and may interact strongly with the support, as in raft structures, and larger particles may all have virtually the same surface crystallographies. Experiments with supported metals having varied average aggregate sizes are not as straightforward or clearly interpreted as experiments with various crystallographic orientations of a single crystal of metal or variation of the number of metal atoms in a molecular metal cluster, but they are widely applicable to a large class of materials and are helpful for elucidation of details of relations between catalyst structure and performance.

Structure-insensitive reactions take place at nearly the same rate on metal aggregates and particles of various sizes and on single crystals with different crystallographic orientations. The best-known examples are hydrogenations, for example, of olefins [63]. Since these reactions are catalyzed in solution by mononuclear metal complexes (the Wilkinson hydrogenation), it has been suggested that a catalytic site for olefin hydrogenation on a metal surface is a single metal atom, which behaves more or less independently of its neighbors. But there are other plausible explanations. For example, the hydrogenation of ethylene takes place on surfaces that are largely covered with carbonaceous residues, and the catalytic site may be better described as part of the residue rather than part of the metal surface. The residues may fuzz out the surfaces, so that they appear almost all alike to an adsorbed reactant.

Structure-insensitive reactions occurring on metal surfaces include those involving the breaking or making of H–H, C–H, or O–H bonds. Some well-documented examples are listed in Table 6-4 [63]. Some other important characteristics of this class of reactions are given in the table: When the metal catalyst is combined with a different and catalytically inactive metal to form an alloy, it leads to a dilution of the catalytic metal in the surface. Consequently, the rate of the structure-insensitive reaction decreases, but not strongly. Similarly, poisons on the surface of the catalytic metal reduce the activity, but only moderately. For these reactions, the variation in activity from one metal to a near neighbor in the periodic table is often not large.

These observations are contrasted to those for reactions classed as structure sensitive [63] (Table 6-4). Not only does the activity vary substantially (at least severalfold, but perhaps by orders of magnitude) from one crystal plane to another, but modification of a surface by alloying or adsorption of poisons exerts a large effect on the activity. The variations in activity from one metal to its near neighbors in the periodic table are large. It is concluded that the catalytic sites for structure-insensitive reactions are quite simple (perhaps single metal atoms), whereas the catalytic sites for structure-sensitive reactions are more complicated, involving a number of metal atoms, with the activity being sensitive to the arrangement of the set of metal atoms (called an ensemble) constituting the site.

Many data are available for a relatively simple structure-sensitive test reaction, the hydrogenolysis of ethane:

$$C_2H_6 + H_2 \rightarrow 2CH_4 \tag{6-45}$$

The data of Fig. 6-45 [98] show the structure sensitivity. The reaction on Ni (100) is markedly faster than the reaction on the Ni (111) surface.

Similarly, there is a significant decrease in activity of silica-supported Ni catalysts as a result of reduction at high temperature to give increasingly large Ni particles [99]. The samples with the largest particles are expected to have largely planar (111) surfaces, whereas the smaller aggregates are expected to

Table 6-4
CLASSIFICATION OF METAL-CATALYZED REACTIONS [63]

Increasing importance of effect →

	Effect of Structure	Effect of Alloying	Effect of Poisons	Effect of Nature of Metal	Kinetics	Nature of Bonds Being Activated	Multiplicity of Site
$CO + \tfrac{1}{2}O_2 \rightarrow CO_2$						O—O	1 or 2 atoms
$H_2 + D_2 \rightarrow 2HD$						H—H	
$C_2H_4 + H_2 \rightarrow C_2H_6$	Structure-insensitive (no effect)	Minor	Minor	Moderate	Based on uniform surface formalism	{H—H, C—H}	
$C_6H_{10} + H_2 \rightarrow C_6H_{12}$						{H—H, C—H}	
$C_6H_6 + 3H_2 \rightarrow C_6H_{12}$						{H—H, C—H}	
$C_3H_9OH \rightarrow C_3H_8CO + H_2$						O—H	
$N_2 + 3H_2 \rightarrow 2NH_3$	Structure-sensitive (moderate effect)	Large	Large	Very large	Based on nonuniform surface, Temkin formalism	N—N	Large multiple site
$C_2H_6 + H_2 \rightarrow 2CH_4$						C—C	

← Increasing importance of effect →

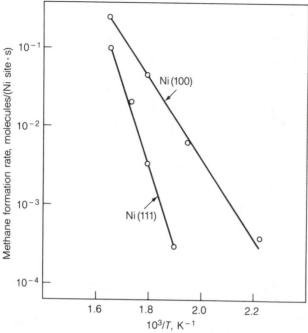

Figure 6-45
Rates of ethane hydrogenolysis catalyzed by two Ni single-crystal surfaces [98]. Data demonstrate that the reaction is structure sensitive.

have more surfaces such as the (100) surface. There is therefore a broad consistency between the results characterizing the single crystals and those characterizing the supported aggregates. However, the reasons for the differences in activity of the different Ni surfaces remain to be explained.

The structure sensitivity of the reaction is consistent with the hypothesis that a catalytic site on the nickel surface consists of more than one Ni atom. Therefore, one might expect that incorporation of a second metal (such as Cu) that is catalytically inactive for the ethane hydrogenolysis would markedly decrease the activity of the surface. Consistent with this expectation, it has been observed (Fig. 6-46) that alloying the catalytically active nickel with Cu reduces the activity by orders of magnitude [100]. The activity for ethane hydrogenolysis also changes by orders of magnitude from one metal to another.

In contrast, the addition of Cu has much less effect on the structure-insensitive cyclohexane dehydrogenation reaction, and addition of Cu in small amounts even increases the activity of the catalyst (Fig. 6-46).

The strong dependence of the activity for hydrogenolysis on the composition of the bimetallic catalyst raises the question of how a relatively small amount of Cu can have such a pronounced effect on the activity of the Ni. It is widely agreed that the catalytically active ensembles for this reaction consist of more than one Ni atom, and some researchers have even suggested that an ensemble consists of 10 or more Ni atoms, although the suggestions are imprecise and still a matter of lively debate [101]. Hydrogen chemisorption data have provided some clarification of the observations. Hydrogen chemisorbs selectively on the Ni atoms in the alloy surface, and hydrogen has therefore been used to determine the fraction of the surface covered with each metal. The results, confirmed by Auger electron spectroscopy [102], show that the surfaces of the Cu–Ni particles are strongly enriched in Cu, which is thermodynamically more favored in the surface positions than in the interior positions.

The use of the term "alloy" here pertains to catalysts used in the form of metal powders having surface areas of about 1 m²/g. Alloys are bulk materials, but the metal entities in supported metal catalysts are often too small to have properties characteristic of the bulk. The dependence of the physical properties on the size of the metal entities provide important opportunities in catalysis. For example, a particular pair of metals in a particular proportion may not be soluble in the bulk (i.e., may not form an alloy) but may be soluble in small aggregates, because surface energetics may dominate in the latter and not the former. Consequently, it is possible to use a second metal in bimetallic clusters [103] to regulate the catalytic properties of a metal. The results illustrated for the Cu–Ni alloy powders in Fig. 6-46 give an indication of how addition of a second metal to a catalyst can change its performance. If one wished to change the catalyst to favor hydrogenation–dehydrogenation reac-

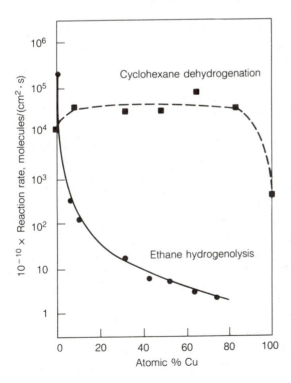

Figure 6-46
Activities of Cu–Ni alloy catalysts at 316°C as a function of the bulk composition [100]. The ethane hydrogenolysis was carried out at an ethane partial pressure of 0.03 atm and a hydrogen partial pressure of 0.20 atm. The cyclohexane dehydrogenation was carried out with a cyclohexane partial pressure of 0.17 atm and a hydrogen partial pressure of 0.83 atm.

tions over hydrogenolysis reactions, one could selectively poison the catalyst (e.g., with sulfur compounds) or incorporate a second metal to reduce its activity for hydrogenolysis while only slightly modifying its performance for the hydrogenation. These ideas are discussed below in the context of a technologically important example, the reforming of naphtha to manufacture lead-free high-octane-number gasoline.

ETHYLENE OXIDATION ON SILVER

One of the most thoroughly investigated reactions catalyzed by supported metals is the partial oxidation of ethylene to give ethylene oxide, accompanied by the undesired oxidation of both ethylene and ethylene oxide to give CO_2. Ethylene oxide is used on a large scale to make ethylene glycol (for antifreeze and other products) and numerous other chemicals. The only surface with a high activity and selectivity for olefin epoxidation is silver, and ethylene is the only olefin that is converted with high selectivity; propylene undergoes combustion to give CO_2.

The reactive species present on the silver under reaction conditions are for the most part unstable under vacuum, and hence the ultrahigh vacuum techniques have shed little direct light on the details of the catalysis. Clean silver surfaces are hardly reactive with ethylene, but silver surfaces partially covered with oxygen are reactive. Surface and subsurface oxides of silver are quite stable and have been investigated extensively by ultrahigh vacuum and other techniques. Various monatomic and diatomic oxygen species exist on the silver surface, apparently even under catalytic reaction conditions, and their complicated chemistry is affected by the presence of other adsorbates such as CO_2 (which improves the selectivity for ethylene oxide formation when present in low concentrations) and by promoters and modifiers added to increase the selectivity to ethylene oxide. Monatomic oxygen species have been implicated in the rate-determining step for the formation of ethylene oxide [104], but some authors contend that dioxygen surface species react with ethylene, an interpretation that is still debated. A schematic representation of an intermediate involving monatomic oxygen on the surface is shown in Fig. 6-47 [104]. The catalytic reaction is structure insensitive, presumably because the oxygen on the surface modifies the surface, nullifying any effect of the underlying silver and making one crystal face about as active as another [105].

Evidence of the structures present on single crystals of silver provides only a beginning toward understanding the nature of the catalysts used in industrial ethylene oxide processes. The industrial catalysts are structurally complex, typically consisting of quite large silver particles (roughly a micron in diameter) that are dispersed on an α-Al_2O_3 support in the presence of several additional components.

The literature of the kinetics is conflicting, and there is a lack of quantitative data for industrial catalysts and reaction conditions.

Catalyst formulations described in patents include alkali metal promoters such as K^+ and Cs^+, which might be present in the form of carbonates under

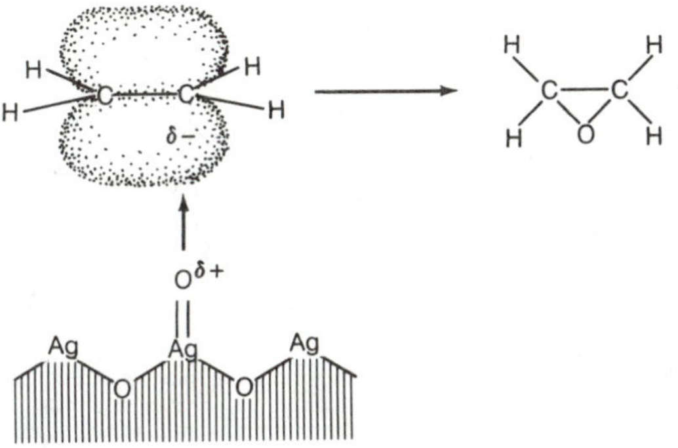

Figure 6-47
Schematic representation of a proposed surface intermediate in the epoxidation of ethylene on an oxidized silver surface [104].

catalytic reaction conditions. These are chemical promoters, affecting the mechanism of the surface-catalyzed oxidation reaction in a way that is not yet understood. Data from a patent [106], shown in Fig. 6-48, indicate how very small amounts of K^+ in the catalyst increase the selectivity for ethylene oxide; the selectivity passes through a maximum with increasing K^+ concentration. This is a common pattern in surface catalysis, and alkali metals are used as promoters in catalysts for a number of different industrial processes [107] (Table 6-1).

The catalyst selectivity is also improved by the continuous addition to the reactor feed of a few parts per million of chlorine-containing organic compounds such as ethylene dichloride. Chlorine on the catalyst surface suppresses the undesired reactions giving CO_2 in a series-parallel reaction network:

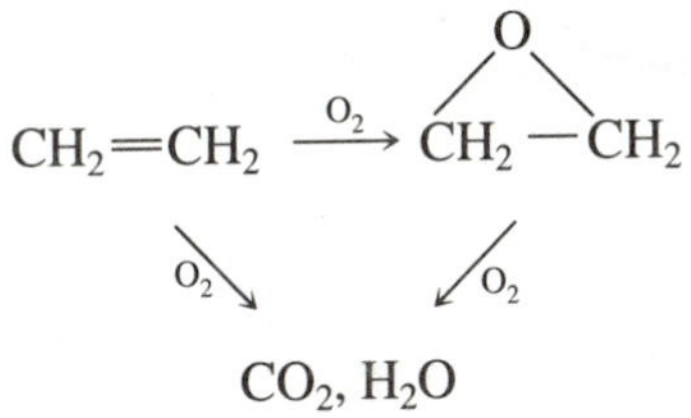

$$CH_2{=}CH_2 \xrightarrow{O_2} \overset{\displaystyle O}{\overset{\displaystyle \diagup\ \diagdown}{CH_2{-}CH_2}}$$

$$\underset{O_2}{\searrow} \qquad \underset{O_2}{\swarrow}$$

$$CO_2,\ H_2O$$

(6-46)

The properties of the catalyst support, α-Al_2O_3, are crucial; it has an unusually low surface area, typically <1 m^2/g. The catalyst pore geometry and particle size (3–10 mm in diameter [108]) are chosen to favor formation of the desired ethylene oxide and to minimize the formation of CO_2. Since the reactions are fast, a significant mass transport resistance within the pore structure would decrease the yield of ethylene oxide because it is an intermediate in a sequential reaction path (see Chapter 4, Section 4.8.1). In an industrial process, as much as 80 percent of the carbon in the ethylene feed is converted into the desired product, with the remainder giving mostly CO_2.

The catalyst support needs to be inert. Reactive supports give less selective catalysts. The point is illustrated by the data of Fig. 6-49 for silver-on-alumina catalysts not containing promoters. As the temperature of preparation of the Al_2O_3 support increased from 300°C to 1800°C, the selectivity increased

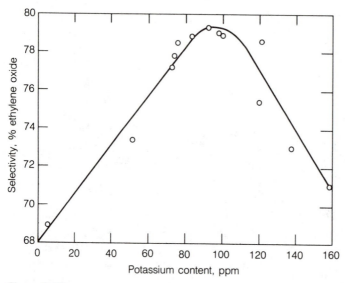

Figure 6-48
Promotion of the ethylene oxidation reaction by potassium [106]. The selectivity (the percentage of the ethylene converted which gives ethylene
oxide) increases through a maximum with increasing promoter concentration on the surface of the supported silver catalyst.

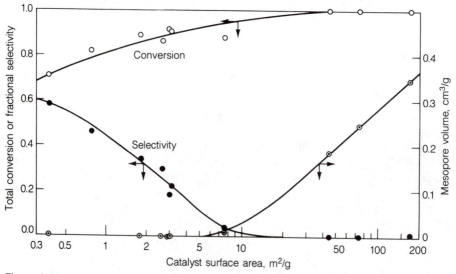

Figure 6-49
Performance of a series of Ag/Al$_2$O$_3$ catalysts for oxidation of ethylene in a
pulse reactor at 250°C [109]. As the temperature of treatment of the support
was increased, the surface area decreased, the activity (measured by the
fractional conversion of ethylene) decreased, and the selectivity to ethylene
oxide increased.

from about 0 percent to 60 percent (which is less than could be obtained under practical conditions) [109]. These results are explained by the conversion of the reactive transition alumina to nearly inert α-Al_2O_3 with increasing temperature of treatment (Fig. 6-7). The conversion to α-Al_2O_3 is accompanied by loss of surface area from about 170 m^2/g to about 0.4 m^2/g and an increase in average pore diameter from 4 nm to roughly 1400 nm, reflecting morphological changes in the Al_2O_3. The undesirable transition alumina is believed to catalyze the isomerization of ethylene oxide to acetaldehyde, which is easily oxidized to give CO_2.

In addition to the chemical promoters, the literature [110] describes textural (or structural) promoters such as barium carbonate. This component may assist in maintaining the dispersion of the silver, although it is not likely that it is used in today's catalysts.

Catalysts described in patents also contain binders to maintain the physical integrity of the support particles. Most of the supports commonly used for metal catalysts, such as transition aluminas, do not require binders because the support materials themselves have reactive functional groups on the surfaces of the primary particles that help bind them together. But the primary particles of the nearly inert α-Al_2O_3 used in ethylene oxide catalysts bind to each other much less strongly, and materials such as clay minerals (e.g., kaolinite) may be added to serve the purpose. After the support particles are formed (e.g., by extrusion), they are heated to remove water (and possibly organic burnout material added to give a pore structure) and then subjected to a high temperature, perhaps about 1500°C, to convert the clay to a glassy (vitrified) state and to set the bond.

The catalyst may be made by impregnation of the support with an aqueous solution containing silver and alkali metal salts. Silver amine complexes have been mentioned in the literature. The water is removed by drying; and when the solid is heated, the silver compound decomposes, giving particles of metallic silver dispersed on the support.

In the industrial process for ethylene oxidation, the catalyst particles are held in bundles of thousands of narrow (20–50 mm i.d.) tubes surrounded by a heat transfer medium [108, 111]. The narrow tubes minimize the temperature rise in the fixed beds, which favors the selectivity. Nonetheless, hot spots unavoidably occur near the reactor inlet, and the catalyst must be formulated so that the highly exothermic combustion reactions do not become so fast that the reaction runs away.

The reactor feeds may contain either air or nearly pure oxygen, in addition to recycled CO_2, water, and ethylene oxide, as well as ethylene and some impurities. The trade-offs between processes using air and oxygen in the feed are similar to those mentioned for the Wacker process in Chapter 2. Space velocities may be as much as about 4500 h^{-1}, temperatures about 200–300°C, pressures about 10–30 atm (the high pressures favor the rates of transport but do not significantly affect the intrinsic kinetics), and per-pass ethylene conversions about 20–50 percent. Selectivity is almost independent of conversion until higher conversions are reached, and then selectivity decreases with increasing conversion [108]. Years of continuing research and development have led to increases in selectivity from <70 percent in the late 1960s to about 80 percent in the late 1980s [104]. The catalysts may last for several years before being regenerated or replaced.

In summary, because of its technological importance, the relative simplicity of the reaction network, and the catalytic activity of silver itself, which can be investigated readily in ultrahigh vacuum experiments, ethylene oxidation has become one of the most thoroughly investigated catalytic processes. But with all the effort, including application of virtually all the methods of ultrahigh vacuum surface science, the details of the reaction mechanism remain elusive. What is clear is that the rate-determining step in the ethylene oxide formation involves some form of oxygen bonded to the silver surface. The industrial catalysts provide an excellent case study; it illustrates how improvements have been made over the years by modification of physical properties to minimize transport influence and by modification of surface composition with promoters and suppressors to somehow change the surface chemistry in ways that lead to increased selectivity.

BIFUNCTIONAL CATALYSIS AND THE REFORMING OF NAPHTHA HYDROCARBONS

In the supported metal catalysts described thus far, the role of the support has been that of a platform, largely inert except for a tendency to help maintain the dispersion of the metal. Sometimes the support is not inert but plays a direct role in the catalysis. The opportunities for bifunctional catalysis involving both the metal and the support are illustrated by Pt/γ-Al_2O_3 and related catalysts that are used for naphtha reforming, the conversion of the gasoline range hydrocarbons into high-octane-number gasoline.

To understand the choice of the components in the catalysts and their complementary roles, consider the processing goals of naphtha reforming: Naphtha (Fig. 2-17) has too low an octane number to be an optimum motor fuel. Reducing the concentration of straight-chain paraffins and increasing the concentration of branched paraffins, naphthenes (cyclic paraffins), and aromatics improves the octane number and thereby the tendency of the fuel to prevent engine knock and the attendant inefficient operation. Reforming of the fuel is preferable to addition of antiknock compounds such as tetraethyl lead as lead emissions from the engine are harmful to the environment.

A desirable set of reactions for the reforming process includes those that do not change the carbon number of the molecules in the naphtha (keeping them in the gasoline boiling range), but cause their rearrangement (reforming) to give molecules with higher octane numbers. The octane numbers (in parentheses) of the following C_7 hydrocarbons give an indication of the desirable processing goals: n-heptane (0), 2-methylhexane (41), 2,2-dimethylpentane (89), 2,2,3-trimethylbutane (113), methylcyclohexane (104), and toluene (124). The obvious processing goals include isomerization to increase the degree of branching of paraffins, dehydrocyclization of paraffins to give naphthenes and aromatics, and dehydrogenation of naphthenes to give aromatics.

This list calls to mind the catalytic properties of platinum from Section 6.4.3. Platinum catalyzes the dehydrogenation and dehydrocyclization reactions rapidly, provided that the hydrogen partial pressure is high enough to prevent the buildup of too much carbonaceous deposit poisoning the metal surface. However, platinum's activity for skeletal isomerization is low. Skeletal isomerization reactions of hydrocarbons are catalyzed by strong acids; these are the reactions proceeding through carbenium ion intermediates and described in Chapters 2 and 5.

Other platinum group metals such as osmium and iridium are also active for the dehydrogenation and dehydrocyclization reactions; but, in contrast to platinum, they are also highly active for hydrogenolysis of paraffins, and this reaction is for the most part a detriment, since it converts gasoline range molecules into gases such as propane that are too light for gasoline. However, it does offer the advantage of increasing the octane number of the product by removal of straight-chain paraffins. In sum, platinum is the metal with the most suitable properties for meeting the goals of naphtha reforming, but by itself it is lacking some of the desirable characteristics of strong acid catalysts.

Consequently, platinum on a nearly inert support (e.g., Pt/SiO$_2$) is not a very good reforming catalyst, being only slightly better than the first known catalysts, for example, Cr$_2$O$_3$/Al$_2$O$_3$ (chromia). A conceptual advance was made by V. Haensel and co-workers in the 1940s when they discovered the advantages of using a support for the platinum that provided a separate catalytic function—acidity—to facilitate the isomerization, as well as other reactions [112]. Such catalysts are now recognized as the prototype bifunctional catalysts.

Haensel and co-workers discovered the benefits of an acidic component of the catalyst; when platinum was supported on less-than-thoroughly-washed Al$_2$O$_3$ made from aluminum chloride rather than on SiO$_2$ or on Al$_2$O$_3$ made from aluminum nitrate, it was found to be more active for the desired reforming reactions. This observation led to the application of Pt/Al$_2$O$_3$ catalysts in which the acidity of the transition alumina was increased by incorporation of chloride. The chloride increases the acid strength of the surface by an inductive effect, roughly as a Cl substituent increases the acid strength of acetic acid.

What is remarkable about the combination of the acidic function with the metal function in the reforming catalyst is that the functions work in concert, and the key reaction intermediates are olefins, as shown in the reaction scheme proposed by Mills et al. [113] (Fig. 6-50). Consider the conversion of n-hexane to benzene. The dehydrocyclization reaction is more facile in the presence of the bifunctional catalyst than when platinum alone is the catalyst, for the following reason: In the first reaction in the pathway, the hexane is rapidly dehydrogenated on the metal surface; the resulting hexene is easily protonated by the acidic sites on the support to give a carbenium ion, which is converted into methylcyclopentane. This is dehydrogenated on the metal surface to give

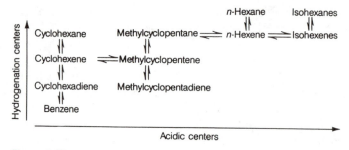

Figure 6-50
Bifunctional catalysis: Reaction network for reforming of C_6 hydrocarbons proposed by Mills et al. [113].

methylcyclopentene, which undergoes an acid-catalyzed rearrangement to give cyclohexene, which is dehydrogenated on the metal surface, ultimately giving benzene.

This sequence of reactions involves an interplay between the two functions of the catalyst and requires transport between them of the intermediate compounds—olefins. The independent action of the two components and the influence of the transport processes on the rate of the reaction were demon-

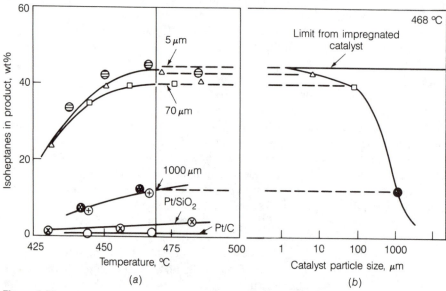

Figure 6-51
Bifunctional catalysis by a mixture of particles of an acidic catalyst (silica–alumina) and a nonacidic supported metal catalyst (Pt/SiO$_2$): demonstration of the transport of intermediates between the particles [114]. Data are for isomerization of n-heptane: (a) conversion as a function of temperature; (b) conversion at 468°C as a function of the diameter of the particles of catalyst. Reaction conditions: 2.5 atm n-heptane partial pressure; 20.0 atm H$_2$ partial pressure; residence time in catalyst bed, 17 s.

strated in experiments with *n*-heptane isomerization catalyzed by a mixture of particles of platinum on a nonacidic support (Pt/SiO₂) and particles of a strong acid (SiO₂–Al₂O₃) [114] (Fig. 6-51). (Can you explain the shape of the curve showing the dependence of rate on catalyst particle size shown in the right-hand part of the figure?)

To recapitulate, the bifunctional catalyst works well for reforming although the individual functions alone do not. The metal alone does not catalyze the branching reactions; they require an acidic function to generate carbenium ions, which undergo the desired branching. The acidic function alone is not sufficient to generate carbenium ions. It is too weakly acidic to protonate paraffins; and, if it were, it would be deactivated very rapidly by carbonaceous deposits. The dehydrogenation activity of the metal function generates olefins, which are rapidly protonated and converted by nearby acid groups.

In practice, the reforming reactions take place rapidly at about 500°C, and the acid-catalyzed reactions are for the most part slower than those catalyzed by the platinum. Since the platinum is expensive, it was imperative in the development of an economical industrial process to minimize its amount and maximize its dispersion. The industrial catalysts have about 1 wt% Pt present in a high dispersion (average cluster size about 1 nm) on high-surface-area η-Al₂O₃ or γ-Al₂O₃ (about 250 m²/g). Platinum is the metal of choice because it is the only one that has activity for the desired reactions without being more than moderately active for undesired reactions such as hydrogenolysis of paraffins. Alumina is the support of choice because it is inexpensive, is easily prepared with the desired physical properties, and interacts strongly enough with the aggregates of platinum to maintain a high metal dispersion. The most economical method for creating the required acidity in the support is to regulate the chloride content of the alumina by including small amounts of chlorine-containing organic compounds in the feed to the reactor, along with traces of water, which maintain the alumina surface in a partially hydroxylated state.

It was a remarkable insight on the part of the industrial researchers who developed the naphtha reforming process to use a reaction environment with a high partial pressure of hydrogen for a process in which the major reactions were dehydrogenations. The reason for the high hydrogen partial pressure was to minimize the rate of formation of carbonaceous deposits (coke), which include polyunsaturated aromatics. Acidic catalysts alone form coke rapidly; the metal function helps slow the coke accumulation, since it is active for hydrogenating the coke precursors.

One of the goals of the reforming process is to maximize the production of high-octane-number compounds such as aromatic hydrocarbons, and these are formed rapidly by dehydrogenation of naphthenes. Consequently, there is a motivation for using low pressure and high temperature in the process, since the dehydrogenation of naphthenes is favored thermodynamically at low hydrogen partial pressures and high temperatures, and the reactions are relatively fast at both low and high pressures. The complicating processing issue is that the higher the temperature and the lower the hydrogen partial pressure (other conditions being unchanged), the higher is the rate of formation of carbonaceous deposits on the catalyst surface and the faster the catalyst deactivation. When partial pressures of hydrogen of as much as 40 or 50 atm are used in industrial processes, the catalyst may last for many months without regeneration. Hydrogen (likely atomic hydrogen formed by dissociation of H_2 on the platinum and spilled over onto the support) reacts with the carbonaceous deposits and their precursors, and the higher the hydrogen partial pressure, the higher are the rates of these desirable reactions. However, the higher the hydrogen partial pressure, the higher also is the rate of the largely undesirable hydrogenolysis reactions.

The incentive to use lower pressures in the process placed a challenge at the doorstep of the chemists and engineers seeking to prepare improved catalysts. A catalyst that would undergo deactivation less rapidly would offer the opportunity for lower-pressure operation and the benefits of higher aromatics yields than those allowed by thermodynamics at the higher pressures. Recognition of this opportunity led to improved reforming catalysts and processes; the innovations centered around the incorporation of a second metal into the catalyst to modify the properties of the platinum and to optimize the performance at low pressures.

A technological breakthrough resulted from the invention of the Re–Pt/Al_2O_3 catalyst by Kluksdahl [115]. This bimetallic catalyst allows operation at pressures of as little as several atmospheres, with high yields of aromatics and a low enough rate of deactivation for economical operation. The rhenium in the catalyst markedly reduces its susceptibility to deactivation by carbonaceous deposits.

Neither the reason for the benefit nor the structure of the catalyst is well understood. The rhenium and platinum under some circumstances form bimetallic alloylike aggregates on the support surface [116], but there is evidence that catalysts prepared in a manner typical of industrial operation may have segregated metals, with the platinum present as highly dispersed aggregates of zerovalent metal and the rhenium present separately in a cationic form on the surface. When the rhenium and the platinum are present on separate particles of support and mixed together in the same reactor, the benefits of the rhenium are still observed [117]. This observation weighs against the involvement of bimetallic aggregates in the catalysis, but it is not sufficient evidence to exclude their existence and possible benefit.

Another bimetallic combination has also found industrial application in catalytic reforming. The catalysts consist of iridium and platinum on the alumina support, and there is evidence from EXAFS spectroscopy and other techniques that the metals are present as small bimetallic aggregates (bimetallic clusters [103]). Now, there is a basis for interpretation of the role of the second metal in minimizing the deactivation of the catalyst. Iridium by itself is not a satisfactory reforming catalyst because it is too active for hydrogenolysis. But hydrogenolysis is a structure-sensitive reaction, requiring an ensemble of metal centers on the catalyst surface. Dilution of the iridium with platinum at the metal surface may minimize the hydrogenolysis markedly and may decrease the activity for dehydrogenation only slightly. Furthermore, the presence of the iridium may be expected to give the catalyst a higher activity for conversion of carbonaceous residues or their precursors, since these reactions have some of the character of hydrogenolysis.

The metal functions in the reforming catalysts are strongly poisoned by sulfur-containing compounds, and it is necessary to remove all but about one part per million of these compounds from the feed stream by processes described in Section 6.4.7, below. However, the sulfur poisoning is not entirely a detriment in the process. When a fresh Re–Pt/η-Al$_2$O$_3$ catalyst is brought on stream, it is first intentionally poisoned with sulfur because sulfur is a selective poison for the structure-sensitive hydrogenolysis reaction; in this respect the role of sulfur is comparable to the role of iridium and possibly the role of rhenium in the bimetallic catalysts. The unpoisoned fresh catalyst has such a high hydrogenolysis activity that this exothermic reaction would cause the reactor temperature to rise disastrously at the start of the operation. The temperature runaway can lead to sintering of the metal and fusion of the catalyst into a nonporous mass. Very high temperatures can also lead to failure of the reactor walls, and the result would be the escape of hot hydrogen into the atmosphere, leading to an explosion and fire.

Since deactivation of the catalysts during operation is unavoidable, they are regenerated. In a typical operation, the catalyst is present in a group of several fixed-bed reactors in series. When the catalyst in one reactor is being regenerated, the others are in operation. The regeneration of a supported platinum or rhenium–platinum catalyst involves burning of the carbonaceous deposits at a low, controlled rate with a stream containing a low partial pressure of oxygen. The metal is redispersed by treatment with O$_2$ and chloride-containing compounds; volatile metal compounds are evidently formed, which are transported through the vapor phase and deposited on the support surface. The resulting highly dispersed platinum is treated in H$_2$ and reduced to the zerovalent state and then treated with sulfur compounds before again being brought on stream in the reforming process. The Ir–Pt catalysts are difficult to regenerate, since volatile iridium oxide may form during burning of the deposits; the regeneration and redispersion requires treatment with HCl and oxygen.

Some reforming processes are carried out at pressures of only a few atmospheres, and the rapid deactivation of the catalyst makes the periodic regeneration strategy described above impractical. Instead, the catalysts are used in moving bed reactors, gradually descending through a screened-in annular space (through which the reactant gases flow radially) and being almost continually removed from the bottom and then transferred to a separate regeneration reactor (also a moving bed) and then back to the reforming reactor. In this respect, the reforming process resembles the cracking process described in Chapter 5, except that the residence time of the catalyst in the reforming reactor is orders of magnitude longer than that of the cracking catalyst in the riser tube reactor.

The classes of reactions of the naphtha compounds occurring in the presence of hydrogen are rather well understood, but since there are hundreds of individual compounds in a naphtha feedstock, there are hundreds of reactions to account for. The complexity of the reacting mixture is too great to allow a full quantitative characterization of the reaction networks and kinetics. The practical alternative is to lump the reactants into compound classes and use a simplified representation of the reaction network whereby conversions of compound classes are represented as though they were true reactions. A representation of the naphtha reforming reactions is summarized in this fashion in Fig. 6-52 [118]. (Can you check the plausibility of this scheme by comparing it with the reaction network for conversion of *n*-hexane into benzene shown in Fig. 6-50?)

The kinetics of conversions of these lumps is represented with Langmuir–Hinshelwood forms of equations familiar from this book for reactions of individual compounds, with terms for competitive inhibition [118], but quantitative results are not available. Process models also account for the deactivation of the catalyst, which affects the various reaction classes differently.

In summary, naphtha reforming is one of the most important of the catalytic processes, and its development provided much of the stimulus for advances in the understanding and further applications of supported metal catalysts. The Pt/Al$_2$O$_3$ reforming catalyst is the prototype supported metal, and the modern reforming catalysts are the prototype supported bimetallic catalysts. The process development led to the discovery of bifunctional catalysts, and the catalysts have evolved to become even more complex, which is the rule in catalytic technology. The functions include the primary metal, Pt (catalyzing reactions such as dehydrogenation and dehydrocyclization), the acidic support (catalyzing hydrocarbon conversions proceeding through carbenium ion intermediates), the chloride (providing greater acid strength), and the second metal, Re (which somehow increases the resistance of the catalyst to deac-

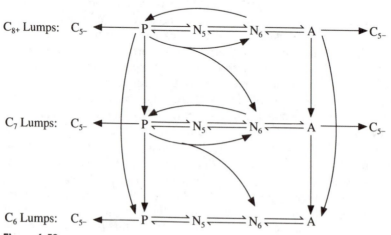

Figure 6-52
Lumped reaction network for catalytic reforming of naphtha [118]. N represents naphthenes; P, paraffins; A, aromatics; and C$_5$− pentane and lighter compounds.

tivation and allows economical operation at low pressures). There is no better process for illustrating concepts in surface catalysis. Unfortunately, since the industrial information is largely proprietary, little can be written about the kinetics and quantitative characterization of the catalytic reactions.

6.4.6 Catalysis by Mixed Metal Oxides: Ammoxidation of Propylene

Mixed metal oxides, those containing more than one kind of metal atom in the bulk, are important catalysts that are used widely for selective oxidation of hydrocarbons. The surface of a mixed metal oxide exposes two different metal ions in addition to O^{2-} ions and —OH groups. Selective oxidation of the hydrocarbon, represented schematically in Fig. 6-53, usually occurs at surface sites having oxygen atoms of limited reactivity (associated with metal M_1 in the figure), and these react with the hydrocarbon to give water and a partially oxidized organic compound rather than the undesired CO_2. These surface sites are then reoxidized indirectly and not by O_2 from the gas phase. The O_2 reacts instead with a surface site associated with the second kind of metal ion (M_2), and oxygen is transported as ions through the bulk of the catalyst from the second site to the first to reoxidize it, with a compensating transport of electrons to complete the cycle.

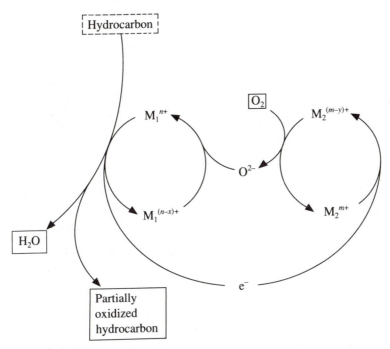

Figure 6-53
Schematic representation of the catalytic cycle for ammoxidation of propylene and related surface-catalyzed partial oxidations. M_1 and M_2 represent the two metals in the mixed metal oxide catalyst.

Essentially this same pattern has already been illustrated by the Wacker oxidation. The organic reactant (ethylene) reacts with an oxide (water), and not directly with O_2, in a step mediated by one metal (palladium). A second metal (copper) reacts with O_2, and it then reoxidizes the first metal. The essential characteristic of this bifunctional catalysis is the protection of the organic reactant from complete oxidation by a cycle whereby it reacts rapidly with an intermediate oxide of limited reactivity but at a negligible rate with O_2 and with other metal oxide groups that might convert it into CO_2 and water.

Metals typically catalyze total oxidation, as illustrated by platinum in the automobile exhaust converter. The epoxidation of ethylene catalyzed by silver (more precisely, by a surface oxide of silver) (Section 6.4.5) is an exception, with the catalyst incorporating only a single kind of metal. Silver is the only good epoxidation catalyst, and ethylene is the only reactant that is epoxidized in high yields.

The more common pattern of surface-catalyzed partial oxidation requires bifunctional surfaces that are most conveniently provided by mixed metal oxides. The reactants usually include olefins and O_2, and the key surface intermediates are allyls, formed by abstraction of hydrogen from the olefin. The most important example in technology is the ammoxidation of propylene to give acrylonitrile:

$$CH_2{=}CHCH_3 + \tfrac{3}{2}O_2 + NH_3 \rightarrow CH_2{=}CHCN + 3H_2O \qquad (6\text{-}47)$$

When ammonia is absent, the oxidation leads to acrolein:

$$CH_2{=}CHCH_3 + O_2 \rightarrow CH_2{=}CHCHO + H_2O \qquad (6\text{-}48)$$

A number of mixed metal oxides have been applied as industrial ammoxidation catalysts, the first and most thoroughly investigated being oxides of bismuth and molybdenum (referred to generically as bismuth molybdates) [119, 120]. Others are uranium antimonates and iron molybdates. The mixed oxides are structurally complex, each typically containing more than one crystalline phase, with the surface compositions and structures being less than well understood.

The propylene oxidation and ammoxidation chemistry is consistent with the schematic representation of the redox cycles of Fig. 6-53. The adsorption of propylene on Bi_2MoO_6 occurs at surface oxygen ions inferred from Raman spectra to bridge bismuth and molybdenum ions [121]. This site is shown schematically in Fig. 6-54 [121].

Adsorption of propylene takes place by abstraction of a hydrogen atom from the methyl group, leading to formation of a surface intermediate inferred to be an allyl. The inference is based on results of experiments with ^{14}C tracers at various positions in the propylene. The α-hydrogen abstraction to form the surface allyl is often rate determining in the selective oxidation.

The reduced surface sites are rapidly reoxidized, the reoxidation occurring by the loss of lattice oxygen from the bulk of Bi_2MoO_6 (or other Bi–Mo

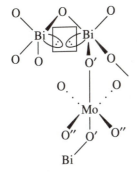

Figure 6-54
Schematic representation of the catalytic site for partial oxidation of propylene on Bi_2MoO_6 [121]. The oxygen designated O' is believed to be responsible for the α-hydrogen abstraction; that designated O", associated with Mo, is believed to be responsible for oxygen insertion. Square designates the center for O_2 reduction and dissociative chemisorption.

oxides) followed by reduction of O_2 from the gas phase and bulk transport of oxide ions to the reduced site [121]. Experiments with $^{18}O_2$ reactant showed that there was a time lag before the appearance of ^{18}O in the oxidized organic product, consistent with diffusion through the bulk of the catalyst [122]. The rapid reoxidation of the site of propylene oxidation under catalytic reaction conditions and the lack of inhibition of this reoxidation by propylene are consistent with the inference that oxygen adsorption and dissociation take place at sites that are spatially and structurally distinct from those where propylene adsorption and oxidation take place [120]. It has been hypothesized that the oxygens inserted into the allylic intermediate to give acrolein (or NH groups inserted to give acrylonitrile) are bound only to Mo ions and that the two lone pairs of electrons associated with Bi–O–Bi surface groups are responsible for reduction of O_2 and reoxidation of the catalyst [121] (Fig. 6-54). Detailed catalytic cycles have been postulated [119]. Numerous other oxidations of olefins are believed to proceed through allylic intermediates on mixed metal oxides.

The industrial catalysts are much more complex than the examples mentioned here. Among the catalysts mentioned in a patent is the following: $Co_6^{2+}Ni_2^{2+}Fe_3^{3+}Bi^{3+}(MoO_4)_{12}$ on 50 wt% SiO_2, with some P and K [123]. Silica is a support in this catalyst. The number of components is an indication that a number of new functions have been incorporated in the catalysts; these are not well understood, but they are likely related in part to increasing the lifetime of the catalyst [124].

The ammoxidation reaction is highly exothermic, and the reaction is carried out in a fluidized-bed reactor to facilitate the heat transfer and allow good temperature control, which is important for maintenance of selectivity. The silica support provides mechanical stability for the catalyst in the abrasive environment of the fluidized bed. Pressures are low (about atmospheric), and temperatures are in the range 400–500°C.

In summary, the essential goals of hydrocarbon oxidation are to give selective conversion to partially oxidized products. The catalysts are complex mixed metal oxides, chosen to have surfaces that react selectively with adsorbed organic reactants at positions where oxygen of only limited reactivity is present. The result is partially oxidized products and a reduced catalytic site; the site is reoxidized by oxygen from the bulk supplied by O_2 from the gas phase, which reacts with the catalyst at positions remote from the catalytic site for hydrocarbon oxidation. The essential goal of the choice of catalyst, reactor, and reaction conditions is to maximize selectivity. The partial oxidation catalysts are among the most complicated and least well understood of all those used in technology.

6.4.7 Catalysis by Metal Sulfides: Hydrodesulfurization and Related Petroleum Hydroprocessing Reactions

Metal sulfides, including MoS_2, WS_2, and many others, are catalytically active for numerous reactions that are also catalyzed by metals [125–130]. The important applications involve small crystallites of metal sulfides dispersed on a high-area porous support, usually γ-Al_2O_3. These catalysts find enormous use in the refining of petroleum to make clean-burning fuels. The most active metal sulfides are typically several orders of magnitude less active than the most active metals. But, in contrast to the metals, they are not poisoned by sulfur compounds and are therefore used as catalysts with feedstocks that contain sulfur. The most important application involves reactions that remove organosulfur compounds from petroleum by reaction with hydrogen to form H_2S and hydrocarbons. Naphtha is subjected to such a process, called hydrodesulfurization, to make hydrocarbon feeds suitable for catalytic reformers with their sulfur-sensitive Re–Pt/Al_2O_3 catalysts.

Petroleum typically contains about 1 wt% sulfur, with wide variations from one crude oil to another, and the sulfur compounds are concentrated in the higher-boiling fractions of the oil (Fig. 2-17). When the heavier oil is burned as a fuel, SO_x, a major air pollutant, is emitted. Excellent progress has been made in abatement of the pollution as a result of catalytic hydrodesulfurization of heavy petroleum fractions, including the residuum, the highest-boiling fraction.

The organosulfur compounds in petroleum include mercaptans (sulfides, RSH), disulfides (RSSR'), and aromatics, including thiophene, benzothiophene, dibenzothiophene, and related compounds; the latter predominate in heavy fuels. The network of reactions of hydrogen with dibenzothiophene, a representative member of this class of compounds, in the presence of a commercial hydrodesulfurization catalyst is shown in Fig. 6-55 [126]. Hydrogenation and hydrogenolysis reactions occur in parallel, the latter being virtually irreversible under practical reaction conditions and leading to formation of H_2S and biphenyl. The catalyst in the near absence of H_2S is highly selective for the hydrogenolysis, which is advantageous since hydrogen is an expensive reactant; but the selectivity decreases sharply as the partial pressure of H_2S in the reactant increases (Fig. 6-55). The reactions take place at high temperatures (about 300–350°C) and pressures (about 50–100 atm), depending on the feedstock. In commercial processes with the heaviest feedstocks, the organosulfur compounds that are most slowly converted are dibenzothiophenes substituted in the 4- and 6-positions (the positions on the rings closest to the

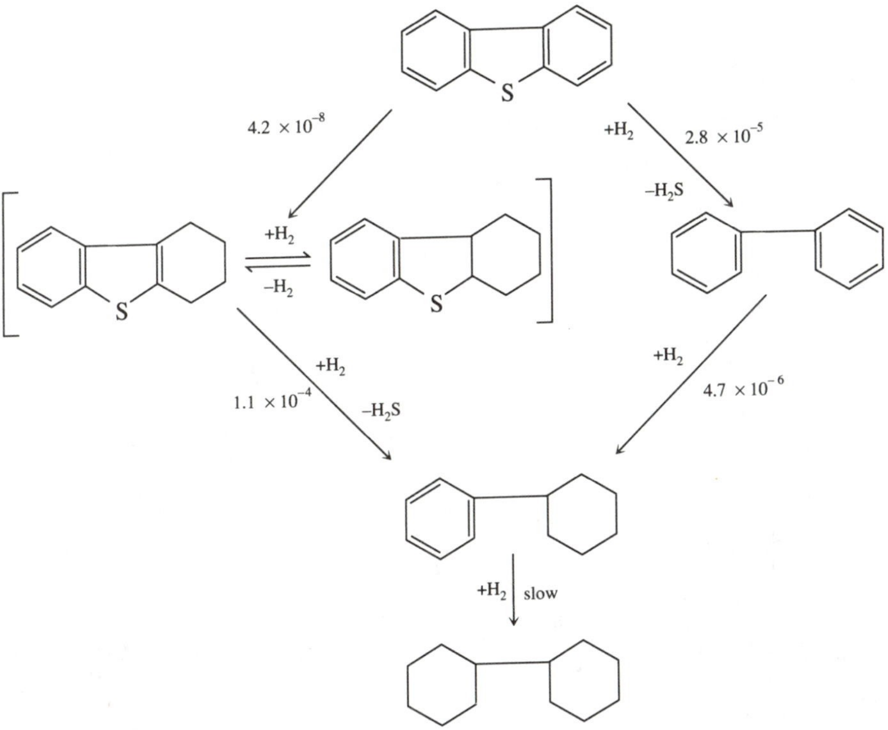

Figure 6-55
Reaction network for hydrodesulfurization and hydrogenation of dibenzothiophene catalyzed by sulfided Co–Mo/Al$_2$O$_3$ at 300°C and 102 atm [126].
Numbers next to the arrows represent the pseudo-first-order rate constants
in units of L/(g of catalyst·s) when the H$_2$S concentration is very small. Addition
of H$_2$S markedly decreases the selectivity for hydrodesulfurization.

sulfur atom). Fractional removals of sulfur are so high that even these unreactive components must be converted.

A number of reactions accompany hydrogenation and hydrodesulfurization, as summarized in the reaction networks for typical reactants shown in
Fig. 6-56. These other hydroprocessing reactions are described below.

Many metal sulfides catalyze the hydroprocessing reactions, MoS$_2$ being
one of the most active. Since MoS$_2$ is expensive, it is used in a highly dispersed
form, supported on a metal oxide, usually γ-Al$_2$O$_3$. There are other catalyst
components, including cobalt or nickel, which are promoters, and sometimes
Ni is used in combination with W rather than Mo, especially when hydrogenation is the principal goal.

Commercial catalysts are prepared by aqueous impregnation of the alumina support with soluble compounds of Mo and the promoter. Much of the
literature refers to preparations with molybdate solutions. Enough molybdenum is used to give about one monolayer on the surface. In acidic solutions,
the molybdate is present almost exclusively in the form of heptamers,

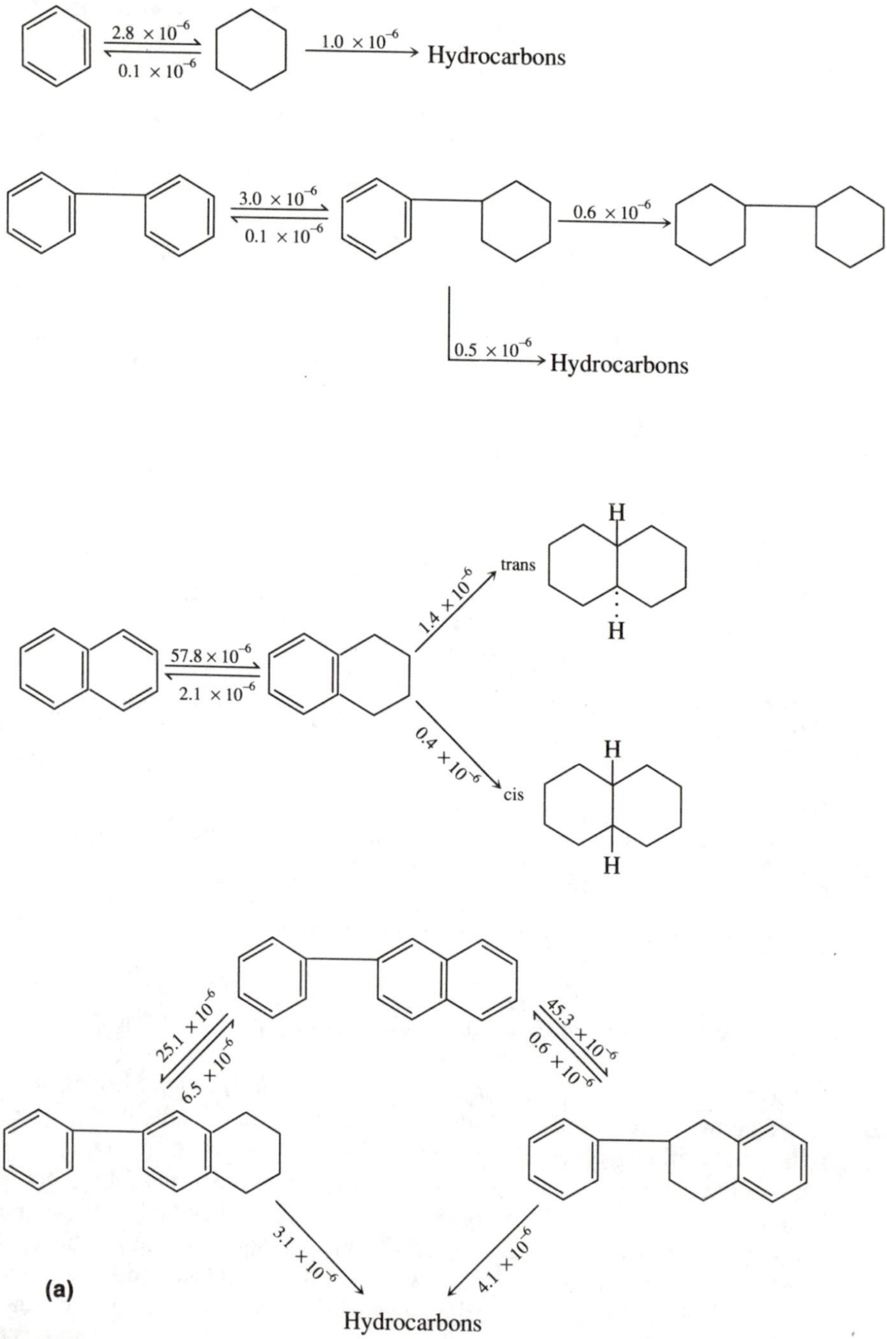

Figure 6-56
(a) (above) Reaction network for the hydrogenation of aromatic hydrocarbons catalyzed by sulfided Co–Mo/γ-Al₂O₃ at 325°C and 75 atm [131]. (b) (opposite page) Reaction network for the hydrogenation and hydrogenolysis of quinoline catalyzed by sulfided Ni–Mo/γ-Al₂O₃ at 350°C and 35 atm [132]. Numbers next to the arrows are pseudo-first-order rate constants in units of L/(g of catalyst·s) when the H₂S concentration is small but sufficient to maintain the catalysts in the sulfided forms.

408

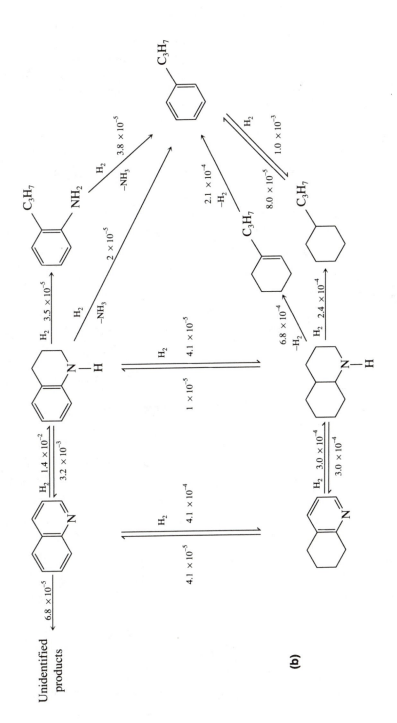

(b)

$[Mo_7O_{24}]^{6-}$, and the resulting surface species are believed to be present in islands containing presumably seven Mo ions, with Mo—O—Al bonds. There are also patches of uncovered support. The promoter Co or Ni is usually added simultaneously, but the chemistry of these preparations is not well known. Some of the promoter ions are present on the surface, but some occupy octahedral and tetrahedral sites in the bulk of the alumina support [127].

Before use, the Mo and some fraction of the Co in the catalyst are converted into the sulfide form, for example, by treatment with H_2S in H_2 or with organosulfur compounds and H_2. The catalyst must be operated in the presence of H_2 and H_2S (or precursors that give H_2S, i.e., organosulfur compounds) to prevent their being reconverted into the oxide form, which is less active.

The surface structures formed in the sulfiding step have been characterized by many spectroscopic methods and have been observed by electron microscopy. They resemble rafts of MoS_2 on the support surface, with a thickness of only a few layers. Bulk MoS_2 has a layer structure represented schematically in Fig. 6-57. The planar surface of sulfur atoms is relatively unreactive, but the edges, with exposed Mo^{2+} ions, are reactive. The cobalt and nickel promoter ions are believed to be present at the edges, perhaps as shown schematically in Fig. 6-58 [128]. This picture of the catalyst structure suggests that the catalytic activity is associated primarily with the promoter nickel more than the molybdenum sites at the MoS_2 crystallite edges; the model is simplified, and the structure is complex but rather well understood.

The promoter increases the activity of the catalyst, as is illustrated by data for hydrodesulfurization of thiophene [129] (Fig. 6-59). The nature of the surface reaction intermediates is essentially unknown, and inferences about the reaction mechanisms are based on presumed analogies to molecular organometallic chemistry [130].

Processes for hydroprocessing of heavy fossil fuels require efficient contacting of the heavy oil with hydrogen. The most efficient reactors are called

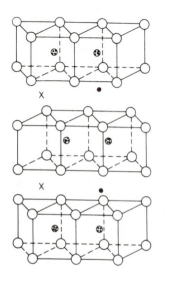

○ S

⊕ Mo

X Octahedral hole

• Tetrahedral hole

Figure 6-57

Layer structure of MoS_2. There are alternating layers of sulfur anions interspersed with layers of molybdenum cations, and there are empty layers between some of the sulfur layers, as shown.

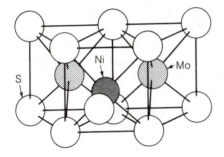

Figure 6-58
Schematic representation of nickel-containing sites at the edges of a MoS$_2$ layer on a carbon support [128].

trickle beds. In a trickle bed, the oil flows downward over particles of catalyst, wetting them but leaving spaces that are not filled with solid or liquid. These void spaces are filled with gas, mostly hydrogen, also flowing downward through the fixed-bed reactor. The hydroprocessing reactions are exothermic, and the reactor usually consists of stages, with interstage cooling of the reactants. Downstream of the reactor, the gas is separated from the oil, scrubbed to remove H$_2$S and other products, and recycled to the reactor. Trickle-bed reactors are also used for numerous other hydrogenation processes.

The hydroprocessing reactions accompanying hydrogenation and hydrodesulfurization depicted in Fig. 6-55 include hydrodenitrogenation, whereby organonitrogen compounds in the feed react with hydrogen to give NH$_3$ and hydrocarbons. Reaction networks for hydrogenation of aromatic hydrocarbons are shown in Fig. 6-56a [131], and a reaction network proposed for the hydro-

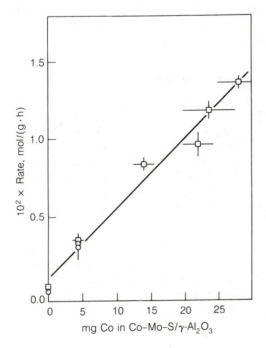

Figure 6-59
Co promotion of thiophene hydrodesulfurization [129].

processing of quinoline is shown in Fig. 6-56*b* [132]; other, slightly different networks have been suggested for the latter. The supported metal sulfide catalyst (a supported Ni–Mo rather than the supported Co–Mo mentioned above) is much less selective for nitrogen removal than for sulfur removal, and the reactions of organonitrogen compounds are typically much slower than the reactions of organosulfur compounds (except when the compounds are present in mixtures, as described in the following example). Removal of the nitrogen, which is present in lower concentrations than sulfur, consumes large amounts of expensive hydrogen.

The nitrogen removal from fossil fuels is motivated in part by the goal of removing sources of NO$_x$ pollution, but there is a more important motivation as well. A downstream process called hydrocracking is used to increase the quality of fuels by increasing the H/C ratio (by hydrogenation) with simultaneous reduction of the molecular weight (cracking). Cracking, described in Chapter 5, requires acidic sites on the catalyst. Some organonitrogen compounds (e.g., quinoline) are strong bases and adsorb strongly on these sites, greatly reducing the cracking activity. Consequently, feeds to hydrocracking reactors are prepared by hydrodenitrogenation to remove most of the inhibitors.

The hydrocracking catalysts are bifunctional, the hydrogenation function being a metal or metal sulfide and the cracking function typically being a zeolite. If the feed contains very little sulfur, the metal component is used; a commonly used catalyst consists of clusters of palladium in the cages of zeolite Y. If the feed contains substantial amounts of sulfur, a metal would be deactivated, and a supported metal sulfide such as the Co–Mo or Ni–Mo sulfides mentioned above is used. The catalyst is then a complicated composite including crystallites of zeolite and MoS$_2$-like structures promoted with Co^{2+} or Ni^{2+} ions and dispersed on a support such as γ-Al$_2$O$_3$. The catalyst has a complexity comparable to that of the cracking catalysts described in Chapter 5.

EXAMPLE 6-7   Competitive Hydroprocessing Reactions

Problem

When organosulfur compounds alone react with hydrogen in the presence of a sulfided Ni–Mo/γ-Al$_2$O$_3$ catalyst, the reactions are substantially faster than the reactions of organonitrogen compounds reacting under the same conditions [133]. However, when mixtures of organosulfur and organonitrogen compounds react with hydrogen, the organonitrogen compounds react faster, with the rates being virtually independent of the concentration of the organosulfur compounds. Explain the observations and suggest how to model the kinetics.

Solution

The results suggest competitive inhibition whereby the various reactants compete for surface catalytic sites. Evidently, the organonitrogen compounds are more strongly adsorbed than the organosulfur compounds and are able to displace them from the catalytic sites. The decrease in reactivity of the organosulfur compounds in the presence of the organonitrogen compounds is an indication of a decreased surface concentration of the former. The greater reactivity of the organonitrogen compounds in a mixture with the organosulfur compounds indicates that they occupy most of the catalytic sites, hindering reaction of the organosulfur compounds by blocking their adsorption on catalytic sites. The pattern is quite general in catalytic hydroprocessing.

The results suggest simple models based on the Langmuir assumptions of an ideal surface. If the only strongly adsorbed species are the organonitrogen compounds, then, to a first approximation, the rate of hydrodesulfurization might be represented as

$$r_{HDS} = \frac{k K_S P_S K_{H_2} P_{H_2}}{1 + K_S P_S + K_N P_N} \tag{6-49}$$

where $K_N P_N \gg (1 + K_S P_S)$. This is one of the simplest forms of equation that can explain all the observations stated above, and it is similar to what has been found to give a good quantitative representation of many data [133]. Similar equations can be written for other reactions. ∎

All the reactions mentioned in this section take place to some degree in any of the hydroprocessing operations, with the selectivity controlled by the catalyst formulation and the reaction conditions. There is another set of reactions that is important, especially in the processing of the heaviest fractions of petroleum (which are being used more and more as oil reserves are depleted). These are reactions that cause catalyst deactivation, including coke formation and hydrodemetallization. The latter is a class of reactions involving hydrogen and the organometallic components of heavy fossil fuels [134]. These components are mostly porphyrinlike structures incorporated in large (typically 5-nm diameter) colloidlike structures, aggregates of large molecules with molecular weights of roughly 1000, called asphaltenes. These are found in the heaviest fraction of petroleum (used to make asphalt). The metals are primarily nickel and vanadium. When these organometallic structures react in the presence of a catalyst in the sulfur-containing atmosphere of a hydroprocessing reactor, they form metal sulfides and organic products that may be further converted.

The vanadium and nickel sulfides form solid deposits in the reactor that partially cover the catalyst surface and deactivate it. Ultimately, they build up sufficiently to block the pores of the catalyst, choking off the transport of the reactant molecules, which is another mode of deactivation. The catalyst is also deactivated by deposits of coke.

Catalyst deactivation is a dominant issue in the application of catalytic hydroprocessing of heavy fossil fuels. Good progress has been made in the formulation of catalysts with optimum physical and chemical properties. The catalyst may have many small pores, with diameters of about 10 nm, to provide high surface area, and some much larger pores, with diameters of the order of 1000 nm, to provide access of the asphaltenes to the particle interior. The issues of intraparticle mass transport introduced in Chapter 4 play a prominent role.

The deactivation of a catalyst used in a pilot plant reactor is usually characterized in an experiment in which the feed flow rate to the reactor is held constant, and the temperature of the reactor is raised just fast enough to maintain a particular product property such as the concentration of sulfur. Results of such an experiment are shown in Fig. 6-60 [135]. The catalyst temperature was raised rather rapidly in the initial period (of probably a few weeks), and then there was a period of slow deactivation characterized by the slow, nearly linear temperature rise, followed by a period of rapid temperature rise. Understanding of the processes of catalyst deactivation has arisen from data characterizing samples of catalyst taken from the reactor at various times. The data of Fig. 6-61 were obtained with an electron microprobe; they show radial profiles of carbon, nickel, and vanadium in a cylindrical catalyst particle (formed by extrusion) as a function of the time of operation. The deposits of coke and of nickel, vanadium, and iron sulfides are nonuniform. The results show that the effectiveness factor for each of the hydrodemetallization reactions and that for coke formation are less than unity; the resistance to intraparticle diffusion of the asphaltenes and perhaps other coke precursors is significant. There is an opportunity to tailor the pore size distribution of the catalyst to optimize its performance. (There is also a motivation to use very small particles of

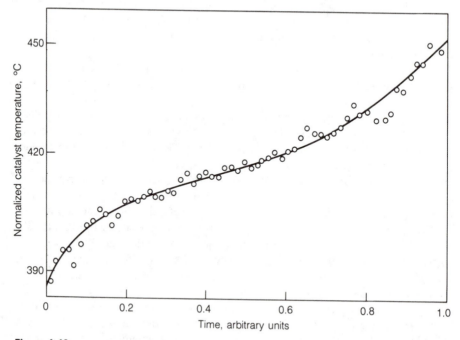

Figure 6-60
Typical deactivation profile for hydrodesulfurization of petroleum residuum [135]. The temperature of the catalyst bed was raised to compensate for the loss of activity. The product sulfur concentration was held constant. Data are incompletely specified, but they were obtained for $\frac{1}{16}$-in. cylindrical catalyst particles.

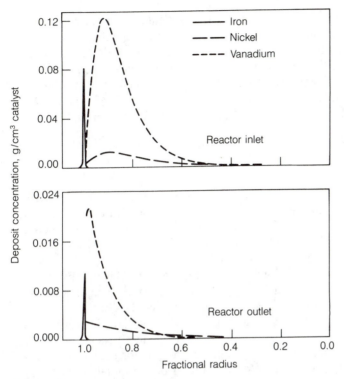

Figure 6-61
Profiles showing the concentrations of deposited Fe, V, and Ni in and
on particles of cylindrical catalyst used for hydroprocessing of petroleum residuum [135]. Analyses were done with an electron microprobe.
The catalyst particle diameter was probably about $\frac{1}{16}$ in.

catalyst. Why are such small particles not used in a trickle-bed reactor? What
reactor design would you suggest as an alternative?)

Mathematical modeling and computer calculations have been reported,
indicating that industrial researchers have been able to design catalysts with
optimum loadings of Co and Mo (or Ni and Mo) and optimum pore geometries
for various feedstocks and applications. The results remain proprietary, as is
typical of catalytic technology, but the issues become clear from a qualitative
discussion.

EXAMPLE 6-8 Lifetime of a Hydroprocessing Catalyst

Problem

To extend the lifetime of a hydroprocessing catalyst, it may be used with
another, upstream reactor (called a guard bed) containing an inexpensive
catalyst specifically chosen to accumulate deposits of metal sulfides and
to minimize the deposits in the expensive downstream catalyst. Assume
that the demetallization catalyst has a particle density of 1.0 g/cm^3 and a
pore volume of 0.5 cm^3/g and that the deposition profiles shown in Fig.

6-61 are roughly representative of a used demetallization catalyst. Use these data to do an approximate design of a demetallization reactor to operate under the conditions applied to generate the pilot plant data of the figure. Assume that the pilot plant was operated with a liquid hourly space velocity of 1.0, removing about 50 percent of the V and 50 percent of the Ni from a heavy oil containing 136 ppm of V and 48 ppm of Ni. For a feed rate of 20,000 bbl/day, what reactor volume is required for the demetallization? If the catalyst particle diameter is $\frac{1}{32}$ in., approximately how long will the catalyst last before its pores are plugged?

Solution

Assume that the reactor will operate with a liquid hourly space velocity (LHSV) of 1.0, as did the pilot plant. This is the flow rate of the liquid (under standard conditions) in units of reactor volumes per hour. The oil feed rate is given as 20,000 barrels/day, which is 833 barrels/h. Therefore, the reactor volume is 833 barrels, which is 132 m³.

To estimate the life of the catalyst before pore plugging by metal sulfide deposits, assume, as a first approximation, that the demetallization catalyst behaves like the catalyst represented by the data of Fig. 6-61. To estimate the mass of metal sulfide deposits, assume that the density of the metal sulfides is about 5 g/cm³ and that the effectiveness factor for demetallization is about 0.2 on average for the reactor (Fig. 6-61). This is small enough that the depth of penetration of the metals is virtually independent of the catalyst particle size. (Can you explain this statement in terms of the Thiele model?) The mass of metal sulfide deposits is, then, the volume of the reactor times the volume fraction filled with catalyst particles (about 0.5) times the volume fraction of the particles that is pores (0.5) times the fraction of the pore volume occupied by deposits (0.2) times the density of the deposits. This works out to be about 3.3×10^7 g, of which about 2.6×10^7 g is assumed to be Ni + V. Since about half of the metal is removed in the reactor, the amount of oil processed before plugging contains twice this amount of Ni + V. From the metal content of the oil (184 ppm), this allows a determination of the amount of oil processed, about 2.8×10^{11} g. If the oil has a specific gravity of about 1, the volume processed is about 1.8×10^6 barrels. Since the flow rate of oil is 20,000 barrels/day, the length of the run is about 1.8×10^6 barrels/ (2×10^4 barrels/day), or 90 days. ∎

In summary, the metal sulfide catalysts are active for a wide range of hydroprocessing reactions, and they are the catalysts of choice for these reactions when the reactants contain sulfur, which poisons metal catalysts. Like the metals, the metal sulfides are used in the form of highly dispersed particles on high-area supports, usually alumina, and their activities are increased by the presence of promoters. Sometimes the supports are inert, as in hydrodesulfurization catalysts, but sometimes they are catalytically active themselves, as in hydrocracking catalysts. The structures of the supported metal sulfide catalysts are complex but fairly well understood. In applications for heavy fossil fuels, the catalyst deactivation phenomena are of dominant importance and dictate the design of the physical properties of the catalyst particles. The designs are based on the principles of diffusion with reaction in porous particles described in Chapter 4.

REFERENCES

1. Somorjai, G. A., *Chemistry in Two Dimensions: Surfaces*, Cornell University Press, Ithaca, New York, 1981.
2. Somorjai, G. A., in L. L. Hegedus, ed., *Catalyst Design, Progress and Perspectives*, Wiley, New York, 1987, p. 11.
3. Binnig, G., Rohrer, H., Gerber, C., and Stoll, E., *Surf. Sci.*, **144**, 321 (1984).
4. Van Hove, M. A., Weinberg, W. H., and Chen, C.-M., *Low Energy Electron Diffraction: Experiment, Theory, and Surface Structure Determination*, Springer, New York, 1986.
5. Weisz, P. B., *CHEMTECH*, **3**, 498 (1973).
6. Lecloux, A. J., in J. R. Anderson and M. Boudart, eds., *Catalysis—Science and Technology*, Vol. 2, Springer, Berlin, 1981, p. 171.
7. Boehm, H.-P., and Knözinger, H., in J. R. Anderson and M. Boudart, eds., *Catalysis—Science and Technology*, Vol. 4, Springer, Berlin, 1983, p. 39.
8. (a) Lamb, H. H., Gates, B. C., and Knözinger, H., *Angew. Chem. Int. Ed. Engl.*, **27**, 1127 (1988); (b) Lamb, H. H., and Gates, B. C., *J. Chem. Soc., Chem. Commun.*, **1990**, 1296.
9. Oberlander, R. K., in B. E. Leach, ed., *Applied Industrial Catalysis*, Academic Press, Vol. 3, New York, 1984, p. 63.
10. Poisson, R., Brunelle, J.-P., and Nortier, P., in A. B. Stiles, ed., *Catalyst Supports and Supported Catalysts*, Butterworths, Boston, 1987, p. 11.
11. Knözinger, H., and Ratnasamy, P., *Catal. Rev.—Sci. Eng.*, **17**, 31 (1978).
12. Ertl, G., and Koch, J., *Z. Naturforsch.*, **25a**, 1096 (1970).
13. Brunauer, S., Emmett, P. H., and Teller, E., *J. Am. Chem. Soc.*, **60**, 309 (1938).
14. Brunauer, S., and Emmett, P. H., *J. Am. Chem. Soc.*, **59**, 2682 (1937).
15. Christmann, K., Behm, R. J., Ertl, G., Van Hove, M. A., and Weinberg, W. H., *J. Chem. Phys.*, **70**, 4168 (1979).
16. Koetzle, T. F., Muller, J., Tipton, D. L., Hart, D. W., and Bau, R., *J. Am. Chem. Soc.*, **101**, 5631 (1979).
17. Behm, R. J., Christmann, K., Ertl, G., and Van Hove, M. A., *J. Chem. Phys.*, **73(6)**, 2984 (1980).
18. Bradshaw, A. M., and Hoffmann, F. M., *Surf. Sci.*, **72**, 513 (1978).
19. Conrad, H., Ertl, G., and Küppers, J., *Surf. Sci.*, **76**, 323 (1978).
20. Engel, T., and Ertl, G., *Adv. Catal.*, **28**, 1 (1979).

21. Koestner, R. J., Frost J. C., Stair, P. C., Van Hove, M. A., and Somorjai, G. A., *Surf. Sci.*, **117**, 491 (1982).
22. Koestner, R. J., Van Hove, M. A., and Somorjai, G. A., *CHEMTECH*, **13**, 376 (1983).
23. Davis, S. M., Zaera, F., and Somorjai, G. A., *J. Catal.*, **77**, 439 (1982).
24. Christmann, K., Ertl, G., and Shimizu, H., *J. Catal.*, **61**, 397 (1980).
25. Shimizu, H., Christmann, K., and Ertl, G., *J. Catal.*, **61**, 412 (1980).
26. Knözinger, H., and Kaerlein, C.-P., *J. Catal.*, **25**, 436 (1972).
27. Hall, W. K., *Acc. Chem. Res.*, **8**, 257 (1975).
28. Burwell, R. L., *CHEMTECH*, **4**, 370 (1974).
29. Iwasawa, Y., *Adv. Catal.*, **35**, 187 (1987).
30. Allum, K. G., Hancock, R. D., McKenzie, S., and Pitkethly, R. C., *Proc. 5th Int. Congr. Catal.* (Palm Beach), Vol. 1, p. 477, 1972.
31. Evnin, A. B., Rabo, J., and Kasai, P. H., *J. Catal.*, **30**, 109 (1973).
32. Kuznetsov, B. N., Startsev, A. N., and Yermakov, Y. I., *J. Mol. Catal.*, **8**, 135 (1980).
33. Yermakov, Y. I., Kuznetsov, B. N., and Zakharov, V. A., *Catalysis by Supported Complexes*, Elsevier, Amsterdam, 1981.
34. McDaniel, M. P., *Adv. Catal.*, **33**, 47 (1985).
35. Karol, F. J., *Catal. Rev.—Sci. Eng.*, **26**, 557 (1984).
36. Ballard, D. G. H., *Adv. Catal.*, **22**, 263 (1973).
37. (a) Kirlin, P. S., van Zon, F. B. M., Koningsberger, D. C., and Gates, B. C., *J. Phys. Chem.*, **94**, 8439 (1990); (b) Chang, J.-R., Gron, L. U., Honji, A., Sanchez, K. M., and Gates, B. C., *ibid.*, to be published.
38. Kirlin, P. S., Knözinger, H., and Gates, B. C., *J. Phys. Chem.*, **94**, 8451 (1990).
39. Pino, P., and Mulhaupt, R., *Angew. Chem. Int. Ed. Engl.*, **19**, 857 (1980).
40. Natta, G., and Danusso, F., eds., *Stereoregular Polymers and Stereospecific Polymerizations*, Vols. 1 and 2, Pergamon, Oxford, 1967.
41. Cossee, P., *J. Catal.*, **3**, 80 (1964); Arlman, E. J., and Cossee, P., *J. Catal.*, **3**, 99 (1964).
42. Cossee, P., in A. D. Ketley, ed., *The Stereochemistry of Macromolecules*, Vol. 1, p. 145, Marcel Dekker, New York, 1967.
43. Rodriguez, L. A. M., and Gabant, J. A., *J. Polym. Sci. Part C*(4), 125 (1963).
44. Goodall, B., paper presented at International Symposium, "Transition-Metal Catalyzed Polymerizations: Unsolved Problems," Michigan, 1981.
45. Burwell, R. L., Jr., Haller, G. L., Taylor, K. C., and Read, J. F., *Adv. Catal.*, **20**, 1 (1969).
46. Taylor, H. S., *Proc. R. Soc.* (London) *A*, **108**, 105 (1925).
47. (a) Fitzharris, W. D., Katzer, J. R., and Manogue, W. H., *J. Catal.*, **76**, 369 (1982); (b) Bartholomew, C. H., Agrawal, P, K., and Katzer, J. R., *Adv. Catal.*, **31**, 135 (1982).
48. Alstrup, I., and Andersen, N. T., *J. Catal.*, **104**, 466 (1987).
49. Ertl, G., in B. C. Gates, L. Guczi, and H. Knözinger, eds., *Metal Clusters in Catalysis*, Elsevier, Amsterdam, 1987, p. 577.

50. Rootsaert, W. J. M., and Sachtler, W. M. H., *Z. Phys. Chem. N.F.*, **26,** 16 (1960).

51. Taylor, K. C., in J. R. Anderson and M. Boudart, eds., *Catalysis— Science and Technology*, Vol. 5, Springer, Berlin, 1984, p. 119.

52. Engel, T., and Ertl, G., in D. P. Woodruff, ed., *The Chemical Physics of Solid Surfaces and Heterogeneous Catalysis*, Vol. 4, Elsevier, Amsterdam, 1982, p. 73.

53. Jennings, R., ed., *Ammonia Synthesis—Theory and Practice*, Pergamon Press, to be published.

54. Topham, S. A., in J. R. Anderson and M. Boudart, eds., *Catalysis— Science and Technology*, Vol. 7, Springer, Berlin, 1985, p. 1.

55. Nielson, A., *Catal. Rev.—Sci. Eng.*, **23,** 17 (1981).

56. Spencer, N. D., Schoonmaker, R. C., and Somorjai, G. A., *J. Catal.*, **74,** 129 (1982).

57. (a) Ertl, G., *Proc. 7th Int. Congr. Catal. (Tokyo)*, Part A, p. 21, 1980; (b) Ertl, G., in ref. 53.

58. Bozso, F., Ertl, G., Grunze, M., and Weiss, M., *J. Catal.*, **49,** 18 (1977).

59. Bozso, F., Ertl, G., and Weiss, M., *J. Catal.*, **50,** 519 (1977).

60. Emmett, P. H., and Brunauer, S., *J. Am. Chem. Soc.*, **56,** 35 (1934).

61. Boudart, M., *Catal. Rev.—Sci. Eng.*, **23,** 1 (1981).

62. Ertl, G., Weiss, M., and Lee, S. B., *Chem. Phys. Lett.*, **60(3),** 391 (1979).

63. Boudart, M., and Djéga-Mariadassou, G., *Kinetics of Heterogeneous Catalytic Reactions*, Chap. 4. Princeton University Press, Princeton, N.J., 1984.

64. Temkin, M. I., and Pyzhev, V., *Acta Physicochim. USSR,* **12,** 217 (1940).

65. Boudart, M., *AIChE J.*, **2,** 62 (1956); *Ind. Eng. Chem. Fundam.*, **25,** 656 (1986); *Catal. Lett.*, **1,** 21 (1988).

66. Stoltze, P., and Norskov, J. K., *Phys. Rev. Lett.*, **55,** 2502 (1985); Stoltze, P., *Phys. Scripta,* **36,** 824 (1987).

67. Bare, S. R., Strongin, D. R., and Somorjai, G. A., *J. Phys. Chem.*, **90,** 4726 (1986); Strongin, D. R., Bare, S. R., and Somorjai, G. A., *J. Catal.*, **103,** 289 (1987); Strongin, D. R., and Somorjai, G. A., *Catal. Lett.*, **1,** 61 (1988).

68. (a) LeBlanc, J. R., Madhavan, S., and Porter, R. E., in *Kirk-Othmer Encyclopedia of Chemical Technology*, 3rd edition, Vol. 2, Wiley, New York, 1978, p. 470; (b) Emmett, P. H., in E. Drauglis and R. I. Jaffee, eds., *The Physical Basis for Heterogeneous Catalysis*, Plenum, New York, 1975, p. 3.

69. Freifelder, M., *Catalytic Hydrogenation in Organic Synthesis*, Wiley, New York, 1978.

70. Sachtler, W. M. H., *Proc. 8th Int. Congr. Catal.* (Berlin), Vol. 1, p. 151, 1984.

71. Bell, A. T., *Proc. 9th Int. Congr. Catal.* (Calgary), Vol. 5, p. 134, 1988.

72. Beebe, T. P., Jr., and Yates, J. T., Jr., *J. Am. Chem. Soc.*, **108,** 663 (1986).

73. Hattori, T., and Burwell, R. L., Jr., *J. Phys. Chem.*, **83,** 241 (1979).

74. Somorjai, G. A., *Proc. 8th Int. Congr. Catal.* (Berlin), Vol. 1, p. 113, 1984.
75. Davis, S. M., Zaera, F., and Somorjai, G. A., *J. Catal.*, **85**, 206 (1984).
76. Gault, F., *Adv. Catal.*, **30**, 1 (1981).
77. Winterbottom, J. M., in *Catalysis* (a Specialist Periodical Report), The Royal Society of Chemistry, London, 1981, p. 141.
78. Knözinger, H., in S. Patai, ed., *The Chemistry of Functional Groups. The Chemistry of the Hydroxyl Group*, Interscience, New York, 1971, p. 642.
79. Knözinger, H., Bühl, H., and Kochloefl, K., *J. Catal.*, **24**, 57 (1972).
80. Kokes, R. J., and Dent, A. L., *Adv. Catal.*, **22**, 1 (1972).
81. Eischens, R. P., Pliskin, W. A., and Low, M. J. D., *J. Catal.*, **1**, 180 (1962).
82. Anderson, J. R., *Structure of Metallic Catalysts*, Academic Press, London, 1975.
83. Chakraborti, S., Datye, A. K., and Long, N. J., *J. Catal.*, **108**, 444 (1987).
84. Logan, A. D., Braunschweig, E. J., and Datye, A. K., *Langmuir*, **4**, 827 (1988).
85. Koningsberger, D. C., and Prins, R., eds., *X-Ray Absorption: Principles, Applications, Techniques of EXAFS, SEXAFS, and XANES*, Wiley, New York, 1988.
86. Yates, D. J. C., Murrell, L. L., and Prestridge, E. B., *J. Catal.*, **57**, 41 (1979).
87. Fung, A. S., Tooley, P. A., Koningsberger, D. C., Kelley, M. J., and Gates, B. C., *J. Phys. Chem.*, **95**, 225 (1991).
88. Cotton, F. A., and Walton, R. A., *Multiple Bonds Between Metal Atoms*, Wiley, New York, 1982.
89. Stevenson, S. A., Dumesic, J. A., Baker, R. T. K., and Ruckenstein, E., eds., *Metal-Support Interactions in Catalysis, Sintering, and Redispersion*, Van Nostrand Reinhold, New York, 1987.
90. Van't Blik, H. F. J., Van Zon, J. B. A. D., Huizinga, T., Vis, J. C., Koningsberger, D. C., and Prins, R., *J. Am. Chem. Soc.*, **107**, 3139 (1985).
91. Haller, G. L., and Resasco, D. E., *Adv. Catal.*, **36**, 173 (1989).
92. Stevenson, S. A., Raupp, G. B., Dumesic, J. A., Tauster, S. J., and Baker, R. T. K., in *Metal-Support Interactions in Catalysis, Sintering, and Redispersion*, S. A. Stevenson, J. A. Dumesic, R. T. K. Baker, and E. Ruckenstein, eds., Van Nostrand Reinhold, New York, 1987, p. 3.
93. Christmann, K., Ertl, G., and Pignet, T., *Surf. Sci.*, **54**, 365 (1976).
94. Via, G. H., Sinfelt, J. H., and Lytle, F. W., *J. Chem. Phys.*, **71**, 690 (1979).
95. Kip, B. J., Duivenvoorden, F. B. M., Koningsberger, D. C., and Prins, R., *J. Catal.*, **105**, 26 (1987).
96. Ruckenstein, E., in S. A. Stevenson, J. A. Dumesic, R. T. K. Baker, and E. Ruckenstein, eds., *Metal-Support Interactions in Catalysis, Sintering, and Redispersion*, Van Nostrand Reinhold, New York, 1987, p. 141.

97. Boudart, M., *J. Mol. Catal.,* **30,** 27 (1985).
98. Goodman, D. W., *Surf. Sci.,* **123,** L679 (1982).
99. Carter, J. T., Cusumano, J. A., and Sinfelt, J. H., *J. Phys. Chem.,* **70,** 2257 (1966).
100. Sinfelt, J. H., Carter, H. L., and Yates, D. J. C., *J. Catal.,* **24,** 283 (1972).
101. Sachtler, W. M. H., and Van Santen, R. A., *Adv. Catal.,* **26,** 69 (1977); Martin, G. A., *Catal. Rev.—Sci. Eng.,* **30,** 519 (1988).
102. Helms, C. R., *J. Catal.,* **36,** 114 (1975).
103. Sinfelt, J. H., *Bimetallic Catalysts, Discoveries, Concepts, and Applications,* Wiley, New York, 1983.
104. Van Santen, R. A., and Kuipers, H. P. C. E., *Adv. Catal.,* **35,** 265 (1987).
105. Sajkowski, D. J., and Boudart, M., *Catal. Rev.—Sci. Eng.,* **29,** 325 (1987).
106. Nielsen, R. P., and La Rochelle, J. H., U.S. Patent 4,356,312 (1982).
107. Mross, W.-D., *Catal. Rev.—Sci. Eng.,* **25,** 591 (1983).
108. Cawse, J. N., Henry, J. P., Swartzlander, M. W., and Wadia, P. H., in *Kirk-Othmer Encyclopedia of Chemical Technology,* Vol. 9, Wiley, New York, 1980, p. 432.
109. Kanoh, H., Nishimura, T., and Ayame, A., *J. Catal.,* **57,** 372 (1979).
110. Spath, H. T., Tomazic, G. S., Wurm, H., and Torkar, K., *J. Catal.,* **26,** 18 (1972).
111. Berty, J. M., in B. E. Leach, ed., *Applied Industrial Catalysis,* Vol. 1, Academic Press, New York, 1983, p. 207.
112. Sterba, M. J., and Haensel, V., *Ind. Eng. Chem. Prod. Res. Dev.,* **15,** 2 (1976).
113. Mills, G. A., Heinemann, H., Milliken, T. H., and Oblad, A. G., *Ind. Eng. Chem.,* **45,** 134 (1953).
114. Weisz, P. B., and Swegler, E. W., *Science,* **126,** 31 (1957).
115. Kluksdahl, H. E., U.S. Patent 3,415,737 (1968).
116. Meitzner, G., Via, G. H., Lytle, F. W., and Sinfelt, J. H., *J. Chem. Phys.,* **87,** 6354 (1987).
117. Bertolacini, R. J., and Pellett, R. J., in B. Delmon and G. F. Froment, eds., *Catalyst Deactivation,* Elsevier, Amsterdam, 1980, p. 72.
118. Ramage, M. P., Graziani, K. R., Schipper, P. H., Krambeck, F. J., and Choi, B. C., *Adv. Chem. Eng.,* **13,** 193 (1987).
119. Grasselli, R. K., and Burrington, J. D., *Adv. Catal.,* **30,** 133 (1981).
120. Gates, B. C., Katzer, J. R., and Schuit, G. C. A., *Chemistry of Catalytic Processes,* McGraw-Hill, New York, 1979, Chapter 4.
121. Glaeser, L. C., Brazdil, J. F., Hazle, M. A., Mehicic, M., and Grasselli, R. K., *J. Chem. Soc. Faraday Trans. I,* **81,** 2903 (1985).
122. Keulks, G. W., and Krenzke, L. D., *Proc. 6th Int. Congr. Catal.* (London), Vol. 2, p. 806, 1977.
123. Grasselli, R. K., Heights, G. H., and Callahan, J. L., U.S. Patent 3,414,631 (1968); Grasselli, R. K., Heights, G. H., and Hardman, H. F., U.S. Patent 3,642,930 (1972).

124. Schuit, G. C. A., and Gates, B. C., *CHEMTECH*, **13**, 693 (1983).
125. Weisser, O., and Landa, S., *Sulphide Catalysts, Their Properties and Applications*, Pergamon, Oxford, 1973.
126. Houalla, M., Nag, N. K., Sapre, A. V., Broderick, D. H., and Gates, B. C., *AIChE J.*, **24**, 1015 (1978).
127. Knözinger, H., *Proc. 9th Int. Congr. Catal.* (Calgary), Vol. 5, p. 20, 1988.
128. Bouwens, S. M. A. M., Koningsberger, D. C., De Beer, V. H. J., Louwers, S. P. A., and Prins, R., *Catal. Lett.*, **5**, 273 (1990).
129. Candia, R. H., Topsøe, H., and Clausen, B. S., *Proc. 9th Iberoamer. Symp. Catal.* (Lisbon), 1984, p. 211.
130. Angelici, R. J., *Acc. Chem. Res.*, **21**, 387 (1988).
131. Sapre, A. V., and Gates, B. C., *Ind. Eng. Chem. Proc. Des. Dev.*, **20**, 68 (1981).
132. Sundaram, K. M., Katzer, J. R., and Bischoff, K. B., unpublished results.
133. Girgis, M. J., and Gates, B. C., *Ind. Eng. Chem. Res.*, in press, 1991.
134. Quann, R. J., Ware, R. A., Hung, C.-W., and Wei, J., *Adv. Chem. Eng.*, **14**, 95 (1988).
135. Tamm, P. W., Bridge, A. G., and Harnsberger, H. F., *Ind. Eng. Chem. Proc. Des. Dev.*, **20**, 262 (1981).
136. Dent, A. L., and Kokes, R. J., *J. Phys. Chem.*, **74**, 3653 (1970).
137. Emmett, P. H., and Brunauer, S., *J. Am. Chem. Soc.*, **59**, 1553 (1937).

FURTHER READING

Ultrahigh vacuum surface science is introduced in Somorjai's book [1]. S. J. Gregg and K. S. W. Singh's *Adsorption, Surface Area, and Porosity*, 2nd edition, Academic Press, London, 1967, is a good introduction. *Physical and Chemical Aspects of Adsorbents and Catalysts*, edited by B. G. Linsen, J. M. H. Fortuin, C. Okkerse, and J. J. Steggerda, Academic Press, London, 1967, is also helpful. Good background material about the preparation of metal-oxide-supported catalysts is given in a review by J. P. Brunelle, *Pure Appl. Chem.*, **50**, 1211 (1978). Kinetics of surface-catalyzed reactions is summarized in the book by Boudart and Djéga-Mariadassou [63]. Olefin polymerization is the subject of R. P. Quirk's *Transition Metal Catalyzed Polymerization*, Cambridge University Press, 1982, and K. Soga and T. Keii's *Catalytic Polymerization of Olefins*, Elsevier, Amsterdam, 1986. Sinfelt's book [103] gives good case studies in the characterization of supported metal catalysts. Related books are cited in the reading list at the end of this volume.

PROBLEMS

6-1 If a particle consisting of porous γ-Al_2O_3 has a void fraction of 0.5 and a surface area of 200 m^2/g, what is the approximate average size of the aggregated primary particles constituting the solid?

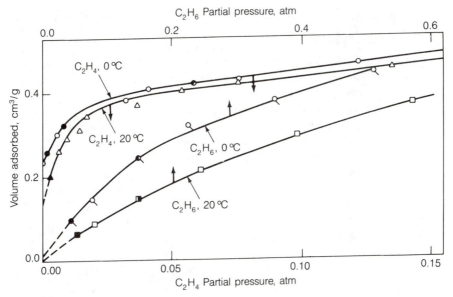

Figure 6-62
Adsorption isotherms for ethylene and ethane on ZnO [136]. The open symbols represent adsorption and the closed symbols desorption.

6-2 Adsorption isotherms for ethylene and ethane on zinc oxide are shown in Fig. 6-62, and enthalpies of adsorption of these compounds are shown in Fig. 6-63 [136]. Explain the shapes of the curves, and use the principles of thermodynamics to show how the data in the second figure are determined from the data in the first.

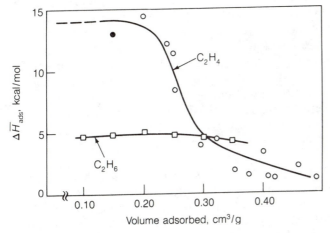

Figure 6-63
Enthalpies of adsorption of ethylene and ethane on ZnO [136].

6-3 Design a metal-oxide-supported bifunctional catalyst for the Aldox process.

6-4 A catalyst referred to as "silicated alumina" has been made by the reaction of $Si(OC_2H_5)_4$ with the surface of γ-Al_2O_3 in the presence of a nonaqueous solvent. The catalyst incorporates a submonolayer coverage of Si-containing groups on the Al_2O_3 surface and has been advocated as useful for the skeletal isomerization of but-1-ene. What are some of the plausible surface structures to account for the catalytic activity? What chemistry explains the synthesis of these surface species?

6-5 The process for synthesis of methanol from CO and H_2 and the process for synthesis of ammonia from N_2 and H_2 have a number of characteristics in common. Considering thermodynamics, kinetics, and other important criteria, list the important common criteria for process design and the important common characteristics of the two process flow sheets. Often, methanol and ammonia plants are located next to each other. Why?

6-6 Samples of MgO that have been exposed to air have infrared spectra indicative of carbonates, but these bands are not characteristic of SiO_2; γ-Al_2O_3 is intermediate in character. Explain.

6-7 What structures do you expect for (a) ethanol adsorbed on partially dehydroxylated MgO; (b) pyridine adsorbed on partially dehydroxylated γ-Al_2O_3; (c) $H_4Os_4(CO)_{12}$ adsorbed on largely dehydroxylated MgO; and (d) CO adsorbed on partially dehydroxylated γ-Al_2O_3?

6-8 Consider the Temkin–Pyzhev equation for the kinetics of ammonia synthesis. Show that the rate of reaction is predicted to be zero at equilibrium. Where does the reaction equilibrium constant appear in this equation?

6-9 When coal or petroleum is used to generate synthesis gas $(CO + H_2)$ as a source of hydrogen for ammonia synthesis, thorough feed preparation is needed to remove sulfur. Why? How can the sulfur be removed?

6-10 Data of Emmett and Brunauer [68b, 137] showing the transient adsorption of nitrogen on a doubly promoted iron catalyst at 1 atm and various temperatures are presented in Fig. 6-64. Use these data to estimate the initial rate of ammonia synthesis from a stoichiometric mixture of H_2 and N_2 at 1 atm and 400°C. Make use of these data for an order-of-magnitude estimate of the initial rate at 100 atm.

6-11 Many of the concepts of surface catalysis and even of chemical equilibrium originated in investigations of ammonia synthesis catalysis. Consult the reviews by Topham [54] and Emmett [68b] and enumerate these concepts.

6-12 Until recently, when selectivities of about 80 percent have been reported with some frequency, many researchers had accepted the working hypothesis that a selectivity for ethylene oxide formation from ethylene and oxygen (defined as the fraction of the carbon in ethylene converted to ethylene oxide) could not exceed $\frac{6}{7}$. This led some researchers to favor a mechanism whereby adsorbed dioxygen species reacted with ethylene

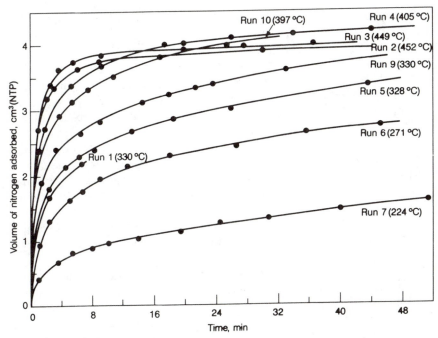

Figure 6-64
Nitrogen adsorption by a doubly promoted iron catalyst at 1 atm and various temperatures [68b].

to give ethylene oxide, leaving monatomic oxygen on the surface that could react with ethylene to give CO_2. Explain the logic of this reasoning.

6-13 According to a rule of thumb, the dispersion D of metal in a supported transition metal catalyst is related approximately to the average metal particle diameter d, as follows: $d = 1/D$, where d is in nm. How good is this approximation for hemispherical particles of Pt? Plot the estimate from this rule of thumb against your estimate from crystallographic data for particles 1–5 nm in diameter.

6-14 As this book was being written, the detrimental impact of freons on the ozone layer protecting the earth's surface from excessive ultraviolet radiation had become clear enough to lead to political action to restrict freons. Research on the manufacture of substitute freons was leading to the discovery of new catalysts. Find out what the substitutes for freon are and how they are made.

6-15 A. B. McEwen, W. F. Maier, R. H. Fleming, and S. M. Baumann [*Nature*, **329**, 531 (1987)] observed that a number of reactions of hydrocarbons with hydrogen, including H–D exchange and olefin hydrogenation, are catalyzed not only by the surface of platinum but also by a surface of platinum covered by a layer of silica, which was apparently free of pores that would allow transport of hydrocarbon molecules to the underlying metal surface. How can you explain this observation? What does it imply about the nature of the catalytic sites?

6-16 Two detrimental effects have been observed when silica-supported catalysts were used in a flow reactor at temperatures exceeding about 600°C: (1) solid deposits formed downstream of the reactor, plugging a heat exchanger, and (2) the catalyst lost much of its physical strength and formed powder. What properties of silica explain these observations? What would you expect the composition of the deposits to be?

6-17 When alumina is made from bauxite by the important Bayer process, which involves dissolution of the ore in strong base to form aluminum hydroxide, the resulting alumina contains substantial amounts of sodium and iron. Why? Suggest a route to alumina synthesis that gives a product free of these impurities.

6-18 Check the physical properties of isotactic and atactic polypropylene and suggest how to separate the undesired atactic from the desired isotactic product.

6-19 Examine Fig. 6-52 and identify the compounds constituting the following lumps and the products of their conversions in the reforming process: C_6 naphthene; C_7 aromatic; C_8 aromatic; and C_6 paraffin. Identify the classes of reactions leading to the formation and to the conversion of these lumps.

6-20 A typical hydrodesulfurization catalyst has about enough molybdenum in it to form a monolayer of the metal oxide on the γ-Al_2O_3 support, which has been estimated to be about 6 Mo atoms/nm^2. If the support surface area is 150 m^2/g, about how much Mo is present in the catalyst? If, upon sulfiding of the catalyst, the Mo is converted to MoS_2 rafts on the support that are estimated to be two layers thick, what fraction of the support surface is covered with the MoS_2?

6-21 Kinetics of the hydrogenation and simultaneous hydrogenolysis of dibenzothiophene (DBT) has been measured at 275–325°C and represented with the following rate equations [Broderick, D. H., and Gates, B. C., *AIChE J.*, **27**, 663 (1981)]:

$$r_{hyd} = \frac{kK_{DBT}K_{H_2}C_{DBT}C_{H_2}}{1 + K_{DBT}C_{DBT}} \qquad (6\text{-}50)$$

where $kK_{H_2} = 3.75 \times 10^{10} \exp(-1.2 \times 10^8/RT)$ mol/(g of catalyst·s) and $K_{DBT} = 2.0 \exp(1430/RT)$L/mol;

$$r_{HDS} = \frac{kK_{DBT}K_{H_2}C_{DBT}C_{H_2}}{(1 + K_{DBT}C_{DBT} + K_{H_2S}C_{H_2S})^2(1 + K_{H_2}C_{H_2})} \qquad (6\text{-}51)$$

where $k = (7.87 \times 10^{11}) \exp(-30,100/RT)$ mol/(g of catalyst·s), $K_{DBT} = (1.8 \times 10^{-1}) \exp(4540/RT)$ L/mol, $K_{H_2} = (4.0 \times 10^{-3}) \exp(8360/RT)$

L/mol, $K_{H_2S} = (7.0 \times 10^{-1})$ exp$(5260/RT)$ L/mol, and T is in K and R in cal/(mol·K).

Use these results to determine the selectivity of the catalyst for hydrodesulfurization and summarize the results in a figure showing the rate of hydrogenolysis divided by the rate of hydrogenation plus the rate of hydrogenolysis as a function of the reactant concentration and the H_2S concentration. Explain the advantage of scrubbing the recycle gas to remove H_2S. How does the temperature affect the selectivity for hydrogenolysis? What are the processing implications?

6-22 Topsøe, H., Topsøe, N., Bohlbro, H., and Dumesic, J. A. [*Proc. 7th Int. Congr. Catal.* (Tokyo), Part A, p. 247, 1980] measured rates of ammonia synthesis catalyzed by Fe/MgO and other iron catalysts. For the range of catalysts, including Fe/MgO with various average Fe particle sizes, the rate of the catalytic reaction was found to be virtually the same for all the catalysts when expressed as a turnover frequency, molecules/(surface site·s), with the number of surface sites determined by chemisorption of N_2 at 300°C. Turnover frequencies based on chemisorption of H_2 and of CO, on the other hand, varied by orders of magnitude. Explain the results and their practical significance.

6-23 In processes including ammonia oxidation (used in the manufacture of nitric acid), the metal catalyst is used in the form of a wire screen (or gauze) and not as supported metal crystallites. Can you tell why the screen is the preferred form?

6-24 Devise a simple, low-cost apparatus for determining the activities of a set of catalysts for ammonia synthesis at atmospheric pressure. Include a device for measurement of the ammonia formation rate.

6-25 The kinetics of hydrogenolysis of ethane catalyzed by metals has been found to be of the form

$$r = kP_{C_2H_6}^m P_{H_2}^n \tag{6-52}$$

where $m \cong 1$ and n has values between -2 and $\frac{1}{2}$, depending on the metal. The observation has been explained in terms of a sequence of reactions involving dehydrogenation of ethane on the surface to form a species C_2H_x followed by the rate-determining hydrogenation of adsorbed C_2H_x [Cimino, A., Boudart, M., and Taylor, H. S., *J. Phys. Chem.*, **58**, 796 (1954)]. Show that the postulated sequence of reactions is consistent with the observed kinetics.

6-26 Knözinger and Stolz [*Ber. Bunsenges. Phys. Chem.*, **75**, 1055 (1971)] observed that 4-methyl pyridine was more strongly bound to Lewis acid sites of alumina than pyridine. However, 2,4,6-trimethyl pyridine was quantitatively displaced by pyridine. Explain these results. What do they imply about the usefulness of the substituted pyridines for probing the acidic sites on surfaces?

6-27 Using data presented in this chapter, estimate the area occupied by each of the following adsorbates in a monolayer on silica gel: argon, CO, CO_2, O_2, and n-C_4H_{10}.

6-28 Successful chemisorption experiments require rigorously clean surfaces. Consider an experiment with a single crystal catalyst having an area of 1 cm^2 in a vacuum chamber. Estimate the time required to give one monolayer of coverage with a strongly adsorbed contaminant for pressures of 10^{-3}, 10^{-6}, 10^{-9}, and 10^{-12} Torr. If the vacuum chamber has a volume of 10 L, what pressure of the contaminant is sufficient for monolayer coverage of the catalyst?

6-29 Small amounts of Pt/Al_2O_3 or Pt on some other amorphous support are included in mixtures with catalysts for cracking of gas oil to make gasoline. The purpose is to reduce emissions of CO. Explain the role of these catalyst components. Find out what components are added to cracking catalysts to minimize the emissions of SO_x; how do they work?

6-30 Ammonia synthesis catalysts are strongly poisoned by water. What consequences does this poisoning have for process design and operation? Consider the possible sources of feed impurities.

6-31 Early catalysts for isomerization of n-butane to give isobutane were combinations of HCl and $AlCl_3$ or supported analogues. More modern catalysts for this reaction incorporate a metal such as platinum on the support and are longer-lived. What is the role of the platinum?

6-32 Conceive a process and choose the catalysts to convert n-butane and methane to methyl t-butyl ether.

6-33 Ni/alumina catalysts are used for reactions including hydrogenation of fatty acids and olefins and for conversion of CO and H_2 to methane. The catalysts used for the former reaction may contain 20 wt% Ni. Assume that 20 g of such a catalyst is to be prepared with $Ni(NO_3)_2$ by the incipient wetness technique, whereby just enough solution is used to fill the pores of the support. If the pore volume of the alumina is 0.6 cm^3/g, how much $Ni(NO_3)_2$ solution should be used and what is the required concentration? (Problem provided by Prof. J. A. Schwarz.)

6-34 A Cu/alumina catalyst was prepared by the incipient wetness method with 10 g of the support and 100 cm^3 of aqueous 0.10-molar $Cu(NO_3)_2$. The support adsorbed 80 percent of the copper ions. What was the copper content of the catalyst? (Problem provided by Prof. J. A. Schwarz.)

6-35 Professor V. N. Gryaznov conceived a catalytic membrane reactor in which butene is dehydrogenated to butadiene on one side of a palladium membrane while toluene undergoes hydrogenolysis to give methane and benzene on the other side. Sketch a schematic diagram of the reactor and describe the patterns for flow of heat and of hydrogen. Explain how the reactor works and why the membrane is made of palladium.

6-36 The supports used in naphtha reforming catalysts are either γ-Al_2O_3 or η-Al_2O_3. The former has the advantage of being more robust than the latter, and the latter has the advantage of being more strongly acidic than the former. For either, a crucial requirement is high purity of the alumina.

Why is this important? What impurities need to be avoided in the manufacture?

6-37 Figure 6-51 shows a reaction of *n*-hexane to give methylcyclopentane. What is the expected mechanism of this reaction?

6-38 Solid superacids have been made from mixed metal oxides incorporating sulfates [Yamaguchi, T., *Appl. Catal.*, **61**, 1 (1990)]. Consult the reference and find an explanation for the role of the sulfates in increasing the acidity of the surfaces.

6-39 A recent patent entitled "Catalyst Composition and Process for Oxidation of Ethylene to Ethylene Oxide" (Bhasin, M. M., Ellgen, P. C., and Hendrix, C. D., U.S. Patent 4,916,243, 1990) reports highly selective supported silver catalysts promoted by Cs^+ in combination with other alkali metals. Consult this patent and suggest the optimum composition of a catalyst for ethylene oxidation. State how to make the catalyst. Make use of the examples in the patent.

6-40 H. Davy in 1818 described a safety lamp for miners in which he placed a spiral of platinum wire over the flame area. When the flame went out because the concentration of mine gas was too high, the lamp glowed. How does the lamp work?

6-41 Design a catalytic stove for backpackers.

6-42 Design a catalytic self-cleaning oven.

Glossary

Activator A group that combines with an enzyme catalyst, modifying the structure of the catalytic site and increasing the activity of the catalyst.

Activity A measure of how fast a chemical reaction is catalyzed. It may be expressed as a rate or a rate constant.

Adsorbate Molecule that is adsorbed on a solid.

Adsorbent Solid on which a molecule is adsorbed.

Adsorption Combination with a surface; see chemisorption and physisorption (physical adsorption).

BET isotherm An isotherm that represents the possibility of multilayer adsorption.

BET surface area The surface area of a porous solid as determined from adsorption of N_2 and interpreted in the context of the BET model of adsorption.

Catalyst A substance that, usually as a result of chemical bonding with one or more reactants, increases the rate of a chemical reaction without being consumed to any significant degree.

Chemisorption Adsorption by chemical bonding (chemical adsorption).

Cocatalyst A substance which by itself does not catalyze a reaction but which, in combination with a catalyst, affects the rate of the reaction without being consumed to a significant degree.

Concerted reaction A reaction in which more than one bond-breaking and/or bond-forming step occurs almost simultaneously (also referred to as a synchronous reaction).

Effectiveness factor Observed rate constant of a catalytic reaction divided by the rate constant that would be observed under the same conditions in the absence of any mass and heat transport resistances.

Eley–Rideal mechanism Mechanism of a catalytic reaction involving the (usually rate-determining) combination of a chemisorbed species with another species from the fluid phase; alternatively, the nonchemisorbed species might be physically adsorbed.

Enzyme A biological catalyst which is usually a protein, that is, a polymer formed from a combination of the 20 naturally occurring amino acids.

General acid catalysis Catalysis in which an undissociated acid donates a proton to a reactant.

Inhibitor A substance that reduces the rate of a catalytic reaction, often as a result of bonding chemically to the catalyst. A competitive inhibitor bonds to the catalyst in competition with reactants.

Langmuir isotherm The simplest relationship between the partial pressure of adsorbate and the equilibrium surface concentration of adsorbate which expresses the idea of saturation of the surface. The surface is assumed to have a fixed number of identical sites, and the maximum coverage of noninteracting molecules corresponds to one monolayer.

Langmuir–Hinshelwood kinetics An expression for the rate of a solid-catalyzed reaction with the rate-determining step involving only adsorbed species. The concentrations of adsorbed species are related to the external phase concentrations through the Langmuir isotherm.

Langmuir–Hinshelwood mechanism Mechanism of a catalytic reaction involving the (usually rate-determining) conversion of reactants which are chemisorbed.

Michaelis–Menten kinetics Kinetics indicating the possibility of saturation of catalytic sites by reactant. In the simplest form, it is expressed by an equation of the form

$$ r = \frac{kK_R C_R}{1 + K_R C_R} $$

where R is the reactant.

Physical adsorption (physisorption) Adsorption resembling condensation, involving weak (van der Waals) interactions between adsorbent and adsorbate.

Poison A substance that combines strongly with a catalyst causing a loss of its activity.

Promoter A substance that affects the performance of a catalyst, for example, making it more active, selective, or stable.

Rate-determining step One step in a sequence or cycle that is the bottleneck; all the other steps achieve virtual equilibrium.

Regenerability Susceptibility of a catalyst to regeneration of its activity, for example, in a treatment such as burning off of carbonaceous deposits.

Selectivity A measure of catalytic activity for one reaction relative to that for other reactions. This term is often used more or less interchangeably with "specificity" and "directivity."

Specific acid catalysis Catalysis in which a hydrated proton (H_3O^+) donates a proton to a reactant.

Specific base catalysis Catalysis in which an OH^- ion accepts a proton from a reactant.

Stability A measure of the rate of loss of catalytic activity (or selectivity).

Steady-state approximation The approximation in chemical kinetics, according to which, after a short induction period, the concentrations of highly reactive intermediates in a chemical reaction all achieve nearly time-invariant values.

Substrate Reactant. Alternatively, "substrate" is sometimes used as a synonym for "support." Because of the possible confusion, this term is largely avoided in the text.

Support A solid, the surface of which incorporates catalytic components. The surface of the support may be involved in catalysis, but instead may be inert.

Turnover Single reaction event or turn of a catalytic cycle. To demonstrate the occurrence of catalysis rather than a stoichiometric reaction, the experimenter must demonstrate that the number of reaction events per catalytic site exceeds unity.

Turnover frequency (turnover number) Reaction rate in units of molecules per catalytic site per second, strictly, only for the case in which the catalytic sites are saturated with reactant, that is, for the case for which the reaction is zero order in reactant. Often the term is used more loosely as the rate of reaction in molecules per second per surface metal atom.

Reading List

BOOKS

Anderson, J. R., *Structure of Metallic Catalysts*, Academic Press, New York, 1975. Description of bulk and supported metal catalysts.

Bond, G. C., *Catalysis by Metals*, Academic Press, London, 1962. Dated but still one of the best compendia.

Bond, G. C., *Heterogeneous Catalysis* (2nd edition), Oxford University Press, Oxford, 1987. Short, very elementary introduction.

Campbell, I. M., *Catalysis at Surfaces*, Chapman and Hall, New York, 1988 (paperbound). Elementary introduction.

Gates, B. C., Katzer, J. R., and Schuit, G. C. A., *Chemistry of Catalytic Processes*, McGraw-Hill, New York, 1979. Petrochemical and petroleum process chemistry.

Hegedus, L. L., ed., *Catalyst Design: Progress and Perspectives*, Wiley–Interscience, New York, 1987. Various viewpoints of catalytic science and engineering and some prospects for catalyst design.

Leach, B. E., ed., *Applied Industrial Catalysis*, Academic Press, New York, 1983–1984. Practical catalysis in depth.

Butt, J. B., and Petersen, E. E., *Activation, Deactivation, and Poisoning of Catalysts,* Academic Press, San Diego, 1988.

Pines, H., *The Chemistry of Catalytic Hydrocarbon Conversions*, Academic Press, New York, 1981. An organic chemist's summary of catalytic reactions in petroleum refining and petrochemical conversion.

Satterfield, C. N., *Heterogeneous Catalysis in Practice*, McGraw-Hill, New York, 1980. Introduction.

Schwab, G.-M., *Catalysis from the Standpoint of Chemical Kinetics* (translated from the German by H. S. Taylor and R. Spence), van Nostrand, New York, 1937. The first unified, quantitative book about catalysis.

433

Stiles, A. B., ed., *Catalyst Supports and Supported Catalysts: Theoretical and Applied Concepts*, Butterworths, Boston, 1987. Practical monograph.

Tamaru, K., *Dynamic Heterogeneous Catalysis*, Academic Press, London, 1978. Concepts: kinetics, adsorption.

Tanabe, K., *Solid Acids and Bases*, Academic Press, New York, 1970. Compendium, neglecting zeolites and organic solids.

Thomas, C. L., *Catalytic Processes and Proven Catalysts*, Academic Press, New York, 1970. Almost a handbook; a good book to consult about what catalysts are used in industry.

Thomas, J. M., and Thomas, W. J., *Introduction to the Principles of Heterogeneous Catalysis*, Academic Press, New York, 1967. Dated; but one of the most complete single-volume statements of principles of surface catalysis.

Twigg, M. V., ed., *Catalyst Handbook*, 2nd edition, CRC Press, Boca Raton, Florida, 1989. Lots of practical information.

Information about catalytic processes appears in the *Kirk-Othmer Encyclopedia of Chemical Technology*, Wiley, New York.

Other recommended general references are given in the section Further Reading at the end of each chapter. A thorough listing of books about catalysis is given in J. T. Richardson's *Principles of Catalyst Development*, Plenum Press, New York, 1989.

MONOGRAPH SERIES

Advances in Catalysis, Academic Press. Volumes appearing yearly include reviews of topics in catalysis.

Catalysis, Reinhold. P. H. Emmett edited this series of seven pre-1960 volumes reviewing the chemistry and technology of catalysis.

Catalysis—Science and Technology, Springer. J. R. Anderson and M. Boudart edit this series, a successor to Emmett's *Catalysis*.

A continuing series of books appears under the heading *Catalysis by Metal Complexes*, edited by R. Ugo and B. R. James and published by Reidel. Ugo also edited the series *Aspects of Homogeneous Catalysis*, published by Reidel.

The continuing series *Studies in Surface Science and Catalysis* is published by Elsevier.

The review series *Catalysis* is published by CRC Press; early volumes were published by the Royal Society in London.

JOURNALS AND SOURCES OF PRIMARY LITERATURE

Applied Catalysis. Elsevier.

Catalysis Letters. J. C. Baltzer. Short communications.

Catalysis Reviews—Science and Engineering. Marcel Dekker.

Journal of Catalysis. Academic Press. Mostly chemistry of catalysis by surfaces.

Journal of Molecular Catalysis. Elsevier. Emphasizing molecular catalysis and relations between solution and surface catalysis.

Kinetics and Catalysis (*Kinetika i. Kataliz*). The USSR's standard journal in the field.

Reaction Kinetics and Catalysis Letters. Akademiai Kiado. Short communications, mostly from Eastern Europe.

Proceedings of the International Congress on Catalysis appear every four years, with various publishers.

Proceedings of the International Zeolite Conference appear regularly and contain many papers on catalysis.

Many less specialized journals include reports of work on catalysis; some of the best are *Journal of the American Chemical Society*, *Journal of Physical Chemistry*, *Journal of the Chemical Society Faraday Transactions I*, and *Industrial and Engineering Chemistry Research*.

Acknowledgments

CHAPTER 1

Figure 1-1: Reprinted with permission from *J. Am. Chem. Soc.*, **92**, 6777. Copyright (1970) American Chemical Society.

CHAPTER 2

Figure 2-7: Reprinted with permission from *J. Chem. Soc.*, p. 2000. Copyright (1957) Royal Society of Chemistry, Cambridge, UK.

Figure 2-9: Reprinted with permission from *J. Am. Chem. Soc.*, **88**, 2502. Copyright (1966) American Chemical Society.

Figure 2-10: Reprinted with permission from *Proc. R. Soc. A 197*, p. 141. Copyright (1949) Royal Society of Chemistry, Cambridge, UK.

Figure 2-12: Reprinted with permission from *Friedel-Crafts Chemistry*, p. 368. Copyright (1973) John Wiley & Sons, Inc., New York.

Figure 2-13: Reprinted with permission from *Chem. Rev.* **57**, 935. Copyright (1957) American Chemical Society.

Figure 2-17: Reprinted with permission from P. B. Venuto and E. T. Habib, Jr. *Fluid Catalytic Cracking with Zeolite Catalysts*, p. 7, 1979, by courtesy of Marcel Dekker, Inc., New York. Sources of data are given in the reference.

Figure 2-20: Reprinted with permission from *Chemistry and Chemical Engineering of Catalytic Processes*, p. 137. Copyright (1980) by Kluwer Academic Publishers.

Figure 2-25: Reprinted with permission from *Adv. Catal.*, **25**, 272. Copyright (1976) Academic Press.

Figure 2-26: Reprinted with permission, *J. Mol. Catal.*, **4**, 243. Copyright (1978) Elsevier Science Publishers.

Figures 2-27, 2-28, 2-29, 2-30: Reprinted with permission from B. C. Gates, et al. *Chemistry of Catalytic Processes:* Figures 2-1 on p. 115, 2-3 on p. 116, and 2-4 on p. 117. Copyright (1979) McGraw-Hill, Inc.

Figure 2-31: Reprinted with permission from University Science Books.

Figures 2-32 and 2-33: Reprinted with permission from *J. Am. Chem. Soc.*, **110**, 3773. Copyright (1988) American Chemical Society.

Figures 2-34, 2-35, 2-36: Reprinted with permission from *Science,* figures 1, 2, and 4, Vol. 217, p. 403, 30 July 1982, *Mechanism and Stereoselectivity of Asymmetric Hydrogenation.* Copyright 1982 by the AAAS.

Figure 2-37: Reprinted from F. A. Cotton and G. Wilkinson: *Advanced Inorganic Chemistry,* 3rd ed. p. 790. Copyright (1972) John Wiley and Sons, Inc.

Figure 2-38: Reprinted with permission from *Hydrocarbon Processing,* April, 1970.

Figure 2-39: Reprinted with permission from *Proc. 9th Int. Congr. Catal.* p. 254. Copyright (1988) The Chemical Institute of Canada.

Figure 2-40: Reprinted with permission from *J. Am. Chem. Soc.,* **98,** 846. Copyright (1976) American Chemical Society.

Figure 2-42: Reprinted with permission from *J. Am. Chem. Soc.,* **112,** 4911. Copyright (1990) American Chemical Society.

Figures 2-43 and 2-44: Reprinted with permission from *Adv. Catal.,* **23,** 263. Copyright (1973) Academic Press.

Figure 2-48: Reprinted with permission from *J. Am. Chem. Soc.,* **96,** 2614. Copyright (1974) American Chemical Society.

Figure 2-49: Reprinted with permission from *J. Chem. Soc., Chem. Commun.* p. 859. Copyright (1975) The Royal Society of Chemistry, London, UK.

Figure 2-50: Reprinted with permission from *J. Am. Chem. Soc.,* **101,** 6110. Copyright (1979) American Chemical Society.

Figure 2-51: Reprinted with permission from *J. Am. Chem. Soc.,* **98,** 1056. Copyright (1976) American Chemical Society.

Figure 2-55: Reprinted with permission from *Catalysis in Micellar and Macromolecular Systems.* Copyright (1975) Academic Press.

Figure 2-56: Reprinted with permission from D. Forster, A. Hershman, and D. E. Morris: *Catal. Rev.-Sci. Eng.,* **23,** 89 (1981) by courtesy of Marcel Dekker, Inc., New York.

Figure 2-57: Reprinted with permission from *J. Am. Chem. Soc.,* **111,** 1123. Copyright (1989) American Chemical Society.

CHAPTER 3

Figure 3-1: Reprinted with permission from *Nature,* Vol. 206, p. 757. Copyright © 1965 Macmillan Magazines, Ltd.

Figure 3-2: Reprinted from Linus Pauling: *The Nature of the Chemical Bond,* Third ed. Copyright © 1939 and 1940, Third edition © 1960 by Cornell University. Used by permission of the publisher, Cornell University Press.

Figure 3-3: Reprinted from A. Fersht: *Enzyme Structure and Mechanism, 2nd edition,* p. 10. Copyright © 1977, 1985 by W. H. Freeman & Co. Reprinted by permission.

Figure 3-4: Reprinted with permission from *Macromolecules,* **12,** 633. Copyright (1979) American Chemical Society.

Figure 3-6: Reprinted with permission from J. E. Bailey and D. F. Ollis: *Biochemical Engineering Fundamentals,* 1E, Figure 1.5 on p. 8. Copyright (1977) McGraw-Hill, Inc.

Figure 3-8: Reprinted with permission from *Nature,* Vol. 266, p. 328. Copyright © 1977 Macmillan Magazines, Ltd.

Figure 3-9: Reprinted with permission from W. S. Bennett and T. A. Steitz: *Proc. Nat. Acad. Sci. of U. S.,* **75,** 4848 (1978).

Figure 3-10: Reprinted with permission from *Biochemistry,* **14,** 1088. Copyright (1975) American Chemical Society.

Figures 3-11 and 3-12: Reprinted with permission from S. Blackburn: *Enzyme Structure and Function,* pp. 455–456 (1976) by courtesy of Marcel Dekker, Inc., NY.

Figure 3-13: Reprinted with permission from C. Cantor and P. R. Schimmel: *Biophysical Chemistry,* p. 930. Copyright (1980) Irving Geis.

Figure 3-14: Reprinted with permission from R. J. P. Williams: *J. Mol. Catal.,* **30,** 1. Copyright (1985) Elsevier Science Publishers.

Figure 3-15: Reprinted with permission from: *Coord. Chem. Rev.,* **54,** 1. Copyright (1984) Elsevier Science Publishers.

CHAPTER 4

Figure 4-3: Reprinted with permission from *J. Catal.,* **31,** 27. Copyright (1973) Academic Press.

Figures 4-6 and 4-7: Figures 7 and 8 reprinted from *Reaction Kinetics and Adsorption Equilibria in the Vapor-Phase Dehydration of Ethanol,* Robert L. Kabel, and Lennart N. Johanson, *AIChEJ.,* Vol. 8, No. 5, pp. 621–628 (1962). Reproduced by permission of the American Institute of Chemical Engineers © 1962 AIChE.

Figure 4-10: Reprinted in modified form with permission from R. Thornton and B. C. Gates: *Proc. 5th Int. Congr. Catal.,* **1,** 357. Copyright (1973) Elsevier Science Publishers.

Figures 4-11 and 4-12: Reprinted with permission from *J. Catal.,* **24,** 315. Copyright (1972) Academic Press.

Figure 4-13: Reprinted from J. Chem. Soc. B, p. 1015. Copyright (1966) Royal Society of Chemistry, Cambridge, U.K.

Figures 4-14 and 4-15: Reprinted with permission from *Proc. 6th Int. Congr. Catal.,* **1,** 499. Copyright (1977) The Chemical Society, London, UK.

Figure 4-18: Figures 2 & 3 reprinted from *Oligomerization of Isobutylene on Cation Exchange Resins,* W. O. Haag, AIChE Symposium Series, Vol. 63, No. 73, pp. 140–147 (1967). Reproduced by permission of the American Institute of Chemical Engineers © AIChE.

Figure 4-23: Reprinted with permission from *Ind. Eng. Chem.,* **10,** 185. Copyright (1971) American Chemical Society.

Figure 4-25: Reprinted with permission from *Ind. Eng. Chem. Fundam.,* **4,** 317. Copyright (1965) American Chemical Society.

Figure 4-26: Reprinted with permission from *Adv. Catal.* **3,** 249. Copyright (1951) Academic Press.

Figure 4-27: Reprinted with permission from *Chem. Eng. Sci.,* **17,** 265, P. B. Weisz and J. S. Hicks, *The behaviour of porous catalyst particles in view of internal mass and heat diffusion effects.* Copyright (1962) Pergamon Press, Inc.

Figure 4-29: Figures 2 & 4 reprinted from *Mass Transport and Reaction in Sulfonic Acid Resin Catalyst: The Dehydration of t-Butyl Alcohol,* H. W. Heath, Jr., B. C. Gates, AIChE J., **187,** No. 2, pp. 322–323 (1972). Reproduced by permission of the American Institute of Chemical Engineers © 1972 AIChE.

Figure 4-30: Reprinted with permission from *Chem. Eng. Sci.,* **45,** 1605, A. Rehfinger and U. Hoffmann *Kinetics of Methyl Tertiary Butyl Ether Liquid Phase Synthesis Catalyzed by Ion Exchange Resin-I. Intrinsic Rate Expression in Liquid Phase Activities.* Copyright (1990) Pergamon Press, Inc.

CHAPTER 5

Figures 5-4 and 5-5: Reprinted with permission from *Chem. Rev.* **79**, 91. Copyright (1979) American Chemical Society.

Figure 5-9: Reprinted by permission from *Nature*, vol. 271, p. 572. Copyright © 1978 Macmillan Magazines, Ltd.

Figures 5-10 and 5-11: Reprinted with permission from C. E. Lyman, P. M. Betteridge, and E. F. Moran: *Intrazeolite Chemistry, A.C.S. Symposium Series,* Stuckey, G. D.; Dwyer, F. G., Eds., **218**, 199. Copyright (1983) American Chemical Society.

Figure 5-17: Reprinted with permission from J. V. Smith, *Molecular Sieve Zeolites, Advances in Chemistry Series,* **101**, 171. Flanigan, E. M.; Sand, L. B., Eds. Copyright (1971) American Chemical Society.

Figure 5-18: Reprinted with permission from *Intercalation Chemistry*, p. 113. Copyright (1982) Academic Press.

Figure 5-19: Reprinted from R. M. Barrer: *Zeolites: Science and Technology,* F. Ribeiro, A. E. Rodrigues, L. D. Rollmann, and C. Naccache, Eds., p. 237. Copyright (1984) Martinus Nijhoff Publishers, The Hague, The Netherlands.

Figure 5-20: Reprinted with permission from *CHEMTECH,* **3**, 498. Copyright (1973) American Chemical Society.

Figure 5-21: Reprinted with permission from *J. Catal.,* **61**, 390. Copyright (1980) Academic Press.

Figure 5-22: Reprinted with permission from *J. Catal.,* **101**, 132. Copyright (1986) Academic Press.

Figure 5-24: Reprinted with permission from *Ind. Eng. Chem. Prod. Res. Dev.,* **8**, 24. Copyright (1969) American Chemical Society.

Figure 5-26: Reprinted with permission from P. B. Venuto, *Catalysis in Organic Synthesis,* p. 72, Smith, G. V. Ed. Copyright (1977) Academic Press.

Figures 5-27 and 5-28: Reprinted with permission from *CHEMTECH,* **3**, 498. Copyright (1973). American Chemical Society.

Figure 5-29: Reprinted with permission from *Ind. Eng. Chem. Fundam.,* **11**, 540. Copyright (1972) American Chemical Society.

Figure 5-30: Reprinted with permission from *Proc. 7th Int. Congr. Catal., Part A,* **1**, 1. Copyright (1981) Kodansha, Ltd., Tokyo, Japan.

Figures 5-31 and 5-32: Reprinted with permission from P. B. Weisz *Proc. 7th Int. Congr. Catal.,* **1**, 1. Copyright (1981) Kodansha, Ltd., Tokyo, Japan.

Figure 5-33: Reprinted with permission from D. H. Olson and W. O. Haag: *Catalytic Materials Relationship Between Structure and Reactivity, ACS Symposium Series 248,* p. 280. Whyte, Jr., T. E.; Dalla Betta, R. A.; Derouane, E. G.; Baker, R. T. K., Eds. Copyright (1984) American Chemical Society.

Figure 5-34: Reprinted with permission from *J. Catal.,* **56**, 139. Copyright (1979) Academic Press.

Figure 5-35: Reprinted with permission from *J. Catal.,* **22**, 371. Copyright (1971) Academic Press.

Figure 5-36: Reprinted with permission from *Inorg. Chem.,* **18**, 2840. Copyright (1979) American Chemical Society.

CHAPTER 6

Figure 6-2: Reprinted with permission from G. A. Somorjai: *Catalyst Design, Progress and Perspectives,* L. L. Hegedus, Ed., p. 11. Copyright (1987) John Wiley & Sons, Inc.

Figure 6-3: Reprinted with permission from *Surface Science,* **144**, 321.

Copyright (1984) Elsevier Science Publishers.

Figure 6-4: Reprinted from A. J. Lecloux: *Catalysis—Science and Technology, Vol. 2,* J. R. Anderson and M. Boudart, Eds., p. 171. Copyright (1981) Springer-Verlag, Berlin, Germany.

Figure 6-6: Reprinted with permission from *Angewandte Chemie Int. Ed. in English,* **27,** 1127 (1988) VCH Verlagsgesellschaft, Germany.

Figure 6-7: Reprinted with permission from A. B. Stiles, *Catalyst Supports and Supported Catalysts,* R. Poisson, J.-P. Brunelle, and P. Nortier, Eds. Copyright (1987) Butterworth-Heinemann, Stoneham, MA.

Figure 6-8: Reprinted with permission from *Angewandte Chemie Int. Ed. in English,* **27,** 1127, (1988) VCH Verlagsgesellschaft, Germany.

Figure 6-9: Reprinted with permission from *Catal. Rev.-Sci. Eng.,* **17,** 31 (1978) by courtesy of Marcel Dekker, Inc.

Figure 6-10: Reprinted from G. Ertl and J. Koch: *Zeitschrift für Naturforschung, 25a,* p. 1908. Copyright (1970) Verlag der Zeitschrift.

Figure 6-11: Reprinted from *J. Am. Chem. Soc.,* **59,** 2682. Copyright (1937) American Chemical Society.

Figure 6-12: Reprinted from *J. Am. Chem. Soc.,* **60,** 309. Copyright (1938) American Chemical Society.

Figure 6-14: Reprinted with permission from *J. Chem. Phys.,* **73** (6), 2984. Copyright (1980) American Institute of Physics.

Figures 6-15 and 6-16: Reprinted with permission from *Surf. Sci.,* **72,** 513. Copyright (1978) Elsevier Science Publishers.

Figure 6-17: Reprinted with permission from *CHEMTECH,* **13,** 376. Copyright (June, 1983) American Chemical Society.

Figure 6-18: Reprinted with permission from *J. Catal.,* **77,** 439. Copyright (1982) Academic Press.

Figure 6-19: Reprinted with permission from *J. Catal.,* **30,** 109. Copyright (1973) Academic Press.

Figure 6-20: Reprinted with permission from *Adv. Catal.,* **35,** 187. Copyright (1987) Academic Press.

Figure 6-21: Reprinted with permission from *J. Phys. Chem.* **94,** 8447. Copyright (1990) American Chemical Society.

Figure 6-22: Reproduced from G. Natta, P. Corradini, and G. Allegra, Atti Acad. Nazl. Lincei, Rend., Classe Sci. Fis. Mat. e Nat., *26,* 155 (1959).

Figure 6-23: Reprinted from P. Crossee: *The Stereochemistry of Macromolecules,* A. D. Ketley, Ed., **1,** 145. Copyright (1967) Marcel Dekker, Inc.

Figure 6-24: Reprinted with permission from L. A. M. Rodriguez and J. A. Gabant, *J. Poly Sci., Part C (4),* p. 125. Copyright © (1963) John Wiley and Sons, Inc.

Figure 6-25: Reprinted with permission from *J. Catal.,* **76,** 369. Copyright (1982) Academic Press.

Figure 6-26: Reprinted with permission from W. J. M. Rootsgaert and W. M. H. Sachtler, *Zeitschrift für Physikalische Chemie (Neue Folge),* **26,** 16 (1960).

Figure 6-27: Reprinted with permission from G. Ertl, *Proc. 7th Int. Congr. Catal., Part A,* p. 31. Copyright (1980) Kodansha, Ltd., Japan.

Figure 6-28: Reprinted with permission from T. Engl, G. Ertl, *The Chemical Physics of Solid Surfaces and Heterogeneous Catalysis,* D. P. Woodruff, Ed., **4,** Fig. 7, p. 80. Copyright (1982) Elsevier Science Publishers.

Figure 6-29: Reprinted from H. Conrad, G. Ertl, and J. Küppers, Surface

Science, **76**, 323. Copyright (1978) Elsevier Science Publishers.

Figure 6-30: Reprinted with permission from *J. Chem. Phys.* **69**, 1267. Copyright (1978) American Institute of Physics.

Figure 6-31: Reprinted from *Proc. 7th Int. Congr. Catal., Part A*, p. 21. Copyright (1981) Kodansha, Ltd., Tokyo, Japan.

Figure 6-32: Reprinted with permission from *Chem. Phys. Lett.* **60**, 391. Copyright (1979) Elsevier Science Publishers.

Figure 6-33: Reprinted with permission from *Proc. 7th Int. Congr. Catal., Part A*, p. 30. Copyright (1980) Kodansha, Ltd., Japan.

Figure 6-34: Reprinted with permission from G. A. Somorjai, *Catalyst Design, Progress and Perspectives*, L. L. Hegedus, Ed., p. 16. Copyright © (1987) John Wiley and Sons, Inc.

Figure 6-35: Reprinted with permission from *J. Catal.*, **24**, 57. Copyright (1972) Academic Press.

Figures 6-36 and 6-37: With permission, *Adv. Catal.*, **22**, 1. Copyright (1972) Academic Press.

Figure 6-40: Reprinted with permission from *J. Phys. Chem.*, **95**, 225. Copyright (1991) American Chemical Society.

Figure 6-41: Reprinted with permission from *J. Am. Chem. Soc.*, **107**, 3139. Copyright (1985) American Chemical Society.

Figure 6-43: Reprinted with permission from *J. Chem. Phys.*, **71**, 690. Copyright (1979) American Institute of Physics.

Figure 6-44: Reprinted with permission from *J. Catal.*, **105**, 26. Copyright (1987) Academic Press.

Figure 6-45: Reprinted with permission from *Surface Science*, **123**, 679. Copyright (1982) Elsevier Science Publishers.

Figure 6-46: Reprinted with permission from *J. Catal.*, **24**, 283. Copyright (1972) Academic Press.

Figure 6-47: With permission, *Adv. Catal.*, **35**, 265. Copyright (1987) Academic Press.

Figure 6-49: Reprinted with permission from *J. Catal.*, **57**, 372. Copyright (1979) Academic Press.

Figure 6-50: Reprinted with permission from *Ind. Eng. Chem.*, **45**, 134. Copyright (1953) American Chemical Society.

Figure 6-51: Reprinted with permission from *Science,* figure 1, Vol. 126, 5 July 1957, p. 31, *Stepwise Reaction on Separate Catalytic Centers: Isomerization of Saturated Hydrocarbons,* Weisz, P. B. Copyright 1957 by the AAAS.

Figure 6-52: Reprinted with permission, *Adv. Chem. Eng.*, **13**, 193. Copyright (1987) Academic Press.

Figure 6-54: Reprinted with permission from *J. Chem. Soc., Faraday Trans. I,* **81**, 2903. Copyright (1985) The Royal Society of Chemistry, Cambridge, UK.

Figure 6-55: Figure 5 reprinted from ''Hydrosulfurization of Dibenzothiophene Catalyzed by Sulfided C_0O-M_0O_3/r-Al_2O_3: The Reaction Network'' M. Houalla, N. K. Nag, A. V. Sapre, D. H. Broderick, and B. C. Gates, AIChEJ., Vol. 24, No. 6, p. 1019 (1978). Reproduced by permission of the American Institute of Chemical Engineers © 1978 AIChE.

Figure 6-56a: Reprinted with permission from *Ind. Eng. Chem. Proc. Des. Dev.,* **20**, 68. Copyright (1981) American Chemical Society.

Figure 6-58: Reprinted with permission from *Catal. Lett.*, **5**, 280. Copyright (1990) J. C. Baltzer Scientific Publishing Company.

Figure 6-59: Reprinted with permission from H. Topsoe, *Proc. 9th*

Iberoamerican Symp. Catal, p. 217, (1984).

Figures 6-60 and 6-61: Reprinted with permission from *Ind. Eng. Chem. Proc. Des. Dev.,* **20,** 262. Copyright (1981) American Chemical Society.

Figures 6-62 and 6-63: Reprinted with permission from *J. Phys. Chem.,* **74,** 3653. Copyright (1970) American Chemical Society.

Figure 6-64: Reprinted from P. H. Emmett, *The Physical Basis for Heterogeneous Catalysis,* E. Drauglis and R. I. Jaffee, Eds., p. 3. Copyright (1975) Plenum Press.

TABLES

Table 2-2: Reprinted with permission of University Science Books.

Table 2-3: Reprinted with permission of University Science Books.

Table 2-5: Reprinted from G. Henrici-Olivé and S. Olivé, *Coordination and Catalysis,* p. 172. Copyright (1977) VCH Verlagsgesellschaft. Source of information: J. Halpern, personal communication to G. Henrici-Olivé and S. Olivé.

Table 4-1: Table 3 reprinted from *Reaction Kinetics and Adsorption Equilibria in the Vapor-Phase Dehydration of Ethanol,* Robert L. Kabel, and Lennart N. Johanson, *AIChE Journal,* Vol. 8, No. 5, pp 621–628 (1962). Reproduced by permission of The American Institute of Chemical Engineers © 1962 AIChE.

Table 5-5: Reprinted from *J. Catal.,* **1,** 307. Copyright (1962) Academic Press.

Table 5-6: Reprinted with permission from *Faraday Disc. of Chem. Soc.,* **71,** 317. Copyright (1981) The Royal Society of Chemistry.

Table 5-8: Reprinted from *J. Catal.,* **88,** 240. Copyright (1984) Academic Press.

Table 5-9: Reprinted from *J. Catal.,* **29,** 292. Copyright (1973) Academic Press.

Table 6-4: Reprinted by permission of *Princeton University Press,* Boudart, Michel and G. Djega-Mariadassou, *Kinetics of Heterogeneous Catalytic Reaction.* Copyright © 1984 by Princeton University Press. Table 5.9, page 190, reprinted by permission.

Index

PHYSICAL AND CHEMICAL CONSTANTS

Avogadro's number		$= 6.022 \times 10^{23}$ mol^{-1}
Speed of light		$= 2.9979 \times 10^{8}$ m/s
Electron mass		$= 9.109 \times 10^{-31}$ kg
Proton mass		$= 1.6726485 \times 10^{-27}$ kg
		$= 1.007276470$ amu
Gas constant	R	$= 8.3144$ J/(mol $\cdot$ K)
		$= 1.9872$ cal/(mol $\cdot$ K)
		$= 0.08206$ L atm/(mol $\cdot$ K)
Ice point		$= 273.15$ K
Molar volume at NTP		$= 22.414 \times 10^{3}$ cm^3/mol
		$= 2.2414 \times 10^{-2}$ m^3/mol
Planck's constant	h	$= 6.626 \times 10^{-34}$ J s
		$= 6.626 \times 10^{-27}$ erg s
Boltzmann's constant		$= 1.38066 \times 10^{-23}$ J/K
Other numbers	π	$= 3.14159$
	e	$= 2.7183$
	$\ln 10$	$= 2.3026$

CONVERSION FACTORS

1 cal	$= 4.184$ joules (J)
1 eV/molecule	$= 23.06$ kcal/mol